Probability in Electrical Engineering
and Computer Science

Jean Walrand

Probability in Electrical Engineering and Computer Science

An Application-Driven Course

 Springer

Jean Walrand
Department of EECS
University of California, Berkeley
Berkeley, CA, USA

The link includes examples of Python demos and also Python labs used at Berkeley:
https://sites.google.com/berkeley.edu/probabilityineecs/home

ISBN 978-3-030-49997-6   ISBN 978-3-030-49995-2   (eBook)
https://doi.org/10.1007/978-3-030-49995-2

This Springer imprint is published by the registered company Springer Nature Switzerland AG
The registered company address is: Gewerbestrasse 11, 6330 Cham, Switzerland

To my wife Annie, my daughters Isabelle and Julie, and my grandchildren Melanie and Benjamin, who will probably never read this book.

Preface

This book is about extracting information from noisy data, making decisions that have uncertain consequences, and mitigating the potentially detrimental effects of uncertainty.

Applications of those ideas are prevalent in computer science and electrical engineering: digital communication, GPS, self-driving cars, voice recognition, natural language processing, face recognition, computational biology, medical tests, radar systems, games of chance, investments, data science, machine learning, artificial intelligence, and countless (in a colloquial sense) others.

This material is truly exciting and fun. I hope you will share my enthusiasm for the ideas.

Berkeley, CA, USA   Jean Walrand
April 2020

The original version of this book was revised as the author would like to acknowledge the funding information. The below mentioned text has been added to the Acknowledgements section of the updated version. "This book is an open access publication. This title is freely available in open access edition with generous support from the Library of the University of California, Berkeley." The correction to the book is available at https://doi.org/10.1007/978-3-030-49995-2_17

The original version of this book was revised as the author would like to acknowledge the source of the Python materials used in this book. The below mentioned text has been added to the copyright page of the updated version. "The link includes examples of Python demos and also Python labs used at Berkeley: https://sites.google.com/berkeley.edu/probabilityineecs/home." The correction to the book is available at https://doi.org/10.1007/978-3-030-49995-2_16

Acknowledgements

I am grateful to my colleagues and students who made this book possible. I thank Professor Ramtin Pedarsani for his careful reading of the manuscript, Sinho Chewi for pointing out typos in the first edition and suggesting improvements of the text, Dr. Abhay Parekh for teaching the course with me, Professors David Aldous, Venkat Anantharam, Tom Courtade, Michael Lustig, John Musacchio, Shyam Parekh, Kannan Ramchandran, Anant Sahai, David Tse, Martin Wainwright, and Avideh Zakhor for their useful comments, Stephan Adams, Kabir Chandrasekher, Dr. Shiang Jiang, Dr. Sudeep Kamath, Dr. Jerome Thai, Professors Antonis Dimakis, Vijay Kamble, and Baosen Zhang for serving as teaching assistants for the course and designing assignments, Professor Longbo Huang for translating the book in Mandarin and providing many valuable suggestions, Professors Pravin Varaiya and Eugene Wong for teaching me Probability, Professor Tsu-Jae King Liu for her support, and the students in EECS126 for their feedback.

This book is an open access publication. This title is freely available in open access edition with generous support from the Library of the University of California, Berkeley.

Finally, I want to thank Professor Takek El-Bawab for making a number of valuable suggestions for the second edition and the Springer editorial team, including Mary James, Zoe Kennedy, Vidhya Hariprasanth, and Lavanya Venkatesan for their help in the preparation of this edition.

Introduction

This book is about applications of probability in electrical engineering and computer science. It is not a survey of all the important applications. That would be too ambitious. Rather, the course describes real, important, and representative applications that make use of a fairly wide range of probability concepts and techniques.

Probabilistic modeling and analysis are essential skills for computer scientists and electrical engineers. These skills are as important as calculus and discrete mathematics. The systems that these scientists and engineers use and/or design are complex and operate in an uncertain environment. Understanding and quantifying the impact of this uncertainty is critical to the design of systems.

The book was written for the upper-division course EECS126 "*Probability in EECS*" in the Department of Electrical Engineering and Computer Sciences of the University of California, Berkeley. The students have taken an elementary course on probability. They know the concepts of event, probability, conditional probability, Bayes' rule, discrete random variables and their expectation. They also have some basic familiarity with matrix operations. The students in this class are smart, hard-working, and interested in clever and sophisticated ideas. After taking this course, the students are familiar with Markov chains, stochastic dynamic programming, detection, and estimation. They have both an intuitive understanding and a working knowledge of these concepts and their methods. Subsequently, many students go on to study artificial intelligence and machine learning. This course provides them with a background that enables them to go beyond blindly using toolboxes.

In contrast to most introductory books on probability, the material is organized by applications. Instead of the usual sequence—probability space, random variables, expectation, detection, estimation, Markov chains—we start each topic with a concrete, real, and important EECS application. We introduce the theory as it is needed to study the applications. We believe that this approach makes the theory more relevant by demonstrating its usefulness as it is introduced. Moreover, an emphasis is on hands-on projects where the students use Python notebooks available from the book website to simulate and calculate. Our colleagues at Berkeley designed these projects carefully to reinforce the intuitive understanding of the concepts and to prepare the students for their own investigations.

The chapters, except for the last one and the appendices, are divided into two parts: A and B. Parts A contain the key ideas that should be accessible to junior-level students. Parts B contain more difficult aspects of the material. It is possible to teach only the appendices and parts A. This would constitute a good junior-level course. One possible approach is to teach parts A in a first course and parts B in a second course. For a more ambitious course, one may teach parts A, then parts B. It is also possible to teach the chapters in order. The last chapter is a collection of more advanced topics that the reader and instructor can choose from.

The appendices should be useful for most readers. Appendix A discusses the elementary notions of probability on simple examples. Students might benefit from a quick read of this chapter.

Appendix B reviews the basic concepts of probability. Depending on the background of the students, it may be recommended to start the course with a review of that appendix.

The theory starts with *models* of uncertain quantities. Let us denote such quantities by $\mathbf{X}$ and $\mathbf{Y}$. A model enables one to calculate the expected value $E(h(\mathbf{X}))$ of a function $h(\mathbf{X})$ of $\mathbf{X}$. For instance, $\mathbf{X}$ might specify the output of a solar panel every day during 1 month and $h(\mathbf{X})$ the total energy that the panel produced. Then $E(h(\mathbf{X}))$ is the average energy that the panel produces per month. Other examples are the average delay of packets in a communication network or the average time a data center takes to complete one job (Fig. 1).

Fig. 1 Evaluation

$$\mathbf{X} \overset{?}{\to} E(h(\mathbf{X}))$$

Estimating $E(h(\mathbf{X}))$ is called *performance evaluation*. In many cases, the system that handles the uncertain quantities has some parameters θ that one can select to tune its operations. For instance, the orientation of the solar panels can be adjusted. Similarly, one may be able to tune the operations of a data center. One may model the effect of the parameters by a function $h(\mathbf{X}, \theta)$ that describes the measure of performance in terms of the uncertain quantities $\mathbf{X}$ and the tuning parameters θ (Fig. 2).

Fig. 2 Optimization

$$\max_{\theta} E(h(\mathbf{X}, \theta))$$

One important problem is then to find the values of the parameters θ that maximize $E(h(\mathbf{X}, \theta))$. This is not a simple problem if one does not have an analytical expression for this average value in terms of θ. We explain such *optimization problems* in the book.

There are many situations where one observes $\mathbf{Y}$ and one is interested in guessing the value of $\mathbf{X}$, which is not observed. As an example, $\mathbf{X}$ may be the signal that a transmitter sends and $\mathbf{Y}$ the signal that the receiver gets (Fig. 3).

Fig. 3 Inference

$$\mathbf{Y} \overset{?}{\to} \mathbf{X}$$

Fig. 4 Control

The problem of guessing X on the basis of Y is an *inference problem*. Examples include detection problems (Is there a fire? Do you have the flu?) and estimation problems (Where is the iPhone given the GPS signal?). Finally, there is a class of problems where one uses the observations to act upon a system that then changes. For instance, a self-driving car uses observations from laser range finders, GPS, and cameras to steer the car. These are *control problems* (Fig. 4).

Thus, the course discusses performance evaluation, optimization, inference, and control problems. Some of these topics are called artificial intelligence in computer science and statistical signal processing in electrical engineering. Probabilists call them examples. Mathematicians may call them particular cases. The techniques used to address these topics are introduced by looking at concrete applications such as web search, multiplexing, digital communication, speech recognition, tracking, route planning, and recommendation systems. Along the way, we will meet some of the giants of the field.

The website

https://www.springer.com/us/book/9783030499945

provides additional resources for this book, such as an Errata, Additional Problems, and Python Labs.

About This Second Edition

This second edition differs from the first in a few aspects. The Matlab exercises have been deleted as most students use Python. Python exercises are not included in the book; they can be found on the website. The appendix on Linear Algebra has been deleted. The relevant results from that theory are introduced in the text when needed. Appendix A is new. It is motivated by the realization that some students are confused by basic notions. The chapters on networks are new. They were requested by some colleagues. Basic statistics are discussed in Chap. 8. Neural networks are explained in Chap. 12.

Contents

PageRank: A

1

Application: Ranking the most relevant pages in web search
Topics: Finite Discrete Time Markov Chains, SLLN

Background:

- probability space (B.1.1);
- conditional probability (B.1.5);
- discrete random variable (B.2.1);
- expectation and conditional expectation for discrete RVs (B.2.2), (B.3.5).

1.1 Model

The World Wide Web is a collection of linked web pages (Fig. 1.1). These pages and their links form a graph. The nodes of the graph are pages $\mathscr{X}$ and there is an arc (a directed edge) from i to j if page i has a link to j.

Intuitively, a page has a high rank if other pages with a high rank point to it. (The actual ordering of search engines results depends also on the presence of the search keywords in the pages and on many other factors, in addition to the rank measure that we discuss here.) Thus, the *rank* $\pi(i)$ of page i is a positive number and

$$\pi(i) = \sum_{j \in \mathscr{X}} \pi(j) P(j, i), i \in \mathscr{X},$$

© The Author(s) 2021
J. Walrand, *Probability in Electrical Engineering and Computer Science*,
https://doi.org/10.1007/978-3-030-49995-2_1

Fig. 1.1 Pages point to one
another in the web. Here,
$P(A, B) = 1/2$ and
$P(D, E) = 1/3$

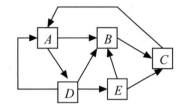

Fig. 1.2 Larry page

Fig. 1.3 Balance equations?

where $P(j, i)$ is the fraction of links in j that point to i and is zero if there is no
such link. In our example, $P(A, B) = 1/2$, $P(D, E) = 1/3$, $P(B, A) = 0$, etc.
(The basic idea of the algorithm is due to Larry Page (Fig. 1.2), hence the name
PageRank. Since it ranks pages, the name is doubly appropriate.)

We can write these equations in matrix notation as

$$\pi = \pi P, \tag{1.1}$$

where we treat π as a row vector with components $\pi(i)$ and P as a square matrix
with entries $P(i, j)$ (Figs. 1.3, 1.4 and 1.5).

Equations (1.1) are called the *balance equations*. Note that if π solves these
equations, then any multiple of π also solves the equations. For convenience, we
normalize the solution so that the ranks of the pages add up to one, i.e.,

$$\sum_{i \in \mathcal{X}} \pi(i) = 1. \tag{1.2}$$

Fig. 1.4 Copy of Fig. 1.2.
Recall that $P(A, B) = 1/2$
and $P(D, E) = 1/3$, etc.

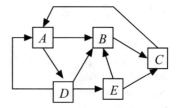

Fig. 1.5 Andrey Markov.
1856–1922

For the example of Fig. 1.4, the balance equations are

$$\pi(A) = \pi(C) + \pi(D)(1/3)$$
$$\pi(B) = \pi(A)(1/2) + \pi(D)(1/3) + \pi(E)(1/2)$$
$$\pi(C) = \pi(B) + \pi(E)(1/2)$$
$$\pi(D) = \pi(A)(1/2)$$
$$\pi(E) = \pi(D)(1/3).$$

Solving these equations with the condition that the numbers add up to one yields

$$\pi = [\pi(A), \pi(B), \pi(C), \pi(D), \pi(E)] = \frac{1}{39}[12, 9, 10, 6, 2].$$

Thus, page A has the highest rank and page E has the smallest. A search engine that uses this method would combine these ranks with other factors to order the pages. Search engines also use variations on this measure of rank.

1.2   Markov Chain

Imagine that you are browsing the web. After viewing a page i, say for one unit of time, you go to another page by clicking one of the links on page i, chosen at random. In this process, you go from page i to page j with probability $P(i, j)$ where $P(i, j)$ is the same as we defined earlier. The resulting sequence of pages that you visit is called a Markov chain, a model due to Andrey Markov (Fig. 1.4).

1.2.1   General Definition

More generally, consider a finite graph with nodes $\mathcal{X} = \{1, 2, \ldots, N\}$ and directed edges. In this graph, some edges can go from a node to itself. To each edge (i, j) one assigns a positive number $P(i, j)$ in a way that the sum of the numbers on the edges out of each node is equal to one. By convention, $P(i, j) = 0$ if there is no edge from i to j.

The corresponding matrix $P = [P(i, j)]$ with nonnegative entries and rows that add up to one is called a *stochastic matrix*. The sequence $\{X(n), n \geq 0\}$ that goes from node i to node j with probability $P(i, j)$, independently of the nodes it visited before, is then called a *Markov chain*. The nodes are called the *states* of the Markov chain and the $P(i, j)$ are called the *transition probabilities*. We say that $X(n)$ is the state of the Markov chain at time n, for $n \geq 0$. Also, $X(0)$ is called the initial state. The graph is the *state transition diagram* of the Markov chain.

Figure 1.6 shows the state transition diagrams of three Markov chains.

Thus, our description corresponds to the following property:

$$P[X(n+1) = j | X(n) = i, X(m), m < n] = P(i, j), \forall i, j \in \mathcal{X}, n \geq 0. \quad (1.3)$$

The probability of moving from i to j does not depend on the previous states. This "amnesia" is called the *Markov property*. It formalizes the fact that $X(n)$ is indeed a "state" in that it contains all the information relevant for predicting the future of the process.

1.2.2   Distribution After n Steps and Invariant Distribution

If the Markov chain is in state j with probability $\pi_n(j)$ at step n for some $n \geq 0$, it is in state i at step $n + 1$ with probability $\pi_{n+1}(i)$ where

$$\pi_{n+1}(i) = \sum_{j \in \mathcal{X}} \pi_n(j) P(j, i), i \in \mathcal{X}. \quad (1.4)$$

Indeed, the event that the Markov chain is in state i at step $n + 1$ is the union over all j of the disjoint events that it is in state j at step n and in state i at step $n + 1$. The probability of a disjoint union of events is the sum of the probabilities of the individual events. Also, the probability that the Markov chain is in state j at step n and in state i at step $n + 1$ is $\pi_n(j) P(j, i)$.

Thus, in matrix notation,

$$\pi_{n+1} = \pi_n P,$$

so that

$$\pi_n = \pi_0 P^n, n \geq 0. \tag{1.5}$$

Observe that $\pi_n(i) = \pi_0(i)$ for all $n \geq 0$ and all $i \in \mathcal{X}$ if and only if π_0 solves the balance equations (1.1). In that case, we say that π_0 is an *invariant distribution*. Thus, an invariant distribution is a nonnegative solution π of (1.1) whose components sum to one.

1.3   Analysis

Natural questions are

- Does there exist an invariant distribution?
- Is it unique?
- Does π_n approach an invariant distribution?

The next sections answer those questions.

1.3.1   Irreducibility and Aperiodicity

We need the following definitions.

Definition 1.1 (Irreducible, Aperiodic, Period)

(a) A Markov chain is *irreducible*, if it can go from any state to any other state, possibly after many steps.
(b) Assume the Markov chain is irreducible and let

$$d(i) := g.c.d.\{n \geq 1 \mid P^n(i, i) > 0\}. \tag{1.6}$$

(If S is a set of positive integers, $g.c.d.(S)$ is the greatest common divisor of these integers.)

Then $d(i)$ has the same value d for all i, as shown in Lemma 2.2. The Markov chain is *aperiodic* if $d = 1$. Otherwise, it is periodic with *period d*.

◇

The Markov chains (a) and (b) in Fig. 1.6 are irreducible and (c) is not. Also, (a) is periodic and (b) is aperiodic.

Fig. 1.6 Three Markov chains with three states $\{1, 2, 3\}$ and different transition probabilities. (**a**) is irreducible, periodic; (**b**) is irreducible, aperiodic; (**c**) is not irreducible

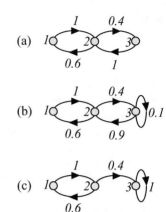

1.3.2 Big Theorem

Simple examples show that the answers to Q2–Q3 can be negative. For instance, every distribution is invariant for a Markov chain that does not move. Also, a Markov chain that alternates between the states 0 and 1 with $\pi_0(0) = 1$ is such that $\pi_n(0) = 1$ when n is even and $\pi_n(0) = 0$ when n is odd, so that π_n does not converge.

However, we have the following key result.

Theorem 1.1 (Big Theorem for Finite Markov Chains)

(*a*) *If the Markov chain is finite and irreducible, it has a unique invariant distribution π and $\pi(i)$ is the long-term fraction of time that $X(n)$ is equal to i.*

(*b*) *If the Markov chain is also aperiodic, then the distribution π_n of $X(n)$ converges to π.* ∎

In this theorem, the *long-term fraction of time* that $X(n)$ is equal to i is defined as the limit

$$\lim_{N \to \infty} \frac{1}{N} \sum_{n=0}^{N-1} 1\{X(n) = i\}.$$

In this expression, $1\{X(n) = i\}$ takes the value 1 if $X(n) = i$ and the value 0 otherwise. Thus, in the expression above, the sum is the total time that the Markov chain is in state i during the first N steps. Dividing by N gives the fraction of time. Taking the limit yields the long-term fraction of time.

The theorem says that, if the Markov chain is irreducible, this limit exists and is equal to $\pi(i)$. In particular, this limit does not depend on the particular *realization* of the random variables. This means that every simulation yields the same limit, as you will verify in Problem 1.8.

1.3.3 Long-Term Fraction of Time

Why should the fraction of time that a Markov chain spends in one state converge? In our browsing example, if we count the time that we spend on page A over n time units and we divide that time by n, it turns out that the ratio converges to $\pi(A)$.

This result is similar to the fact that, when we flip a fair coin repeatedly, the fraction of "heads" converges to 50%. Thus, even though the coin has no memory, it makes sure that the fraction of heads approaches 50%. How does it do it?

These convergence results are examples of the Law of Large Numbers. This law is at the core of our intuitive understanding of probability and it captures our notion of statistical regularity. Even though outcomes are uncertain, one can make predictions. Here is a statement of the result. We discuss it in Chap. 2.

Theorem 1.2 (Strong Law of Large Numbers) *Let $\{X(n), n \geq 1\}$ be a sequence of i.i.d. random variables with mean μ. Then*

$$\frac{X(1) + \cdots + X(n)}{n} \rightarrow \mu \text{ as } n \rightarrow \infty, \text{ with probability } 1. \qquad \blacksquare$$

Thus, the *sample mean values* $Y(n) := (X(1) + \cdots + X(n))/n$ converge to the expected value, with probability 1. (See Fig. 1.7.) Note that the sample mean values $Y(n)$ are random variables: for each n, the value of $Y(n)$ depends on the particular realization of the random variables $X(m)$; if you repeat the experiment, the values will probably be different. However, the limit is always μ, with probability 1. We say that the convergence is *almost sure*.[1]

Fig. 1.7 When rolling a balanced die, the sample mean converges to 3.5

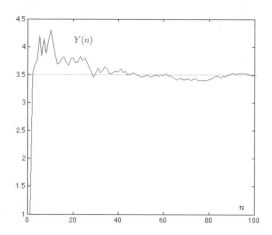

---

[1] "Almost sure" is a somewhat confusing technical expression. It means that, although there are outcomes for which the convergence does not happen, all these outcomes have probability zero. For instance, if you flip a fair coin, the outcome where the coin flips keep on yielding tails is such that the fraction of tails does not converge to 0.5. The same is true for the outcome $H, H, T, H, H, T, \ldots$. So, almost sure means that it happens with probability 1, but not for a set of outcomes that has probability zero.

1.4   Illustrations

We illustrate Theorem 1.1 for the Markov chains in Fig. 1.6. The three situations are
different and quite representative. We explore them one by one.

Figures 1.8, 1.9 and 1.10 correspond to each of the three Markov chains in
Fig. 1.6, as shown on top of each figure. The top graph of each figure shows
the successive values of X_n for $n = 0, 1, \ldots, 100$. The middle graph of the
figure shows, for $n = 0, \ldots, 100$, the fraction of time that X_m is equal to the
different states during $\{0, 1, \ldots, n\}$. The bottom graph of the figure shows, for
$n = 0, \ldots, 100$, the probability that X_n is equal to each of the states.

In Fig. 1.8, the fraction of time that the Markov chain is equal to each of the
states $\{1, 2, 3\}$ converges to positive values. This is the case because the Markov
chain is irreducible. (See Theorem 1.1(a).) However, the probability of being in a
given state does not converge. This is because the Markov chain is periodic. (See
Theorem 1.1(b).)

For the Markov chain in Fig. 1.9, the probabilities converge, because the Markov
chain is aperiodic. (See again Theorem 1.1.)

Finally, for the Markov chain in Fig. 1.10, eventually $X_n = 3$; the fraction of
time in state 3 converges to one and so does the probability of being in state 3. What
happens in this case is that state 3 is *absorbing*: once the Markov chain gets there,
it cannot leave.

Fig. 1.8 Markov chain (a) in
Fig. 1.6

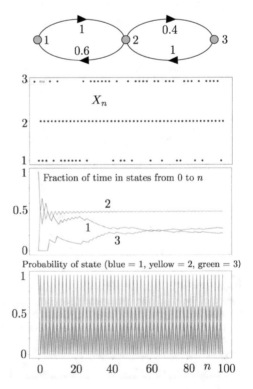

Fig. 1.9 Markov chain (b) in Fig. 1.6

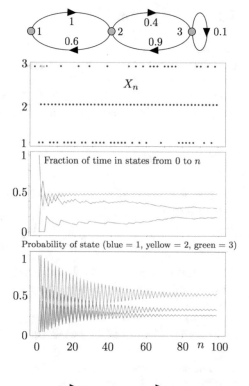

Fig. 1.10 Markov chain (c) in Fig. 1.6

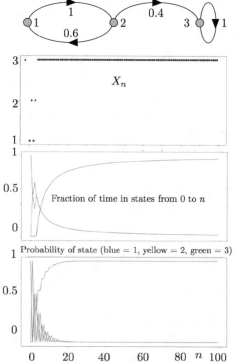

1.5   Hitting Time

Say that you start in page A in Fig. 1.2 and that, at every step, you follow each outgoing link of the page where you are with equal probabilities. How many steps does it take to reach page E? This time is called the *hitting time*, or *first passage time*, of page E and we designate it by T_E. As we can see from the figure, T_E can be as small as 2, but it has a good chance of being much larger than 2 (Fig. 1.11).

1.5.1   Mean Hitting Time

Our goal is to calculate the average value of T_E starting from $X_0 = A$. That is, we want to calculate

$$\beta(A) := E[T_E \mid X_0 = A].$$

The key idea to perform this calculation is to in fact calculate the mean hitting time for all possible initial pages. That is, we will calculate $\beta(i)$ for $i = A, B, C, D, E$ where

$$\beta(i) := E[T_E \mid X_0 = i].$$

The reason for considering these different values is that the mean time to hit E starting from A is clearly related to the mean hitting time starting from B and from D. These in turn are related to the mean hitting time starting from C. We claim that

$$\beta(A) = 1 + \frac{1}{2}\beta(B) + \frac{1}{2}\beta(D). \tag{1.7}$$

To see this, note that, starting from A, after one step, the Markov chain is in state B with probability $1/2$ and it is in state D with probability $1/2$. Thus, after one step, the average time to hit E is the average time starting from B, with probability $1/2$, and it is the average time starting from D, with probability $1/2$.

Fig. 1.11 This is NOT what we mean by *hitting time*!

This situation is similar to the following one. You flip a fair coin. If the outcome is heads you get a random amount of money equal to X and if it is tails you get a random amount Y. On average, you get

$$\frac{1}{2}E(X) + \frac{1}{2}E(Y).$$

Similarly, we can see that

$$\beta(B) = 1 + \beta(C)$$
$$\beta(C) = 1 + \beta(A)$$
$$\beta(D) = 1 + \frac{1}{3}\beta(A) + \frac{1}{3}\beta(B) + \frac{1}{3}\beta(E)$$
$$\beta(E) = 0.$$

These equations, together with (1.7), are called the *first step equations* (FSE). Solving them, we find

$$\beta(A) = 17, \beta(B) = 19, \beta(C) = 18, \beta(D) = 13 \text{ and } \beta(E) = 0.$$

1.5.2 Probability of Hitting a State Before Another

Consider once again the same situation but say that we are interested in the probability that starting from A we visit state C before E. We write this probability as

$$\alpha(A) = P[T_C < T_E \mid X_0 = A].$$

As in the previous case, it turns out that we need to calculate $\alpha(i)$ for $i = A, B, C, D, E$. We claim that

$$\alpha(A) = \frac{1}{2}\alpha(B) + \frac{1}{2}\alpha(D). \tag{1.8}$$

To see this, note that, starting from A, after one step you are in state B with probability $1/2$ and you will then visit C before E with probability $\alpha(B)$. Also, with probability $1/2$, you will be in state D after one step and you will then visit C before E with probability $\alpha(D)$. Thus, the event that you visit C before E starting from A is the union of two disjoint events: either you do that by first going to B or by first going to D. Adding the probabilities of these two events, we get (1.8).

Similarly, one finds that

$$\alpha(B) = \alpha(C)$$

$$\alpha(C) = 1$$

$$\alpha(D) = \frac{1}{3}\alpha(A) + \frac{1}{3}\alpha(B) + \frac{1}{3}\alpha(E)$$

$$\alpha(E) = 0.$$

These equations, together with (1.8), are also called the first step equations. Solving them, we find

$$\alpha(A) = \frac{4}{5}, \alpha(B) = 1, \alpha(C) = 1, \alpha(D) = \frac{3}{5}, \alpha(E) = 0.$$

1.5.3   FSE for Markov Chain

Let us generalize this example to the case of a Markov chain on $\mathcal{X} = \{1, 2, \ldots, N\}$ with transition probability matrix P. Let T_i be the hitting time of state i. For a set $A \subset \mathcal{X}$ of states, let $T_A = \min\{n \geq 0 \mid X(n) \in A\}$ be the hitting time of the set A.

First, we consider the mean value of T_A. Let

$$\beta(i) = E[T_A \mid X_0 = i], i \in \mathcal{X}.$$

The FSE are

$$\beta(i) = \begin{cases} 1 + \sum_j P(i, j)\beta(j), & \text{if } i \notin A \\ 0, & \text{if } i \in A. \end{cases}$$

Second, we study the probability of hitting a set A before a set B, where $A, B \subset \mathcal{X}$ and $A \cap B = \emptyset$. Let

$$\alpha(i) = P[T_A < T_B \mid X_0 = i], i \in \mathcal{X}.$$

The FSE are

$$\alpha(i) = \begin{cases} \sum_j P(i, j)\alpha(j), & \text{if } i \notin A \cup B \\ 1, & \text{if } i \in A \\ 0, & \text{if } i \in B. \end{cases}$$

Third, we explore the value of

$$Y = \sum_{n=0}^{T_A} h(X(n)).$$

That is, you collect an amount $h(i)$ every time you visit state i, until you enter set A. Let

$$\gamma(i) = E[Y \mid X_0 = i], i \in \mathcal{X}.$$

The FSE are

$$\gamma(i) = \begin{cases} h(i) + \sum_j P(i, j)\gamma(j), & \text{if } i \notin A \\ h(i), & \text{if } i \in A. \end{cases} \tag{1.9}$$

Fourth, we consider the value of

$$Z = \sum_{n=0}^{T_A} \beta^n h(X(n)),$$

where β can be thought of as a discount factor. Let

$$\delta(i) = E[Z \mid X_0 = i].$$

The FSE are

$$\delta(i) = \begin{cases} h(i) + \beta \sum_j P(i, j)\delta(j), & \text{if } i \notin A \\ h(i), & \text{if } i \in A. \end{cases}$$

Hopefully these examples give you a sense of the variety of questions that can be answered for finite Markov chains. This is very fortunate, because Markov chains can be used to model a broad range of engineering and natural systems.

1.6 Summary

- Markov Chains: states, transition probabilities, irreducible, aperiodic, invariant distribution, hitting times;
- Strong Law of Large Numbers;
- Big Theorem: irreducible implies unique invariant distribution equal to the long-term fraction of time in the states; convergence to invariant distribution if irreducible and aperiodic;
- Hitting Times: first step equations.

1.6.1   Key Equations and Formulas

Definition of MC	$P[X(n+1) = j \mid X(n) = i, X(m), m < n] = P(i, j)$	(1.3)
P.m.f. of X_n	$\pi_n = \pi_0 P^n$	(1.5)
Balance Equations	$\pi P = \pi$	(1.1)
First Step Equations	$\gamma(i) = h(i) + \sum_j P(i, j)\gamma(j)$	(1.9)

1.7   References

There are many excellent books on Markov chains. Some of my favorites are Grimmett and Stirzaker (2001) and Bertsekas and Tsitsiklis (2008). The original patent on PageRank is Page (2001). The online book Easley and Kleinberg (2012) is an inspiring discussion of social networks. Chapter 14 of that reference discusses PageRank.

1.8   Problems

Problem 1.1 Construct a Markov chain that is not irreducible but whose distribution converges to its unique invariant distribution.

Problem 1.2 Show a Markov chain whose distribution converges to a limit that depends on the initial distribution.

Problem 1.3 Can you find a finite irreducible aperiodic Markov chain whose distribution does not converge?

Problem 1.4 Show a finite irreducible aperiodic Markov chain that converges very slowly to its invariant distribution.

Problem 1.5 Show that a function $Y(n) = g(X(n))$ of a Markov chain $X(n)$ may not be a Markov chain.

Problem 1.6 Construct a Markov chain that is a sequence of i.i.d. random variables. Is it irreducible and aperiodic?

Fig. 1.12 Markov chain for
Problem 1.7

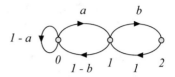

Problem 1.7 Consider the Markov chain $X(n)$ with the state diagram shown in
Fig. 1.12 where $a, b \in (0, 1)$.

(a) Show that this Markov chain is aperiodic;
(b) Calculate $P[X(1) = 1, X(2) = 0, X(3) = 0, X(4) = 1 \mid X(0) = 0]$;
(c) Calculate the invariant distribution;
(d) Let $T_i = \min\{n \geq 0 \mid X(n) = i\}$. Calculate $E[T_2 \mid X(0) = 1]$.

Problem 1.8 Use Python to write a simulator for a Markov chain $\{X(n), n \geq 1\}$
with K states, initial distribution π, and transition probability matrix P. The
program should be able to do the following:

1. Plot $\{X(n), n = 1, \ldots, N\}$;
2. Plot the fraction of time that $X(n)$ is in some chosen states during $\{1, 2, \ldots, m\}$
   as a function of m, for $m = 1, \ldots, N$;
3. Plot the probability that $X(n)$ is equal to some chosen states, for $n = 1, \ldots, N$;
4. Use this program to simulate a periodic Markov chain with five states;
5. Use the program to simulate an aperiodic Markov chain with five states.

Problem 1.9 Use your simulator to simulate the Markov chains of Figs. 1.2 and 1.6.

Problem 1.10 Find the invariant distribution for the Markov chains of Fig. 1.6.

Problem 1.11 Calculate $d(1), d(2)$, and $d(3)$, defined in (1.6), for the Markov
chains of Fig. 1.6.

Problem 1.12 Calculate $d(A)$, defined in (1.6), for the Markov chain of Fig. 1.2.

Problem 1.13 Let $\{X_n, n \geq 0\}$ be a finite Markov chain. Assume that it has
a unique invariant distribution π and that π_n converges to π for every initial
distribution π_0. Then (choose the correct answers, if any)

- X_n is irreducible;
- X_n is periodic;
- X_n is aperiodic;
- X_n might not be irreducible.

Problem 1.14 Consider the Markov chain $\{X_n, n \geq 0\}$ on $\{0, 1\}$ with $P(0, 1) = 0.1$ and $P(1, 0) = 0.3$. Then (choose the correct answers, if any)

- The invariant distribution of the Markov chain is $[0.75, 0.25]$;
- Let $T_1 = \min\{n \geq 0 | X_n = 1\}$. Then $E[T_1 | X_0 = 0] = 1.2$;
- $E[X_1 + X_2 | X_0 = 0] = 0.8$.

Problem 1.15 Consider the MC with the state transition diagram shown in Fig. 1.13.

(a) What is the period of this MC? Explain.
(b) Find all the invariant distributions for this MC.
(c) Does π_n, the distribution of X_n, converge as $n \to \infty$? Explain.
(d) Do the fractions of time the MC spends in the states converge? If so, what is the limit?

Problem 1.16 Consider the MC with the state transition diagram shown in Fig. 1.14.

(a) Find all the invariant distributions of this MC.
(b) Assume $\pi_0(3) = 1$. Find $\lim_n \pi_n$.

Problem 1.17 Consider the MC with the state transition diagram shown in Fig. 1.15.

(a) Find all the invariant distributions of this MC.
(b) Does π_n converge as $n \to \infty$? If it does, prove it.
(c) Do the fractions of time the MC spends in the states converge? Prove it.

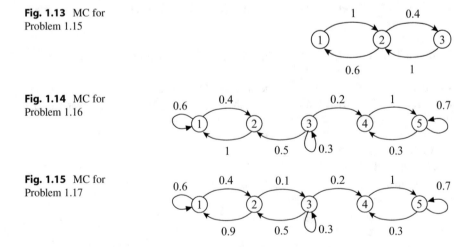

Fig. 1.13 MC for Problem 1.15

Fig. 1.14 MC for Problem 1.16

Fig. 1.15 MC for Problem 1.17

Fig. 1.16 MC for
Problem 1.18

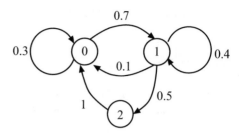

Fig. 1.17 MC for
Problem 1.19

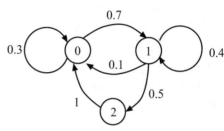

Problem 1.18 Consider the MC shown in Fig. 1.16.

(a) Find the invariant distribution π of this Markov chain.
(b) Calculate the expected time from 0 to 2.
(c) Use Python to plot the probability that, starting from 0, the MC has not reached 2 after n steps.
(d) Use Python to simulate the MC and plot the fraction of time that it spends in the different states after n steps.
(e) Use Python to plot π_n.

Problem 1.19 For the Markov chain $\{X_n, n \geq 0\}$ with transition diagram shown in Fig. 1.17, assume that $X_0 = 0$. Find the probability that X_n hits 2 before it hits 1 twice.

Problem 1.20 Draw an irreducible aperiodic MC with six states and choose the transition probabilities. Simulate the MC in Python. Plot the fraction of time in the six states. Assume you start in state 1. Plot the probability of being in each of the six states.

Problem 1.21 Repeat Problem 1.20, but with a periodic MC.

Problem 1.22 How would you trick the PageRank algorithm into believing that your home page should be given a high rank?

Hint Try adding another page with suitable links.

Problem 1.23 Show that the holding time of a state is geometrically distributed.

Problem 1.24 You roll a die until the sum of the last two rolls is exactly 10. How many times do you have to roll, on average?

Problem 1.25 You roll a die until the sum of the last three rolls is at least 15. How many times do you have to roll, on average?

Problem 1.26 A doubly stochastic matrix is a nonnegative matrix whose rows and columns add up to one. Show that the invariant distribution is uniform for such a transition matrix.

Problem 1.27 Assume that the Markov chain (c) of Fig. 1.6 starts in state 1. Calculate the average number of times it visits state 1 before being absorbed in state 3.

Problem 1.28 A man tries to go up a ladder that has N rungs. Every step he makes, he has a probability p of dropping back to the ground and he goes up one rung otherwise. Use the first step equations to calculate analytically the average time he takes to reach the top, for $N = 1, \ldots, 20$ and $p = 0.05, 0.1$, and 0.2. Use Python to plot the corresponding graphs.

Problem 1.29 Let $\{X_n, n \geq 0\}$ be a finite irreducible Markov chain with transition probability matrix P and invariant distribution π. Show that, for all i, j,

$$\frac{1}{N} \sum_{n=0}^{N-1} 1\{X_n = i, X_{n+1} = j\} \to \pi(i)P(i, j), \text{ w.p. 1 as } N \to \infty.$$

Problem 1.30 Show that a Markov chain $\{X_n, n \geq 0\}$ can be written as

$$X_{n+1} = f(X_n, V_n), n \geq 0,$$

where the V_n are i.i.d. random variables independent of X_0.

Problem 1.31 Let P and $\tilde{P}$ be two stochastic matrices and π a pmf on the finite set $\mathcal{X}$. Assume that

$$\pi(i)P(i, j) = \pi(j)\tilde{P}(j, i), \forall i, j \in \mathcal{X}.$$

Show that π is invariant for P.

Problem 1.32 Let X_n be a Markov chain on a finite set $\mathcal{X}$. Assume that the transition diagram of the Markov chain is a tree, as shown in Fig. 1.18. Show that if

Fig. 1.18 A transition
diagram that is a tree

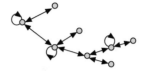

π is invariant and if P is the transition matrix, then it satisfies the following *detailed balance equations*:

$$\pi(i)P(i, j) = \pi(j)P(j, i), \forall i, j.$$

Problem 1.33 Let X_n be a Markov chain such that X_0 has the invariant distribution π and the detailed balance equations are satisfied. Show that

$$P(X_0=x_0, X_1=x_1, \ldots, X_n=x_n)=P(X_N = x_0, X_{N-1} = x_1, \ldots, X_{N-n} = x_n)$$

for all n, all $N \geq n$, and all $x_0, \ldots, x_n$. Thus, the evolution of the Markov chain in *reverse time* $(N, N-1, N-2, \ldots, N-n)$ cannot be distinguished from its evolution in forward time $(0, 1, \ldots, n)$. One says that the Markov chain is *time-reversible*.

Problem 1.34 Let $\{X_n, n \geq 0\}$ be a Markov chain on $\{-1, 1\}$ with $P(-1, 1) = P(1, -1) = a$ for a given $a \in (0, 1)$. Define

$$Y_n = X_0 + \cdots + X_n, n \geq 0.$$

(a) Is $\{Y_n, n \geq 0\}$ a Markov chain? Prove or disprove.
(b) How would you calculate

$$E[\tau|Y_0 = 1] \text{ where } \tau = \min\{n > 0 \mid Y_n = -5 \text{ or } Y_n = 30\}?$$

Problem 1.35 You flip a fair coin repeatedly, forever. Show that the probability that the number of heads is always ahead of the number of tails is zero.

PageRank: B

2

Topics: Sample Space, Trajectories; Laws of Large Numbers: WLLN, SLLN; Proof of Big Theorem.

Background:

- Borel–Cantelli (B.1.2);
- monotonicity of expectation (B.2);
- convergence of expectation (B.8)–(B.9);
- properties of variance: (B.3) and Theorem B.4.

2.1 Sample Space

Let us connect the definition of $\mathbf{X} = \{X_n, n \geq 0\}$ of a Markov chain with the general framework of Sect. B.1. (We write X_n or $X(n)$.) In that section, we explained that a random experiment is described by a sample space. The elements of the sample space are the possible outcomes of the experiment. A probability is defined on subsets, called events, of that sample space. Random variables are real-valued functions of the outcome of the experiment.

To clarify these concepts, consider the case where the X_n are i.i.d. Bernoulli random variables with $P(X_n = 1) = P(X_n = 0) = 0.5$. These random variables describe flips of a fair coin. The random experiment is to flip the coin repeatedly, forever. Thus, one possible outcome of this experiment is an infinite sequence of

© The Author(s) 2021

J. Walrand, *Probability in Electrical Engineering and Computer Science*,

https://doi.org/10.1007/978-3-030-49995-2_2

0's and 1's. Note that an outcome is *not* 0 or 1: it is an infinite sequence since the outcome specifies what happens when we flip the coin forever. Thus, the set Ω of outcomes is the set $\{0, 1\}^\infty$ of infinite sequences of 0's and 1's. If ω is one such sequence, we have $\omega = (\omega_0, \omega_1, \ldots)$ where $\omega_n \in \{0, 1\}$. It is then natural to define $X_n(\omega) = \omega_n$, which simply says that X_n is the outcome of flip n, for $n \geq 0$. Hence $X_n(\omega) \in \Re$ for all $\omega \in \Omega$ and we see that each X_n is a real-valued function defined on Ω. For instance, $X_0(1101001\ldots) = 1$ since $\omega_0 = 1$ when $\omega = 1101001\ldots$. Similarly, $X_1(1101001\ldots) = 1$ and $X_2(1101001\ldots) = 0$. To specify the random experiment, it remains to define the probability on Ω. The simplest way is to say that

$$P(\{\omega | \omega_0 = a, \omega_1 = b, \ldots, \omega_n = z\})$$

$$= P(X_0 = a, \ldots, X_n = z) = 1/2^{n+1}$$

for all $n \geq 0$ and $a, b, \ldots, z \in \{0, 1\}$. For instance,

$$P(\{\omega | \omega_0 = 1\}) = P(X_0 = 1) = 1/2.$$

Similarly,

$$P(\{\omega | \omega_0 = 1, \omega_1 = 0\}) = P(X_0 = 1, X_1 = 0) = 1/4.$$

Observe that we define the probability of a set of outcomes, or event, $\{\omega | \omega_0 = a, \omega_1 = b, \ldots, \omega_n = z\}$ instead of specifying the probability of each outcome ω. The reason is that the probability that we observe a specific infinite sequence of 0's and 1's is zero. That is, $P(\{\omega\}) = 0$ for all $\omega \in \Omega$. Such a description does not tell us much about the coin flips! For instance, it does not specify the bias of the coin, or the fact that successive flips are independent. Hence, the correct way to proceed is to specify the probability of *events*, that are sets of outcomes, instead of the probability of individual outcomes.

For a Markov chain, there is some sample space Ω and each X_n is a function $X_n(\omega)$ of the outcome ω that takes values in $\mathscr{X}$. A probability is defined on subsets of Ω.

In this example, one can choose Ω to be the set of possible infinite sequences of symbols in $\mathscr{X}$. That is, $\Omega = \mathscr{X}^\infty$ and an element $\omega \in \Omega$ is $\omega = (\omega_0, \omega_1, \ldots)$ with $\omega_n \in \mathscr{X}$ for $n \geq 0$. With this choice, one has $X_n(\omega) = \omega_n$ for $n \geq 0$ and $\omega \in \Omega$, as shown in Fig. 2.1. This choice of Ω, similar to what we did for the coin flips, is called the *canonical sample space*. Thus, an outcome is the actual sequence of values of the Markov chain, called the *trajectory*, or *realization* of the Markov chain. It remains to specify the probability of event in Ω. The trick here is that the probability that the Markov chain follows a specific infinite sequence is 0, similarly to the probability that coin flips follow a specific infinite sequence such as all heads. Thus, one should specify the probability of subsets of Ω, not of individual outcomes. One specifies that

Fig. 2.1 In the *canonical* sample space, the outcome ω is the *trajectory* of the Markov chain

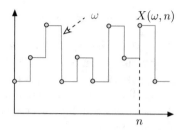

$$P(X_0 = i_0, X_1 = i_1, \ldots, X_n = i_n)$$
$$= \pi_0(i_0) P(i_0, i_1) \times \cdots \times P(i_{n-1}, i_n), \tag{2.1}$$

for all $n \geq 0$ and $i_0, i_1, \ldots, i_n$ in $\mathcal{X}$. Here, $\pi_0(i_0)$ is the probability that the Markov chain starts in state i_0.

This identity is equivalent to (1.3). Indeed, if we let

$$A_n = \{X_0 = i_0, X_1 = i_1, \ldots, X_n = i_n\}$$

and

$$A_{n-1} = \{X_0 = i_0, X_1 = i_1, \ldots, X_{n-1} = i_{n-1}\},$$

then

$$P(A_n) = P[A_n|A_{n-1}]P(A_{n-1}) = P(A_{n-1})P(i_{n-1}, i_n),$$

by (1.3), so that (2.1) holds by induction on n.

Thus, one has defined the probability of events characterized by the first $n + 1$ values of the Markov chain. It turns out that there is one probability on Ω that is consistent with these values.

2.2   Laws of Large Numbers for Coin Flips

Before we discuss the case of Markov chains, let us consider the simpler example of coin flips. Let then $\{X_n, n \geq 0\}$ be i.i.d. Bernoulli random variables with $P(X_n = 0) = P(X_n = 1) = 0.5$, as in the previous section. We think of $X_n = 1$ if flip n yields heads and $X_n = 0$ if it yields tails. We want to show that, as we keep flipping the coin, the fraction of heads approaches 50%. There are two statements that make this idea precise.

2.2.1   Convergence in Probability

The first statement, called the *Weak Law of Large Numbers* (WLLN), says that it is very unlikely that the fraction of heads in n coin flips differs from 50% by even a small amount, say 1%, if n is large. For instance, let $n = 10^5$. We want to show that the likelihood that the fraction of heads among 10^5 flips is more than 51% or less than 49% is small. Moreover, this likelihood can be made as small as we wish if we flip the coin more times.

To show this, let

$$Y_n = \frac{X_0 + \cdots + X_{n-1}}{n}$$

be the fraction of heads in the first n flips. We claim that

$$P(|Y_n - E(Y_n)| \geq \epsilon) \leq \frac{\text{var}(Y_n)}{\epsilon^2}. \tag{2.2}$$

This result is called *Chebyshev's inequality* (Fig. 2.2).

To see (2.2), observe that[1]

$$1\{|Y_n - E(Y_n)| \geq \epsilon\} \leq \frac{(Y_n - E(Y_n))^2}{\epsilon^2}. \tag{2.3}$$

Indeed, if $|Y_n - E(Y_n)| \geq \epsilon$, then $(Y_n - E(Y_n))^2 \geq \epsilon^2$, so that if the left-hand side of inequality (2.3) is one, the right-hand side is at least equal to one. Also, if the left-hand side is zero, it is less than or equal to the right-hand side. Thus, (2.3) holds and (2.2) follows by taking the expected values in (2.3), since $E(1_A) = P(A)$ and $E((Y_n - E(Y_n))^2) = \text{var}(Y_n)$ and since expectation is monotone (B.2).

Fig. 2.2 Pafnuty Chebyshev.
1821–1884

---

[1]By definition, $1\{C\}$ takes the value 1 if the condition C holds and the value 0 otherwise.

Now, $E(Y_n) = 0.5$ and

$$\text{var}(Y_n) = \frac{\text{var}(X_0 + \cdots + X_{n-1})}{n^2} = \frac{n\,\text{var}(X_0)}{n^2}.$$

To see this, recall that if one multiplies a random variable by a, its variance is multiplied by a^2 (see (B.3)). Also, the variance of a sum of independent random variables is the sum of their variances (see Theorem B.4). Hence,

$$P(|Y_n - 0.5| \geq \epsilon) \leq \frac{\text{var}(X_0)}{n\epsilon^2}.$$

Since $X_0 =_D B(0.5)$, we find that

$$\text{var}(X_0) = E\left(X_0^2\right) - (E(X_0))^2$$
$$= E(X_0) - (E(X_0))^2 = 0.5 - 0.25 = 0.25.$$

Thus,

$$P(|Y_n - 0.5| \geq \epsilon) \leq \frac{1}{4n\epsilon^2}.$$

In particular, if we choose $\epsilon = 1\% = 0.01$, we find

$$P(|Y_n - 0.5| \geq 1\%) \leq \frac{2,500}{n} = 0.025 \text{ with } n = 10^5.$$

More generally, we have shown that

$$P(|Y_n - 0.5| \geq \epsilon) \to 0 \text{ as } n \to \infty, \forall \epsilon > 0.$$

This is the WLLN.

2.2.2 Almost Sure Convergence

The second statement is the *Strong Law of Large Numbers* (SLLN). It says that, for all the sequences of coin flips we will ever observe, the fraction Y_n actually converges to 50% as we keep on flipping the coin.

There are many sequences of coin flips for which the fraction of heads does not approach 50%. For instance, the sequence that yields heads for every flip is such that $Y_n = 1$ for all n and thus Y_n does not converge to 50%. Similarly, the sequence 001001001001001 ... is such that Y_n approaches $1/3$ and not 50%. What the SLLN implies is that all those sequences such that Y_n does not converge to 50% have probability 0: they will never be observed.

Thus, this statement is very deep because there are so many sequences to rule out. Keeping track of all of them seems rather formidable. Indeed, the proof of this statement is quite clever. Here is how it proceeds. Note that

$$P(|Y_n - 0.5| \geq \epsilon) \leq E\left(\frac{|Y_n - 0.5|^4}{\epsilon^4}\right), \forall n, \epsilon > 0.$$

Indeed,

$$1\{|Y_n - 0.5| \geq \epsilon\} \leq \frac{|Y_n - 0.5|^4}{\epsilon^4}$$

and the previous inequality follows by taking expectations. Now,

$$E\left(|Y_n - 0.5|^4\right) = E\left(\frac{((X_0 - 0.5) + \cdots + (X_{n-1} - 0.5))^4}{n^4}\right).$$

Also, with $Z_m = X_m - 0.5$, one has

$$E((X_0 - 0.5) + \cdots + (X_{n-1} - 0.5))^4) = E\left(\left(\sum_{m=0}^{n-1} Z_m\right)^4\right)$$

$$= E\left(\sum_{a,b,c,d} Z_a Z_b Z_c Z_d\right),$$

where the sum is over all $a, b, c, d \in \{0, 1, \ldots, n - 1\}$. This sum consists of n terms Z_a^4, $n(n - 1)$ terms $Z_a^2 Z_b^2$ with $a \neq b$ and other terms where at least a factor Z_a is not repeated. The latter terms have zero-mean since $E(Z_a Z_b Z_c Z_d) = E(Z_a)E(Z_b Z_c Z_d) = 0$, by independence, whenever b, c, and d are all different from a. Consequently,

$$E\left(\sum_{a,b,c,d} Z_a Z_b Z_c Z_d\right) = nE\left(Z_0^4\right) + n(n - 1)E\left(Z_0^2 Z_1^2\right) = n\alpha + n(n - 1)\beta$$

with $\alpha = E(Z_0^4)$ and $\beta = E(Z_0^2 Z_1^2)$. Hence, substituting the result of this calculation in the previous expressions, we find that

$$P(|Y_n - 0.5| \geq \epsilon) \leq \frac{n\alpha + n(n - 1)\beta}{n^4 \epsilon^4} \leq \frac{n^2(\alpha + \beta)}{n^4 \epsilon^4} = \frac{\alpha + \beta}{n^2 \epsilon^4}.$$

This inequality implies that[2]

$$\sum_{n\geq 1} P(|Y_n - 0.5| \geq \epsilon) < \infty.$$

This expression shows that the events $A_n := \{|Y_n - 0.5| \geq \epsilon\}$ have probabilities that add up to a finite number. From the Borel–Cantelli Theorem B.1, we conclude that

$$P(A_n, \text{ i.o.}) = 0.$$

This result says that, with probability one, ω belongs only to finitely many A_n's. Hence,[3] with probability one, there is some $n(\omega)$ so that $\omega \notin A_n$ for $n \geq n(\omega)$. That is,

$$|Y_n(\omega) - 0.5| \leq \epsilon, \forall n \geq n(\omega).$$

Since this property holds for an arbitrary $\epsilon > 0$, we conclude that, with probability one,

$$Y_n(\omega) \to 0.5 \text{ as } n \to \infty.$$

Indeed, if $Y_n(\omega)$ does not converge to 50%, there must be some $\epsilon > 0$ so that $|Y_n - 0.5| > \epsilon$ for infinitely many n's and we have seen that this is not the case.

2.3   Laws of Large Numbers for i.i.d. RVs

The results that we proved for coin flips extend to i.i.d. random variables $\{X_n, n \geq 0\}$ to show that

$$Y_n := \frac{X_0 + \cdots + X_{n-1}}{n}$$

approaches $E(X_0)$ as $n \to \infty$. As for coin flips, there are two ways of making that statement precise.

---

[2]Recall that

$$\sum_n \frac{1}{n^2} < \infty.$$

[3]Let $n(\omega) - 1$ be the largest n such that $\omega \in A_n$.

2.3.1   Weak Law of Large Numbers

We need a definition.

Definition 2.1 (Convergence in Probability)   Let $X_n, n \geq 0$ and X be random variables defined on a common probability space. One says that X_n converges in probability to X, and one writes $X_n \xrightarrow{p} X$ if, for all $\epsilon > 0$,

$$P(|X_n - X| \geq \epsilon) \to 0 \text{ as } n \to \infty.$$

$\diamond$

The Weak Law of Large Numbers (WLLN) is the following result.

Theorem 2.1 (Weak Law of Large Numbers)   *Let $\{X_n, n \geq 0\}$ be a sequence of i.i.d. random variables with mean μ. Then*

$$Y_n = \frac{X_0 + \cdots + X_{n-1}}{n} \xrightarrow{p} \mu. \tag{2.4}$$

∎

Proof   Assume that $E(X_n^2) < \infty$. The proof is then the same as for coin flips and is left as an exercise. For the general case, see Theorem 15.14. $\square$

The first result of this type was proved by Jacob Bernoulli (Fig. 2.3).

2.3.2   Strong Law of Large Numbers

We again need a definition.

Definition 2.2 (Almost Sure Convergence)   Let $X_n, n \geq 0$ and X be random variables defined on a common probability space. One says that X_n converges almost surely to X as $n \to \infty$, and one writes $X_n \to X$, a.s. if

Fig. 2.3 Jacob Bernoulli.
1655–1705

Fig. 2.4 When rolling a balanced die, the sample mean converges to 3.5

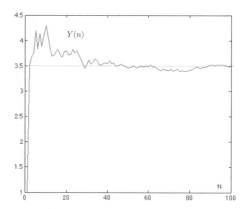

$$P\left(\lim_{n\to\infty} X_n(\omega) = X(\omega)\right) = 1.$$

◇

Thus, this convergence means that the sequence of real numbers $X_n(\omega)$ converges to the real number $X(\omega)$ as $n \to \infty$, with probability one.

Let $\{X_n, n \geq 0\}$ be as in the statement of Theorem 2.1. We have the following result.[4]

Theorem 2.2 (Strong Law of Large Numbers)   *Let $\{X_n, n \geq 0\}$ be a sequence of i.i.d. random variables with mean μ. Then*

$$\frac{X_0 + \cdots + X_{n-1}}{n} \to \mu \text{ as } n \to \infty, \text{ with probability } 1.$$

■

Thus, the *sample mean values* $Y_n := (X_0 + \cdots + X_{n-1})/n$ converge to the expected value, with probability 1. (See Fig. 2.4.)

Proof Assume that

$$E\left(X_n^4\right) < \infty.$$

The proof is then the same as for coin flips and is left as an exercise. The proof of the SLLN in the general case is given in Theorem 15.14.   □

Figure 2.5 illustrates the SLLN and WLLN. The SLLN states that the sample means of i.i.d. random variables converge to the mean, with probability one. The

---

[4]Almost sure convergence implies convergence in probability, so SLLN is stronger than WLLN. See Problem 2.5.

Fig. 2.5 SLLN and WLLN
for i.i.d. $U[0, 1]$ random
variables

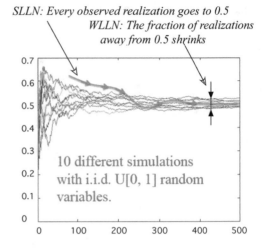

WLLN says that as the number of samples increases, the fraction of realizations where the sample mean differs from the mean by some amount gets small.

2.4   Law of Large Numbers for Markov Chains

The long-term fraction of time that a finite irreducible Markov chain spends in a given state is the invariant probability of that state. For instance, a Markov chain $X(n)$ on $\{0, 1\}$ with $P(0, 1) = a = P(1, 0)$ with $a \in (0, 1]$ spends half of the time in state 0, in the long term. The Markov chain in Fig. 1.2 spends a fraction $12/39$ of the time in state A, in the long term.

To understand this property, one should look at the returns to state i, as shown in Fig. 2.6. The figure shows a particular sequence of values of $X(n)$ and it decomposes this sequence into cycles between successive returns to a given state i. A new cycle starts when the Markov chain comes back to i. The durations of these successive cycles, $T_1, T_2, T_3, \ldots$, are independent and identically distributed, because the Markov chains start afresh from state i at each time T_n, independently of the previous states. This is a consequence of the Markov property for any given value k of T_n and of the fact that the distribution of the evolution starting from state i at time k does not depend on k.

It is easy to see that these random times have a finite mean. Indeed, fix one state i. Then, starting from any given state j, there is some minimum number M_j of steps required to go to state i. Also, there is some probability p_j that the Markov chain will go from j to i in M_j steps. Let then $M = \max_j M_j$ and $p = \min_j p_j$. We can then argue that, starting from any state at time 0, there is at least a probability p that the Markov chain visits state i after at most M steps. If it does not, we repeat the argument starting at time M. We conclude that $T_i \leq M\tau$ where τ is a geometric

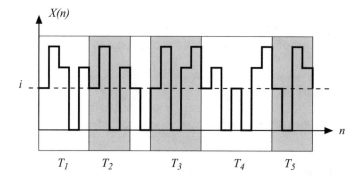

Fig. 2.6 The cycles between returns to state i are i.i.d. The law of large numbers explains the convergence of the long-term fraction of time to a constant

random with parameter p. Hence $E(T_i) \leq ME(\tau) = M/p < \infty$, as claimed. Note also that $E(T_i^4) \leq M^4 E(\tau^4) < \infty$.

The Strong Law of Large Numbers states that

$$\frac{T_1 + T_2 + \cdots + T_k}{k} \to E(T_1), \text{ as } k \to \infty, \text{ with probability 1.} \qquad (2.5)$$

Thus, the long-term fraction of time that the Markov chain spends in state i is given by

$$\lim_{k \to \infty} \frac{k}{T_1 + T_2 + \cdots + T_k} = \frac{1}{E(T_1)}, \text{ with probability 1.} \qquad (2.6)$$

Let us clarify why (2.6) implies that the fraction of time in state i converges to $1/E(T_1)$. Let $A(n)$ be the number of visits to state i by time n. We want to show that $A(n)/n$ converges to $1/E(T_1)$. Then,

$$\frac{k}{T_1 + \cdots + T_{k+1}} < \frac{A(n)}{n} = \frac{k}{n} \leq \frac{k}{T_1 + \cdots + T_k}$$

whenever $T_1 + \cdots + T_k \leq n < T_1 + \cdots + T_{k+1}$. If we believe that $T_{k+1}/k \to 0$ as $k \to \infty$, the inequality above shows that

$$\frac{A(n)}{n} \to \frac{1}{E(T_1)},$$

as claimed. To see why T_{k+1}/k goes to zero, note that

$$P\left(\frac{T_{k+1}}{k} > \epsilon\right) \leq P\left(\frac{M\tau}{k} > \epsilon\right) \leq P(\tau > \alpha k) \leq (1-p)^{\alpha k}$$

with $\alpha = \epsilon/M$.

Thus, by Borel–Cantelli Theorem B.1, the event $T_{k+1}/k > \epsilon$ occurs only for finitely many values of k, which proves the convergence to zero.

2.5   Proof of Big Theorem

This section presents the proof of the main result about Markov chains.

2.5.1   Proof of Theorem 1.1 (a)

Let m_j be the expected return time to state j. That is,

$$m_j = E[T_j | X(0) = j] \text{ with } T_j = \min\{n > 0 | X(n) = j\}.$$

We show that $\pi(j) = 1/m_j$, $j = 1, \ldots, N$ is the unique invariant distribution if the Markov chain is irreducible.

During $n = 1, \ldots, N$ where $N \gg 1$, the Markov chain visits state j a fraction $1/m_j$ of the times. A fraction $P(j, i)$ of those times, it visits state i just after visiting state j. Thus, a fraction $(1/m_j)P(j, i)$ of the times, the Markov chain visits j then i in successive steps. By summing over j, we find the fraction of the times that the Markov chain visits i. Thus,

$$\sum_j \frac{1}{m_j} P(j, i) = \frac{1}{m_i}.$$

Hence, there is one invariant distribution π and it is given by $\pi_i = 1/m_i$, which is the fraction of time that the Markov chain spends in state i.

To show that the invariant distribution is unique, assume that there is another one, say $\phi(i)$. Start the Markov chain with that distribution. Then

$$\frac{1}{N} \sum_{n=0}^{N-1} 1\{X(n) = i\} \to \pi(i).$$

However, taking expectation, we find that the left-hand side is equal to $\phi(i)$. Thus, $\phi = \pi$ and the invariant distribution is unique.[5]

---

[5]Indeed,

$$E(1\{X(n) = i\}) = P(X(n) = i) = \phi(i).$$

Fig. 2.7 An aperiodic
Markov chain

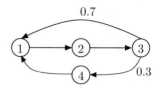

2.5.2 Proof of Theorem 1.1 (b)

If the Markov chain is irreducible but not aperiodic, then π_n may not converge to
the invariant distribution π. For instance, if the Markov chain alternates between 0
and 1 and starts from 0, then $\pi_n = [1, 0]$ for n even and $\pi_n = [0, 1]$ for n odd, so
that π_n does not converge to $\pi = [0.5, 0.5]$.

If the Markov chain is aperiodic, $\pi_n \rightarrow \pi$. Moreover, the convergence is
geometric. We first illustrate the argument on a simple example shown in Fig. 2.7.
Consider the number of steps to go from 1 to 1. Note that

$$\{n > 0 | P^n(1, 1) > 0\} = \{3, 4, 6, 7, 8, 9, 10, \ldots\}.$$

Thus, $P^n(1, 1) > 0$ if $n \geq 6$. Now, $P[X(2) = 1 | X(0) = 2] > 0$, so that
$P[X(n) = 1 | X(0) = 2] > 0$ for $n \geq 8$. Indeed, if $n \geq 8$, then X can go from 2 to
1 in two steps and then from 1 to 1 in $n - 2$ steps. The argument is similar for the
other states and we find that there is some $M > 0$ and some $p > 0$ such that

$$P[X(M) = 1 | X(0) = i] \geq p, i = 1, 2, 3, 4.$$

Now, consider two copies of the Markov chain: $\{X(n), n \geq 0\}$ and $\{Y(n), n \geq 0\}$.
One chooses $X(0)$ with distribution π_0 and $Y(0)$ with the invariant distribution π.
The two Markov chains evolve independently initially. We define

$$\tau = \min\{n > 0 | X(n) = Y(n)\}.$$

In view of the observation above,

$$P(X(M) = 1 \text{ and } Y(M) = 1) \geq p^2.$$

Thus, $P(\tau > M) \leq 1 - p^2$. If $\tau > M$, then the two Markov chains have not met yet
by time M. Using the same argument as before, we see that they have a probability
at least p^2 of meeting in the next M steps. Thus,

$$P(\tau > kM) \leq \left(1 - p^2\right)^k.$$

Now, modify $X(n)$ by gluing it to $Y(n)$ after time τ. This *coupling* operation does
not change the fact that $X(n)$ still evolves according to the transition matrix P, so
that $P(X(n) = i) = \pi_n(i)$ where $\pi_n = \pi_0 P^n$.

Now,

$$\sum_i |P(X(n) = i) - P(Y(n) = i)| \le 2P(X(n) \ne Y(n)) \le 2P(\tau > n).$$

Hence,

$$\sum_i |\pi_n(i) - \pi(i)| \le 2P(\tau > n),$$

and this implies that

$$\sum_i |\pi_n(i) - \pi(i)| \le 2(1 - p^2)^k \text{ if } n > kM.$$

To extend this argument to a general aperiodic Markov chain, we need the fact that for each state i there is some integer n_i such that $P^n(i, i) > 0$ for all $n \ge n_i$. We prove that fact as Lemma 2.3 in the following section.

2.5.3   Periodicity

We start with a property of the set of return times of an irreducible Markov chain.

Lemma 2.1 *Fix a state i and let $S := \{n > 0 | P^n(i, i) > 0\}$ and $d = g.c.d.(S)$. There must be two integers n and $n + d$ in the set S.*

Proof The trick is clever. We first illustrate it on an example. Assume $S = \{9, 15, 21, \ldots\}$ with $d = g.c.d.(S) = 3$. There must be $a, b \in S$ with $g.c.d.\{a, b\} = 3$. Otherwise, the gcd of S would not be 3. Here, we can choose $a = 15$ and $b = 21$. Now, consider the following operations:

$$(a, b) = (15, 21) \to (6, 15) \to (6, 9) \to (3, 6) \to (3, 3).$$

At each step, we go from (x, y) with $x \le y$ to the ordered pair of $\{x, y - x\}$. Note that at each step, each term in the pair (x, y) is an integer linear combination of a and b. For instance, $(6, 15) = (b - a, a)$. Then, $(6, 9) = (b - a, a - (b - a)) = (b - a, 2a - b)$, and so on. Eventually, we must get to $(3, 3)$. Indeed, the terms are always decreasing until we get to zero. Assume we get to (x, x) with $x \ne 3$. At the previous step, we had $(x, 2x)$. The step before must have been $(x, 3x)$, and so on. Going back all the way to (a, b), we see that a and b are both multiples of x. But then, $g.c.d.\{a, b\} = x$, a contradiction.

From this construction, since at each step the terms are integer linear combinations of a and b, we see that

$$3 = ma + nb$$

for some integers m and n. Thus,

$$3 = m^+a + n^+b - m^-a - n^-b,$$

where $m^+ = \max\{m, 0\}$ and $m = m^+ - m^-$, and similarly for n^+ and n^-. Now we can choose

$$N = m^-a + n^-b \text{ and } N + 3 = m^+a + n^+b.$$

The last step of the argument is to notice that if $a, b \in S$, then $\alpha a + \beta b \in S$ for any integers α and β that are not both zero. This fact follows from the definition of S as the return times from i to i. Hence, both N and $N + 3$ are in S.

The proof for a general set S with gcd equal to d is identical.   □

This result enables us to show that the period of a Markov chain is well-defined.

Lemma 2.2 *For an irreducible Markov chain, $d(i)$ defined in (1.6) has the same value for all states.*

Proof Pick $j \neq i$. We show that $d(j) \leq d(i)$. This suffices to prove the lemma, since by symmetry one also has $d(i) \leq d(j)$.

By irreducibility, $P^m(j, i) > 0$ for some m and $P^n(i, j) > 0$ for some n. Now, by definition of $d(i)$ and by the previous lemma, there is some integer N such that $P^N(i, i) > 0$ and $P^{N+d(i)}(i, i) > 0$. But then,

$$P^{m+N+n}(j, j) > 0 \text{ and } P^{m+N+d(i)+n}(j, j) > 0.$$

This implies that the integers $K := n + N + m$ and $K + d(i)$ are both in $S := \{n > 0 | P^n(j, j) > 0\}$. Clearly, this shows that

$$d(j) := g.c.d.(S) \leq d(i).$$
                                                                                        □

The following fact then suffices for our proof of convergence, as we explained in the example.

Lemma 2.3 *Let X be an irreducible aperiodic Markov chain. Let $S = \{n > 0 | P^n(i, i) > 0\}$. Then, there is some n_i such that $n \in S$, for all $n \geq n_i$.*

Proof We know from Lemma 2.1 that there is some integer N such that $N, N+1 \in S$. We claim that

$$n \in S, \forall n > N^2.$$

To see this, first note that for $m > N - 1$ one has

$$mN + 0 = mN,$$
$$mN + 1 = (m - 1)N + (N + 1),$$
$$mN + 2 = (m - 2)N + 2(N + 1),$$
$$\cdots,$$
$$mN + N - 1 = (m - N + 1)N + (N - 1)(N + 1).$$

Now, for $n > N^2$ one can write

$$n = mN + k$$

for some $k \in \{0, 1, \ldots, N - 1\}$ and $m > N - 1$. Thus, n is an integer linear combination of N and $N + 1$ that are both in S, so that $n \in S$. □

2.6　Summary

- Sample Space;
- Laws of Large Numbers: SLLN and WLLN;
- WLLN from Chebyshev's Inequality;
- SLLN from Borel–Cantelli and fourth moment bound;
- SLLN for Markov chains using the i.i.d. return times to a state;
- Proof of Big Theorem.

2.6.1　Key Equations and Formulas

SLLN	$(X_1 + \cdots + X_n)/n \to E(X_1)$, w.p. 1	T.2.2
Chebyshev	$P(|(X_1 + \cdots + X_n)/n - \mu| \geq \epsilon) \leq \text{var}(X_1)/\epsilon^2$	(2.2)
Convergence in Prob.	$P(|X_n - X| \geq \epsilon) \to 0$	D.2.1
Borel–Cantelli	$\sum P(A_n) < \infty \Rightarrow P(A_n, \text{ i.o.}) = 0$	T.B.1
SLLN for MC	$(1\{X_1 = i\} + \cdots + 1\{X_n = i\})/n \to \pi(i)$ w.p. 1	T.1.1

2.7 References

An excellent text on Markov Chains is Chung (1967). A more advanced text on probability theory is Billingsley (2012).

2.8 Problems

Problem 2.1 Consider a Markov chain X_n that takes values in $\{0, 1\}$. Explain why $\{0, 1\}$ is not its sample space.

Problem 2.2 Consider again a Markov chain that takes values in $\{0, 1\}$ with $P(0, 1) = a$ and $P(1, 0) = b$. Exhibit two different sample spaces and the probability on them for that Markov chain.

Problem 2.3 Draw the smallest periodic Markov chain. Show that the fraction of time in the states converges but the probability of being in a state at time n does not converge.

Problem 2.4 For the Markov chain in Problem 2.2, calculate the eigenvalues and use them to get a bound on the distance between the distribution at time n and the invariant distribution.

Problem 2.5 Why does the strong law imply the weak law? More concretely, let X_n, X be random variables such that $X_n \rightarrow X$ almost surely. Show that $X_n \rightarrow X$ in probability.

Hint Fix $\epsilon > 0$ and define $Z_n = 1\{|X_n - X| \geq \epsilon\}$. Use DCT to show that $E(Z_n) \rightarrow 0$ as $n \rightarrow \infty$ if $X_n \rightarrow X$ almost surely.

Problem 2.6 Draw a Markov chain with four states that is irreducible and aperiodic. Consider two independent versions of the Markov chain: one that starts in state 1, the other in state 2. Explain what they will meet after a finite time.

Problem 2.7 Consider the Markov chain of Fig. 1.2. Use Python to calculate the eigenvalues of P. Let λ be the largest absolute value of the eigenvalues other than 1. Use Python to calculate

$$d(n) := \sum_i |\pi(i) - \pi_n(i)|,$$

where $\pi_0(A) = 1$. Plot $d(n)$ and λ^n as functions of n.

Problem 2.8 You flip a fair coin. If the outcome is "head," you get a random amount of money equal to X and if it is" tail," you get a random amount Y. Prove formally that on average, you get

$$\frac{1}{2}E(X) + \frac{1}{2}E(Y).$$

Problem 2.9 Can you find random variables that converge to 0 almost surely, but not in probability?

Problem 2.10 Let $\{X_n, n \geq 1\}$ be i.i.d. zero-mean random variables with variance σ^2. Show that $X_n/n \to 0$ with probability one as $n \to \infty$.

Hint Borel–Cantelli.

Problem 2.11 Let X_n be a finite irreducible Markov chain on $\mathscr{X}$ with invariant distribution π and $f : \mathscr{X} \to \mathfrak{R}$ some function. Show that

$$\frac{1}{N} \sum_{n=0}^{N-1} f(X_n) \to \sum_{i \in \mathscr{X}} \pi(i)f(i) \text{ w.p. 1, as } N \to \infty.$$

Multiplexing: A

3

Application: Sharing Links, Multiple Access, Buffers
Topics: Central Limit Theorem, Confidence Intervals, Queueing, Randomized Protocols

Background:

- General RV (B.4)

3.1   Sharing Links

One essential idea in communication networks is to have different users share common links.

For instance, many users are attached to the same coaxial cable; a large number of cell phones use the same base station; a WiFi access point serves many devices; the high-speed links that connect buildings or cities transport data from many users at any given time (Figs. 3.1 and 3.2).

Networks implement this sharing of physical resources by transmitting bits that carry information of different users on common physical media such as cables, wires, optical fibers, or radio channels. This general method is called *multiplexing*. Multiplexing greatly reduces the cost of the communication systems. In this chapter, we explain statistical aspects of multiplexing.

© The Author(s) 2021
J. Walrand, *Probability in Electrical Engineering and Computer Science*,
https://doi.org/10.1007/978-3-030-49995-2_3

Fig. 3.1 Shared coaxial
cable for internet access, TV,
and telephone

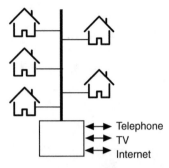

Telephone
TV
Internet

Fig. 3.2 Cellular base
station antennas

Fig. 3.3 A random number ν
of connections share a link
with rate C

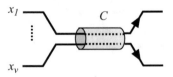

In the internet, at any given time, a number of packet flows share links. For instance, 20 users may be downloading web pages or video files and use the same coaxial cable of their service provider.

The *transmission control protocol (TCP)* arranges for these different flows to share the links as equally as possible (at least, in principle).

We focus our attention on a single link, as shown in Fig. 3.3. The link transmits bits at rate C bps. If ν connections are active at a given time, they each get a rate C/ν. We want to study the typical rate that a connection gets. The nontrivial aspect of the problem is that ν is a random variable.

As a simple model, assume that there are $N \gg 1$ users who can potentially use that link. Assume also that the users are active independently, with probability p. Thus, the number ν of active users is $Binomial(N, p)$ that we also write as $B(N, p)$. (See Sect. B.2.8.)

Figure 3.4 shows the probability mass function for $N = 100$ and $p = 0.1, 0.2$, and 0.5. To be specific, assume that $N = 100$ and $p = 0.2$. The number ν of active users is $B(100, 0.2)$ that we also write as $Binomial(100, 0.2)$. On average, there

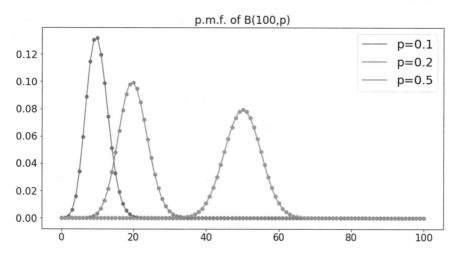

Fig. 3.4 The probability mass function of the *Binomial*(100, *p*) distribution, for *p* = 0.1, 0.2 and 0.5

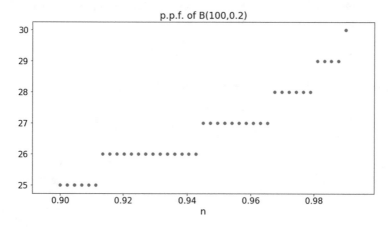

Fig. 3.5 The Python tool "ppf" shows (3.1)

are $Np = 20$ active users. However, there is some probability that a few more than 20 users are active. We want to find a number m so that the likelihood that there are more than m active users is negligible, say 5%. Given that value, we know that each active user gets at least a rate C/m, with probability 95%.

Thus, we can dimension the links, or provision the network, based on that value m. Intuitively, m should be slightly larger than the mean. Looking at the actual distribution, for instance, by using Python's "ppf" as in Fig. 3.5, we find that

$$P(\nu \leq 27) = 0.966 > 95\% \text{ and } P(\nu \leq 26) = 0.944 < 95\%. \tag{3.1}$$

Thus, the smallest value of m such that $P(\nu \leq m) \geq 95\%$ is $m = 27$.

To avoid having to use distribution tables or computation tools, we use the fact that the binomial distribution is well approximated by a *Gaussian* random variable that we discuss next.

3.2   Gaussian Random Variable and CLT

Definition 3.1 (Gaussian Random Variable)

(a) A random variable W is *Gaussian*, or *normal*, with mean 0 and variance 1, and one writes $W =_D \mathcal{N}(0, 1)$, if its probability density function (pdf) is f_W where

$$f_W(x) = \frac{1}{\sqrt{2\pi}} \exp\left\{-\frac{x^2}{2}\right\}, x \in \Re.$$

One also says that W is a *standard normal random variable*, or a *standard Gaussian random variable* (Named after C.F. Gauss, see Fig. 3.6).

(b) A random variable X is Gaussian, or normal, with mean μ and variance σ^2, and we write $X =_D \mathcal{N}(\mu, \sigma^2)$, if

$$X = \mu + \sigma W,$$

where $W =_D \mathcal{N}(0, 1)$. Equivalently,[1] the pdf of X is given by

$$f_X(x) = \frac{1}{\sqrt{2\pi\sigma^2}} \exp\left\{-\frac{(x - \mu)^2}{2\sigma^2}\right\}.$$

◇

Figure 3.7 shows the pdf of a $\mathcal{N}(0, 1)$ random variable W. Note in particular that

Fig. 3.6 Carl Friedrich Gauss. 1777–1855

---

[1] See (B.9).

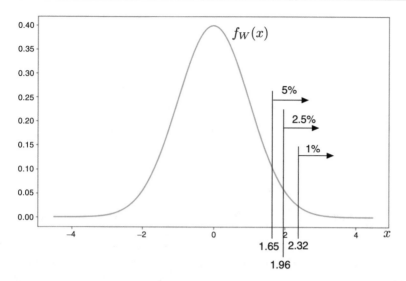

Fig. 3.7 The pdf of a $\mathcal{N}(0, 1)$ random variable

$$P(W > 1.65) \approx 5\%, \ P(W > 1.96) \approx 2.5\% \text{ and } P(W > 2.32) \approx 1\%. \tag{3.2}$$

The Central Limit Theorem states that the sum of many small independent random variables is approximately Gaussian. This result explains that thermal noise, due to the agitation of many electrons, is Gaussian. Many other natural phenomena exhibit a Gaussian distribution when they are caused by a superposition of many independent effects.

Theorem 3.1 (Central Limit Theorem) *Let $\{X(n), n \geq 1\}$ be i.i.d. random variables with mean $E(X(n)) = \mu$ and variance $var(X(n)) = \sigma^2$. Then, as $n \to \infty$,*

$$\frac{X(1) + \cdots + X(n) - n\mu}{\sigma\sqrt{n}} \Rightarrow \mathcal{N}(0, 1). \tag{3.3}$$

∎

In (3.3), the symbol $\Rightarrow$ means *convergence in distribution*. Specifically, if $\{Y(n), n \geq 1\}$ are random variables, then $Y(n) \Rightarrow \mathcal{N}(0, 1)$ means that

$$P(Y(n) \leq x) \to P(W \leq x), \forall x \in \Re,$$

where W is a $\mathcal{N}(0, 1)$ random variable. We prove this result in the next chapter.

More generally, one has the following definition.

Definition 3.2 (Convergence in Distribution) Let $\{X(n), n \geq 1\}$ and X be random variables. One says that $X(n)$ converges in distribution to X, and one writes $X(n) \Rightarrow X$, if

$$P(X(n) \leq x) \to P(X \leq x), \text{ for all } x \text{ s.t. } P(X = x) = 0. \tag{3.4}$$

◇

As an example, let $X(n) = 3 + 1/n$ for $n \geq 1$ and $X = 3$. It is intuitively clear that the distribution of $X(n)$ converges to that of X. However,

$$P(X(n) \leq 3) = 0 \not\to P(X \leq 3) = 1.$$

But,

$$P(X(n) \leq x) \to P(X \leq x), \forall x \neq 3.$$

This example explains why the definition (3.4) requires convergence of $P(X(n) \leq x)$ to $P(X \leq x)$ only for x such that $P(X = x) = 0$.

How does this notion of convergence relate to convergence in probability and almost sure convergence? First note that convergence in distribution is defined even if the random variables $X(n)$ and X are not on the same probability space, since it involves only the distributions of the individual random variables. One can show[2] that

$$X(n) \overset{a.s.}{\to} X \text{ implies } X(n) \overset{P}{\to} X \text{ implies } X(n) \Rightarrow X.$$

Thus, convergence in distribution is the weakest form of convergence.

Also, a fact that I find very comforting is that if $X(n) \Rightarrow X$, then one can construct random variables $Y(n)$ and Y on the *same* probability space so that $Y(n) =_D X(n)$ and $Y =_D X$ and

$$Y(n) \to Y, \text{ with probability } 1.$$

This may seem mysterious but is in fact quite obvious. First note that a random variable with cdf $F(\cdot)$ can be constructed by choosing a random variable $Z =_D U[0, 1]$ and defining (see Fig. 3.8)

$$X(Z) = \inf\{x \in \Re \mid F(x) \geq Z\}.$$

Indeed, one then has $P(X(Z) \leq a) = F(a)$ since $X(Z) \leq a$ if and only if $Z \in [0, F(a)]$, which has probability $F(a)$ since $Z =_D U[0, 1]$. But then, if $X(n) \Rightarrow X$, we have $F_{X_n}(x) \to F_X(x)$ whenever $P(X = x) = 0$, and this implies that

$$X_n(z) = \inf\{x \in \Re \mid F_{X_n}(x) \geq z\} \to X(z) = \inf\{x \in \Re \mid F(x) \geq z\},$$

for all z.

---

[2]See Problems 2.5 and 3.9.

Fig. 3.8 If $Z =_D U[0, 1]$, then cdf of $X(Z)$ is F

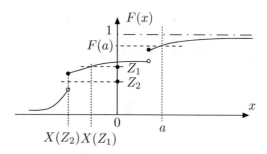

Fig. 3.9 Comparing $Binomial(100, 0.2)$ with $Gaussian(20, 16)$

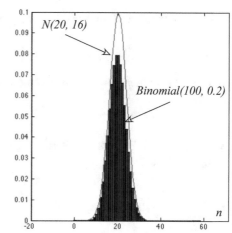

3.2.1 Binomial and Gaussian

Figure 3.9 compares the binomial and Gaussian distributions.

To see why these distributions are similar, note that if $X =_D B(N, p)$, then one can write

$$X = Y_1 + \cdots + Y_N,$$

where the random variables Y_n are i.i.d. and Bernoulli with parameter p. Thus, by the CLT,

$$\frac{X - Np}{\sqrt{N}} \approx \mathcal{N}\left(0, \sigma^2\right),$$

where $\sigma^2 = \text{var}(Y_1) = E(Y_1^2) - (E(Y_1))^2 = p(1 - p)$. Hence, one can argue that

$$B(N, p) \approx_D \mathcal{N}\left(Np, N\sigma^2\right) =_D \mathcal{N}(Np, Np(1 - p)). \tag{3.5}$$

For $p = 0.2$ and $N = 100$, one concludes that $B(100, 0.2) \approx \mathcal{N}(20, 16)$, which is confirmed by Fig. 3.9.

3.2.2 Multiplexing and Gaussian

We now apply the Gaussian approximation of a binomial distribution to multiplexing. Recall that we were looking for the smallest value of m such that $P(B(N, p) > m) \leq 5\%$. The ideas are as follows. From (3.5) and (3.2), we have

(1) $B(N, p) \approx \mathcal{N}(Np, Np(1 - p))$, for $N \gg 1$;

(2) $P(N(\mu, \sigma^2) > \mu + 1.65\sigma) \approx 5\%$.

Combining these facts, we see that, for $N \gg 1$,

$$P(B(N, p) > Np + 1.65\sqrt{Np(1 - p)}) \approx 5\%.$$

Thus, the value of m that we are looking for is

$$m = Np + 1.65\sqrt{Np(1 - p)} = 20 + 1.65\sqrt{16} \approx 27.$$

A look at Fig. 3.9 shows that it is indeed unlikely that ν is larger than 27 when $\nu =_D B(100, 0.2)$.

3.2.3 Confidence Intervals

One can invert the calculation that we did in the previous section and try to guess p from the observed fraction $Y(N)$ of active users out of $N \gg 1$. From the ideas (1) and (2) above, together with the symmetry of the Gaussian distribution around its mean, we see that the events

$$A_1 = \{B(N, p) \geq Np + 1.65\sqrt{Np(1 - p)}\}$$

and

$$A_2 = \{B(N, p) \leq Np - 1.65\sqrt{Np(1 - p)}\}$$

each have a probability close to 5%. With $Y(N) =_D B(N, p)/N$, we see that

$$A_1 = \left\{ Y(N) \geq p + 1.65\sqrt{\frac{p(1 - p)}{N}} \right\}$$

and

$$A_2 = \left\{ Y(N) \leq p - 1.65\sqrt{\frac{p(1 - p)}{N}} \right\}.$$

Hence, the event $A_1 \cup A_2$ has probability close to 10%, so that its complement has probability close to 90%. Consequently,

$$P\left(Y(N) - 1.65\sqrt{\frac{p(1-p)}{N}} \le p \le Y(N) + 1.65\sqrt{\frac{p(1-p)}{N}}\right) \approx 90\%.$$

We do not know p, but $p(1-p) \le 1/4$. Hence, we find

$$P\left(Y(N) - 0.83\frac{1}{\sqrt{N}} \le p \le Y(N) + 0.83\frac{1}{\sqrt{N}}\right) \ge 90\%.$$

For $N = 100$, this gives

$$P(Y(N) - 0.08 \le p \le Y(N) + 0.08) \ge 90\%.$$

For instance, if we observe that 30% of the 100 users are active, then we guess that p is between 0.22 and 0.38, with probability 90%. In other words, $[Y(N) - 0.08, Y(N) + 0.08]$ is a 90%-*confidence interval* for p.

Figure 3.7 shows that we can get a 5%-confidence interval by replacing 1.65 by 2. Thus, we see that

$$\left[Y(N) - \frac{1}{\sqrt{N}}, Y(N) + \frac{1}{\sqrt{N}}\right] \tag{3.6}$$

is a 95%-confidence interval for p.

How large should N be to have a good estimate of p? Let us say that we would like to know p plus or minus 0.03 with 95% confidence. Using (3.6), we see that we need

$$\frac{1}{\sqrt{N}} = 3\%, \text{ i.e., } N = 1,089.$$

Thus, $Y(1,089)$ is an estimate of p with an error less than 0.03, with probability 95%. Such results form the basis for the design of public opinion surveys.

In many cases, one does not know a bound on the variance. In such situations, one replaces the standard deviation by the sample standard deviation. That is, for i.i.d. random variables $\{X(n), n \ge 1\}$ with mean μ, the confidence intervals for μ are as follows:

$$\left[\mu_n - 1.65\frac{\sigma_n}{\sqrt{n}}, \mu_n + 1.65\frac{\sigma_n}{\sqrt{n}}\right] = 90\% - \text{Confidence Interval}$$

$$\left[\mu_n - 2\frac{\sigma_n}{\sqrt{n}}, \mu_n + 2\frac{\sigma_n}{\sqrt{n}}\right] = 95\% - \text{Confidence Interval},$$

where

$$\mu_n = \frac{X(1) + \cdots + X(n)}{n}$$

and

$$\sigma_n^2 = \frac{\sum_{m=1}^{n}(X(m) - \mu_n)^2}{n-1} = \frac{n}{n-1}\left\{\frac{\sum_{m=1}^{n} X(m)^2}{n} - \mu_n^2\right\}.$$

What's up with this $n - 1$ denominator? You probably expected the sample variance to be the arithmetic mean of the squares of the deviations from the sample mean, i.e., a denominator n in the first expression for σ_n^2. It turns out that to make the estimator such that $E(\sigma_n^2) = \sigma^2$, i.e., to make the estimator *unbiased*, one should divide by $n - 1$ instead of n. The difference is negligible for large n, obviously. Nevertheless, let us see why this is so.

For simplicity of notation, assume that $E(X(n)) = 0$ and let $\sigma^2 = \text{var}(X(n)) = E(X(n)^2)$. Note that

$$n^2 E\big((X(1) - \mu_n)^2\big)$$
$$= E\big((nX(1) - X(1) - X(2) - \cdots - X(n))^2\big)$$
$$= E\big((n-1)^2 X(1)^2\big) + E\big(X(2)^2\big) + \cdots + E\big(X(n)^2\big)$$
$$= (n-1)^2\sigma^2 + (n-1)\sigma^2 = n(n-1)\sigma^2.$$

For the second equality, note that the cross-terms $E(X(i)X(j))$ for $i \neq j$ vanish because the random variables are independent and zero-mean.

Hence,

$$E\big((X(1) - \mu_n)^2\big) = \frac{n-1}{n}\sigma^2 \text{ and } \sum_{m=1}^{n} E\big((X(m) - \mu_n)^2\big) = (n-1)\sigma^2.$$

Consequently, an unbiased estimate of σ^2 is

$$\sigma_n^2 := \frac{1}{n-1}\sum_{m=1}^{n} E\big((X(m) - \mu_n)^2\big).$$

3.3   Buffers

The internet is a packet-switched network. A packet is a group of bits of data together with some control information such as a source and destination address,

Fig. 3.10 A switch with multiple input and output ports

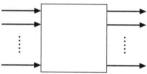

somewhat like an envelope you send by regular mail (if you remember that). A host (e.g., a computer, a smartphone, or a web cam) sends packets to a switch. The switch has multiple input and output ports, as shown in Fig. 3.10.

The switch stores the packets as they arrive and sends them out on the appropriate output port, based on the destination address of the packets. The packets arrive at random times at the switch and, occasionally, packets that must go out on a specific output port arrive faster than the switch can send them out. When this happens, packets accumulate in a buffer. Consequently, packets may face a queueing[3] delay before they leave the switch. We study a simple model of such a system.

3.3.1 Markov Chain Model of Buffer

We focus on packets destined to one particular output port. Our model is in discrete time. We assume that one packet destined for that output port arrives with probability $\lambda \in [0, 1]$ at each time instant, independently of previous arrivals. The packets have random sizes, so that they take random times to be transmitted. We assume that the time to transmit a packet is geometrically distributed with parameter μ and all the transmission times are independent. Let X_n be the number of packets in the output buffer at time n, for $n \geq 0$. At time n, a transmission completes with probability μ and a new packet arrives with probability λ, independently of the past. Thus, X_n is a Markov chain with the state transition diagram shown in Fig. 3.11.

---

[3]Queueing and queuing are alternative spellings; queueing tends to be preferred by researchers and has the peculiar feature of having five vowels in a row, somewhat appropriately.

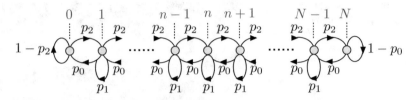

Fig. 3.11 The transition probabilities for the buffer occupancy for one of the output ports

In this diagram,

$$p_2 = \lambda(1 - \mu)$$
$$p_0 = \mu(1 - \lambda)$$
$$p_1 = 1 - p_0 - p_2.$$

For instance, p_2 is the probability that one new packet arrives and that the transmission of a previous does not complete, so that the number of packets in the buffer increases by one.

3.3.2   Invariant Distribution

The balance equations are

$$\pi(0) = (1 - p_2)\pi(0) + p_0\pi(1)$$
$$\pi(n) = p_2\pi(n - 1) + p_1\pi(n) + p_0\pi(n + 1), \ 1 \le n \le N - 1$$
$$\pi(N) = p_2\pi(N - 1) + (1 - p_0)\pi(N).$$

You can verify that the solution is given by

$$\pi(i) = \pi(0)\rho^i, i = 0, 1, \dots, N \text{ where } \rho := \frac{p_2}{p_0}.$$

Since the probabilities add up to one, we find that

$$\pi(0) = \left[\sum_{i=0}^{N} \rho^i \right]^{-1} = \frac{1 - \rho}{1 - \rho^{N+1}}.$$

In particular, the average value of X under the invariant distribution is

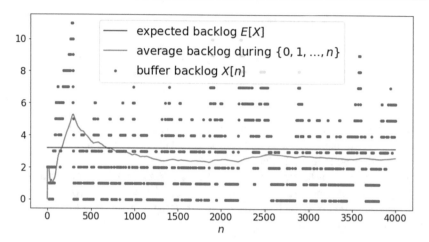

Fig. 3.12 A simulation of the queue with $\lambda = 0.16$, $\mu = 0.20$, and $N = 20$

$$E(X) = \sum_{i=0}^{N} i\pi(i) = \pi(0) \sum_{i=0}^{N} i\rho^i$$

$$= \rho \frac{N\rho^{N+1} - (N+1)\rho^N + 1}{(1-\rho)(1-\rho^{N+1})}$$

$$\approx \frac{\rho}{1-\rho} = \frac{p_2}{p_0 - p_2} = \frac{\lambda(1-\mu)}{\mu - \lambda},$$

where the approximation is valid if $\rho < 1$, i.e., $\lambda < \mu$, and $N \gg 1$ so that $N\rho^N \ll 1$.

Figure 3.12 shows a simulation of this queue when $\lambda = 0.16$, $\mu = 0.20$, and $N = 20$. It also shows the average queue length over n steps and we see that it approaches $\lambda(1-\mu)/(\mu - \lambda) = 3.2$. Note that this queue is almost never full, which explains that one can let $N \to \infty$ in the expression for $E(X)$.

3.3.3 Average Delay

How long do packets stay in the switch? Consider a packet that arrives when there are k packets already in the buffer. That packet then leaves after $k + 1$ packet transmissions. Since each packet transmission takes $1/\mu$ steps, on average, the expected time that the packet spends in the switch is $(k + 1)/\mu$. Thus, to find the expected time a packet stays in the switch, we need to calculate the probability $\phi(k)$ that an arriving packet finds k packets already in the buffer. Then, the expected time W that a packet stays in the switch is given by

$$W = \sum_{k \geq 0} \frac{k+1}{\mu} \phi(k).$$

The result of the calculation is given in the next theorem.

Theorem 3.2 *If* $\lambda < \mu$, *one has*

$$W = \frac{1}{\lambda}E(X) = \frac{1-\mu}{\lambda - \mu}.$$

■

Proof The calculation is a bit lengthy and the details may not be that interesting, except that they explain how to calculate $\phi(k)$ and that they show that the simplicity of the result is quite remarkable.

Recall that $\phi(k)$ is the probability that there are $k+1$ packets in the buffer after a given packet arrives at time n. Thus, $\phi(k) = P[X(n+1) = k+1 \mid A(n) = 1]$ where $A(n)$ is the number of arrivals at time n. Now, if $D(n)$ is the number of transmission completions at time n,

$$\phi(k) = P[X(n) = k + 1, D(n) = 1 \mid A(n) = 1]$$
$$+ P[X(n) = k, D(n) = 0 \mid A(n) = 1].$$

Also,

$$P[X(n) = k + 1, D(n) = 1 \mid A(n) = 1] = \frac{P[X(n) = k + 1, D(n) = 1, A(n) = 1]}{P(A(n) = 1)}$$

$$= \frac{1}{\lambda}P(X(n) = k + 1)P[D(n) = 1, A(n) = 1 \mid X(n) = k + 1]$$

$$= \frac{1}{\lambda}\pi(k + 1)\lambda\mu = \pi(k + 1)\mu.$$

Similarly,

$$P[X(n) = k, D(n) = 0 \mid A(n) = 1] = \frac{P[X(n) = k, D(n) = 0, A(n) = 1]}{P(A(n) = 1)}$$

$$= \frac{1}{\lambda}P(X(n) = k)P[D(n) = 0, A(n) = 1 \mid X(n) = k]$$

$$= \frac{1}{\lambda}\pi(k)\lambda(1 - \mu 1\{k > 0\}) = \pi(k)(1 - \mu 1\{k > 0\}).$$

Hence,

$$\phi(k) = \pi(k)(1 - \mu 1\{k > 0\}) + \pi(k + 1)\mu.$$

Consequently, the expected time W that a packet spends in the switch is given by

$$
\begin{aligned}
W &= \sum_{k \geq 0} \frac{k+1}{\mu} \phi(k) = \frac{1}{\mu} + \frac{1}{\mu} \sum_{k \geq 0} k\pi(k)(1 - \mu 1\{k = 0\}) + \sum_{k \geq 0} k\pi(k+1) \\
&= \frac{1}{\mu} + \frac{1}{\mu} \sum_{k \geq 0} k\pi(k)(1 - \mu) + \sum_{k \geq 1} (k - 1)\pi(k) \\
&= \frac{1}{\mu} + \frac{1-\mu}{\mu} E(X) + E(X) - 1 = \frac{1}{\mu} + \frac{1}{\mu} E(X) - 1 \\
&= \frac{1}{\mu} + \frac{\lambda(1-\mu)}{\mu(\mu - \lambda)} - 1 = \frac{1-\mu}{\mu - \lambda} = \frac{1}{\lambda} E(X).
\end{aligned}
$$

$\square$

3.3.4 A Note About Arrivals

Since the arrivals are independent of the backlog in the buffer, it is tempting to conclude that the probability that a packet finds k packet in the buffer upon its arrival is $\pi(k)$. An argument in favor of this conclusion looks as follows:

$$
\begin{aligned}
P[X_{n+1} = k + 1 \mid A_n = 1] &= P[X_n = k \mid A_n = 1] \\
&= P[X_n = k] = \pi(k),
\end{aligned}
$$

where the second identity comes from the independence of the arrivals A_n and the backlog X_n. However, the first identity does not hold since it is possible that $X_{n+1} = k$, $X_n = k$, and $A_n = 1$. Indeed, one may have $D_n = 1$.

If one assumes that $\lambda < \mu \ll 1$, then the probability that $A_n = 1$ and $D_n = 1$ is negligible and it is then the case that $\pi(k) \approx \pi(k)$. We encounter that situation in Sect. 5.6.

3.3.5 Little's Law

The previous result is a particular case of *Little's Law* (Little 1961) (Fig. 3.13).

Theorem 3.3 (Little's Law) *Under weak assumptions,*

$$
L = \lambda W,
$$

where L is the average number of customers in a system, λ is the average arrival rate of customers, and W is the average time that a customer spends in the system.

∎

Fig. 3.13 John D. C. Little.
b. 1928

One way to understand this law is to consider a packet that leaves the switch after having spent T time units. During its stay, λT packets arrive, on average. So the average backlog in the switch should be λT.

It turns out that Little's law applies to very general systems, even those that do not serve the packets in their order of arrival.

One way to see this is to think that each packet pays the switch one unit of money per unit of time it spends in the switch. If a packet spends T time units, on average, in the switch, then each packet pays T, on average. Thus, the switch collects money at the rate of λT per unit of time, since λ packets go through the switch per unit of time and each pays an average of T. Another way to look at the rate at which the switch is getting paid is to realize that if there are L packets in the switch at any given time, on average, then the switch collects money at rate L, since each packet pays one unit per unit time. Thus, $L = \lambda T$.

3.4   Multiple Access

Imagine a number of smartphones sharing a *WiFi access point*, as illustrated in Fig. 3.14. They want to transmit packets.

If multiple smartphones transmit at the same time, the transmissions garble one another, and we say that they collide. We discuss a simple scheme to regulate the transmissions and achieve a large rate of success. We consider a discrete time model of the situation.

There are N devices. At time $n \geq 0$, each device transmits with probability p, independently of the others. This scheme, called *randomized multiple access*, was proposed by Norman Abramson in the late 1960s for his Aloha network (Abramson 1970) (Fig. 3.15).

The number $X(n)$ of transmissions at time n is then $B(N, p)$ (see (B.4)). In particular, the fraction of time that exactly one device transmits is

$$P(X(n) = 1) = Np(1 - p)^{N-1}.$$

The maximum over p of this success rate occurs for $p = 1/N$ and it is λ^* where

Fig. 3.14 A number of smartphones share a WiFi access point

Fig. 3.15 Norman Abramson, b. 1932

$$\lambda^* = \left(1 - \frac{1}{N}\right)^{N-1} \approx \frac{1}{e} \approx 0.36.$$

In this derivation, we use the fact that

$$\left(1 - \frac{a}{N}\right)^{N} \approx e^{-a} \text{ for } N \gg 1. \tag{3.7}$$

Thus, this scheme achieves a transmission rate of about 36%. However, it requires selecting $p = 1/N$, which means that the devices need to know how many other devices are active (i.e., try to transmit). We discuss an adaptive scheme in the next chapter that does not require that information.

3.5   Summary

- Gaussian random variable $\mathcal{N}(\mu, \sigma^2)$;
- CLT;
- Confidence Intervals;
- Buffers: average backlog and delay; Little's Law;
- Multiple Access Protocol.

3.5.1   Key Equations and Formulas

Definition of $\mathcal{N}(\mu, \sigma^2)$	$f_X(x) = (2\pi\sigma^2)^{-1/2} \exp\{-(x-\mu)^2/(2\sigma^2)\}$	D.3.1
CLT	$(X_1 + \cdots + X_n - n\mu)/\sqrt{n} \Rightarrow \mathcal{N}(0, \sigma^2)$	T.3.1
95%-Confidence Interval	$(X_1 + \cdots + X_n)/n \pm 2\sigma$	S.3.2.3
Little's Law	$L = \lambda W$	T.3.3
Exponential Approximation	$(1 - a/n)^n \approx \exp\{-a\}$	(3.7)

3.6   References

The buffering analysis is a simple example of queueing theory. See Kleinrock (1975–6) for a discussion of queueing models of computer and communication systems.

3.7   Problems

Problem 3.1 Write a Python code to compute the number of people to poll in a public opinion survey to estimate the fraction of the population that will vote in favor of a proposition within α percent, with probability at least $1 - \beta$. Use an upper bound on the variance. Assume that we know that $p \in [0.4, 0.7]$.

Problem 3.2 We are conducting a public opinion poll to determine the fraction p of people who will vote for Mr. Whatshisname as the next president. We ask N_1 college-educated and N_2 non-college-educated people. We assume that the votes in each of the two groups are i.i.d. $B(p_1)$ and $B(p_2)$, respectively, in favor of Whatshisname. In the general population, the percentage of college-educated people is known to be q.

(a) What is a 95%-confidence interval for p, using an upper bound for the variance.
(b) How do we choose N_1 and N_2 subject to $N_1 + N_2 = N$ to minimize the width of that interval?

Problem 3.3 You flip a fair coin 10,000 times. The probability that there are more than 5085 heads is approximately (choose the correct answer)

☐ 15%;
☐ 10%;
☐ 5%;
☐ 2.5%;
☐ 1%.

Problem 3.4 Write a Python simulation of a buffer where packets arrive as a Bernoulli process with rate λ and geometric service times with rate μ. Plot the simulation and calculate the long-term average backlog.

Problem 3.5 Consider a buffer that can transmit up to M packets in parallel. That is, when there are m packets in the buffer, $\min\{m, M\}$ of these packets are being transmitted. Also, each of these packets completes transmission independently in the next time slot with probability μ. At each time step, a packet arrives with probability λ.

(a) What are the transition probabilities of the corresponding Markov chain?
(b) For what values of λ, M, and μ do you expect the system to be stable?
(c) Write a Python simulation of this system.

Problem 3.6 In order to estimate the probability of head in a coin flip, p, you flip a coin n times, and count the number of heads, S_n. You use the estimator $\hat{p} = S_n/n$. You choose the sample size n to have a guarantee

$$P(|S_n/n - p| \geq \epsilon) \leq \delta.$$

(a) What is the value of n suggested by Chebyshev's inequality? (Use a bound on the variance.)
(b) How does this value change when ϵ is reduced to half of its original value?
(c) How does it change when δ is reduced to half of its original value?
(d) Compare this value of n with that given by the CLT.

Problem 3.7 Let $\{X_n, n \geq 1\}$ be i.i.d. $U[0, 1]$ and $Z_n = X_1 + \cdots + X_n$. What is $P(Z_n > n)$? What would the estimate be of the same probability obtained from the Central Limit Theorem?

Problem 3.8 Consider one buffer where packets arrive one by one every 2 s and take 1 s to transmit. What is the average delay through the queue per packet? Repeat the problem assuming that the packets arrive ten at a time every 20 s. This example shows that the delay depends on how "bursty" the traffic is.

Problem 3.9 Show that if $X(n) \xrightarrow{p} X$, then $X(n) \Rightarrow X$.

Hint Assume that $P(X = x) = 0$. To show that $P(X(n) \leq x) \to P(X \leq x)$, note that if $|X(n) - X| \leq \epsilon$ and $X \leq x$, then $X(n) \leq X + \epsilon$.

Multiplexing: B

Topics: Characteristic Functions, Proof of Central Limit Theorem, Adaptive
CSMA

4.1   Characteristic Functions

Before we explain the proof of the CLT, we have to describe the use of characteristic
functions.

Definition 4.1 *Characteristic Function* The characteristic function of a random
variable X is defined as

$$\phi_X(u) = E(e^{iuX}), u \in \Re.$$

In this expression, $i := \sqrt{-1}$.

◇

Note that

$$\phi_X(u) = \int_{-\infty}^{\infty} e^{iux} f_X(x)dx,$$

© The Author(s) 2021
J. Walrand, *Probability in Electrical Engineering and Computer Science*,
https://doi.org/10.1007/978-3-030-49995-2_4

so that $\phi_X(u)$ is the Fourier transform of $f_X(x)$. As such, the characteristic function determines the pdf uniquely.

As an important example, we have the following result.

Theorem 4.1 (Characteristic Function of $\mathcal{N}(0, 1)$) *Let $X =_D \mathcal{N}(0, 1)$. Then,*

$$\phi_X(u) = e^{-\frac{u^2}{2}}. \tag{4.1}$$

■

Proof One has

$$\phi_X(u) = \int_{-\infty}^{\infty} e^{iux} \frac{1}{\sqrt{2\pi}} e^{-\frac{x^2}{2}} dx,$$

so that

$$\frac{d}{du} \phi_X(u) = \int_{-\infty}^{\infty} ix e^{iux} \frac{1}{\sqrt{2\pi}} e^{-\frac{x^2}{2}} dx = -\int_{-\infty}^{\infty} i e^{iux} \frac{1}{\sqrt{2\pi}} de^{-\frac{x^2}{2}}$$

$$= \int_{-\infty}^{\infty} i \frac{1}{\sqrt{2\pi}} e^{-\frac{x^2}{2}} de^{iux} = -u \int_{-\infty}^{\infty} e^{iux} \frac{1}{\sqrt{2\pi}} e^{-\frac{x^2}{2}} dx$$

$$= -u \phi_X(u).$$

(The third equation follows by integration by parts.) Thus,

$$\frac{d}{du} \log(\phi_X(u)) = -u = -\frac{d}{du} \frac{u^2}{2},$$

which implies that

$$\phi_X(u) = A e^{-\frac{u^2}{2}}.$$

Since $\phi_X(0) = E(e^{i0X}) = 1$, we see that $A = 1$, and this proves the result (4.1). □

We are now ready to prove the CLT.

4.2 Proof of CLT (Sketch)

The technique to analyze sums of independent random variables is to calculate the characteristic function. Let then

$$Y(n) = \frac{X(1) + \cdots + X(n) - n\mu}{\sigma \sqrt{n}}, n \geq 1.$$

We have

$$\phi_{Y(n)}(u) = E\left(e^{iuY(n)}\right) = E\left(\Pi_{m=1}^{n}\exp\left\{\frac{iu(X(m) - \mu)}{\sigma\sqrt{n}}\right\}\right)$$

$$= \left[E\left(\exp\left\{\frac{iu(X(1) - \mu)}{\sigma\sqrt{n}}\right\}\right)\right]^{n}$$

$$= \left[E\left(1 + \frac{iu(X(1) - \mu)}{\sigma\sqrt{n}} - \frac{u^2(X(1) - \mu)^2}{2\sigma^2 n} + o(1/n)\right)\right]^{n}$$

$$= \left[1 - u^2/(2n) + o(1/n)\right]^{n} \rightarrow \exp\left\{-u^2/2\right\}, \text{ as } n \to \infty.$$

The third equality holds because the $X(m)$ are i.i.d. and the fourth one follows from the Taylor expansion of the exponential:

$$e^a \approx 1 + a + \frac{1}{2}a^2.$$

Thus, the characteristic function of $Y(n)$ converges to that of a $\mathcal{N}(0, 1)$ random variable. This suggests that the inverse Fourier transform, i.e., the density of $Y(n)$ converges to that of a $\mathcal{N}(0, 1)$ random variable. This last step can be shown formally, but we will not do it here. □

4.3 Moments of $\mathcal{N}(0, 1)$

We can use the characteristic function of a $\mathcal{N}(0, 1)$ random variable X to calculate its moments. This is how. First we note that, by using the Taylor expansion of the exponential,

$$\phi_X(u) = E\left(e^{iuX}\right) = E\left(\sum_{n=0}^{\infty}\frac{1}{n!}(iuX)^n\right)$$

$$= \sum_{n=0}^{\infty}\frac{1}{n!}(iu)^n E(X^n).$$

Second, again using the expansion of the exponential,

$$\phi_X(u) = e^{-u^2/2} = \sum_{m=0}^{\infty}\frac{1}{m!}\left(-\frac{u^2}{2}\right)^m.$$

Third, we match the coefficients of u^{2m} in these two expressions and we find that

$$\frac{1}{(2m)!}i^{2m}E\left(X^{2m}\right) = \frac{1}{m!}\left(-\frac{1}{2}\right)^m,$$

This gives[1]

$$E\left(X^{2m}\right) = \frac{(2m)!}{m!2^m}.$$ (4.2)

For instance,

$$E(X^2) = \frac{2!}{1!2^1} = 1, \ E\left(X^4\right) = \frac{4!}{2!2^2} = 3.$$

Finally, we note that the coefficients of odd powers of u must be zero, so that

$$E\left(X^{2m+1}\right) = 0, \ \text{for } m = 0, 1, 2, \ldots.$$

(This should be obvious from the symmetry of $f_X(x)$.) In particular,

$$\text{var}(X) = E\left(X^2\right) - E(X)^2 = 1.$$

4.4   Sum of Squares of 2 i.i.d. $\mathcal{N}(0, 1)$

Let X, Y be two i.i.d. $\mathcal{N}(0, 1)$ random variables. The claim is that

$$Z = X^2 + Y^2 =_D Exp(1/2).$$

Let θ be the angle of the vector (X, Y) and $R^2 = X^2 + Y^2$. Thus (see Fig. 4.1)

$$dxdy = rdrd\theta.$$

Note that $E(Z) = E(X^2) + E(Y^2) = 2$, so that if Z is exponentially distributed, its rate must be $1/2$. Let us prove that it is exponential. One has

$$f_{X,Y}(x, y)dxdy = f_{X,Y}(x, y)rdrd\theta = \frac{1}{2\pi} \exp\left\{-\frac{x^2+y^2}{2}\right\} rdrd\theta$$

$$= \frac{1}{2\pi} \exp\left\{-\frac{r^2}{2}\right\} rdrd\theta = \frac{d\theta}{2\pi} \times \exp\left\{-\frac{r^2}{2}\right\} rdr$$

$$=: f_\theta(\theta)d\theta \times f_R(r)dr,$$

---

[1] We used the fact that

$$i^{2m} = (-1)^m.$$

Fig. 4.1 Under the change of variables $x = r\cos(\theta)$ and $y = r\sin(\theta)$, we see that $dxdy = rdrd\theta$. That is, $[r, r + dr] \times [\theta, \theta + d\theta]$ covers an area $rdrd\theta$ in the (x, y) plane

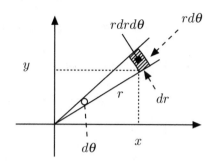

where

$$f_\theta(\theta) = \frac{1}{2\pi}1\{0 < \theta < 2\pi\} \text{ and } f_R(r) = r\exp\left\{-\frac{r^2}{2}\right\}1\{r \geq 0\}.$$

Thus, the angle θ of (X, Y) and the norm $R = \sqrt{X^2 + Y^2}$ are independent and have the indicated distributions. But then, if $V = R^2 =: g(R)$, we find that, for $v \geq 0$,

$$f_V(v) = \frac{1}{|g'(R)|}f_R(r) = \frac{1}{2r}r\exp\left\{-\frac{r^2}{2}\right\} = \frac{1}{2}\exp\left\{-\frac{v}{2}\right\}$$

which shows that the angle θ and $V = X^2 + Y^2$ are independent, the former being uniformly distributed in $[0, 2\pi]$ and the latter being exponentially distributed with mean 2.

4.5   Two Applications of Characteristic Functions

We have used characteristic functions to prove the CLT. Here are two other cute applications.

4.5.1   Poisson as a Limit of Binomial

A Poisson random variable X with mean λ can be viewed as a limit of a $B(n, \lambda/n)$ random variable X_n as $n \to \infty$. To see this, note that

$$E(\exp\{iuX_n\}) = E(\exp\{iu(Z_n(1) + \ldots + Z_n(n))\}),$$

where the random variables $\{Z_n(1), \ldots, Z_n(n)\}$ are i.i.d. Bernoulli with mean λ/n. Hence,

$$E(\exp\{iuX_n\}) = \left[E(\exp\{iu(Z_n(1)\}))\right]^n = \left[1 + \frac{\lambda}{n}(e^{iu} - 1)\right]^n.$$

For the second identity, we use the fact that if $Z =_D B(p)$, then

$$E(\exp\{iuZ\}) = (1 - p)e^0 + pe^{iu} = 1 + p(e^{iu} - 1).$$

Also, since

$$P(X = m) = \frac{\lambda^m}{m!}e^{-\lambda} \text{ and } e^a = \sum_{m=0}^{\infty} \frac{a^m}{m!},$$

we find that

$$E(\exp\{iuX\}) = \sum_{m=0}^{\infty} \frac{\lambda^m}{m!} \exp\{-\lambda\}e^{ium} = \exp\{\lambda(e^{iu} - 1)\}.$$

The result then follows from the fact that

$$\left(1 + \frac{a}{n}\right)^n \to e^a, \text{ as } n \to \infty.$$

4.5.2   Exponential as Limit of Geometric

An exponential random variable can be viewed as a limit of scaled geometric random variables. Let $X =_D Exp(\lambda)$ and $X_n = G(\lambda/n)$. Then

$$\frac{1}{n}X_n \to X, \text{ in distribution.}$$

To see this, recall that

$$f_X(x) = \lambda e^{-\lambda x} 1\{x \geq 0\}.$$

Also,

$$\int_0^\infty e^{-\beta x} = \frac{1}{\beta}$$

if the real part of β is positive.
    Hence,

$$E\left(e^{iuX}\right) = \int_0^\infty e^{iux} \lambda e^{-\lambda x} dx = \frac{\lambda}{\lambda - iu}.$$

Moreover, since

$$P(X_n = m) = (1 - p)^m p, \, m \geq 0,$$

we find that, with $p = \frac{\lambda}{n}$,

$$E\left(\exp\left\{iu\frac{1}{n}X_n\right\}\right) = \sum_{m=0}^{\infty} (1 - p)^m p \exp\{ium/n\}$$

$$= p \sum_{m=0}^{\infty} [(1 - p)exp\{iu/n\}]^m = \frac{p}{1 - (1 - p)exp\{iu/n\}}$$

$$= \frac{\lambda/n}{1 - (1 - \lambda/n)\exp\{iu/n\}}$$

$$= \frac{\lambda}{n(1 - (1 - \lambda/n)(1 + iu/n + o(1/n)))}$$

$$= \frac{\lambda}{\lambda - iu + o(1/n)},$$

where $o(1/n) \to 0$ as $n \to \infty$. This proves the result.

4.6   Error Function

In the calculation of confidence intervals, one uses estimates of

$$Q(x) := P(X > x) \text{ where } X =_D \mathcal{N}(0, 1).$$

The function $Q(x)$ is called the *error function*. With Python or the appropriate smart phone app, you can get the value of $Q(x)$. Nevertheless, the following bounds (see Fig. 4.2) may be useful.

Theorem 4.2 (Bounds on Error Function) *One has*

$$\frac{x}{1 + x^2}\frac{1}{\sqrt{2\pi}}\exp\left\{-\frac{x^2}{2}\right\} \leq Q(x) \leq \frac{1}{x\sqrt{2\pi}}\exp\left\{-\frac{x^2}{2}\right\}, \, \forall x > 0.$$

∎

Proof Here is a derivation of the upper bound. For $x > 0$, one has

$$Q(x) = \int_x^{\infty} f_X(y)dy = \int_x^{\infty} \frac{1}{\sqrt{2\pi}}e^{-\frac{y^2}{2}}dy = \frac{1}{\sqrt{2\pi}}\int_x^{\infty} \frac{y}{y}e^{-\frac{y^2}{2}}dy$$

Fig. 4.2 The error function $Q(x)$ and its bounds

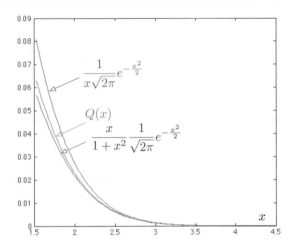

$$\leq \frac{1}{\sqrt{2\pi}} \int_x^\infty \frac{y}{x} e^{-\frac{y^2}{2}} dy = \frac{1}{x\sqrt{2\pi}} \int_x^\infty y e^{-\frac{y^2}{2}} dy$$

$$= -\frac{1}{x\sqrt{2\pi}} \int_x^\infty d e^{-\frac{y^2}{2}} = \frac{1}{x\sqrt{2\pi}} e^{-\frac{x^2}{2}}.$$

For the lower bound, one uses the following calculation, again with $x > 0$:

$$\left(1 + \frac{1}{x^2}\right) \int_x^\infty e^{-\frac{y^2}{2}} dy \geq \int_x^\infty \left(1 + \frac{1}{y^2}\right) e^{-\frac{y^2}{2}} dy$$

$$= -\int_x^\infty d\left(\frac{1}{y} e^{-\frac{y^2}{2}}\right) = \frac{1}{x} e^{-\frac{x^2}{2}}.$$

□

4.7   Adaptive Multiple Access

In Sect. 3.4, we explained a randomized multiple access scheme. In this scheme, there are N active station and each station attempts to transmit with probability $1/N$ in each time slot. This scheme results in a success rate of about $1/e \approx 36\%$. However, it requires that each station knows how many other stations are active.

To make the scheme *adaptive* to the number of active devices, say that the devices adjust the probability $p(n)$ with which they transmit at time n as follows:

$$p(n+1) = \begin{cases} p(n), & \text{if } X(n) = 1; \\ ap(n), & \text{if } X(n) > 1; \\ \min\{bp(n), 1\}, & \text{if } X(n) = 0. \end{cases}$$

Fig. 4.3 Bruce Hajek

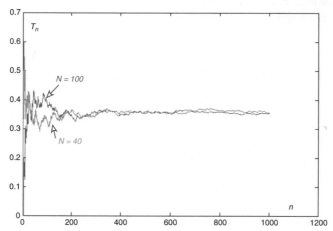

Fig. 4.4 Throughput of the adaptive multiple access scheme

In these update rules, a and b are constants with $a \in (0, 1)$ and $b > 1$. The idea is to increase $p(n)$ if no device transmitted and to decrease it after a collision. This scheme is due to Hajek and Van Loon (1982) (Fig. 4.3).

Figure 4.4 shows the evolution over time of the success rate T_n. Here,

$$T_n = \frac{1}{n} \sum_{m=0}^{n-1} 1\{X(m) = 1\}.$$

The figure uses $a = 0.8$ and $b = 1.2$. We see that the throughput approaches the optimal value for $N = 40$ and for $N = 100$. Thus, the scheme adapts automatically to the number of active devices.

4.8   Summary

- Characteristics Function;
- Proof of CLT;
- Moments of Gaussian;
- Sum of Squares of Gaussians;
- Poisson as limit of Binomial;
- Exponential as limit of Geometric;
- Adaptive Multiple Access Protocol.

4.8.1   Key Equations and Formulas

Characteristic Function	$\phi_X(u) = E(\exp\{iuX\})$	D. 4.1
For $\mathcal{N}(0, 1)$	$\exp\{-u^2/2\}$	T. 4.1
Moments of $\mathcal{N}(0, 1)$	$E(X^{2m}) = (2m)!/(m!2^m)$	(4.2)
Error Function $P(\mathcal{N}(0, 1) > x)$	Bounds	T. 4.2

4.9   References

The CLT is a classical result, see Bertsekas and Tsitsiklis (2008), Grimmett and Stirzaker (2001) or Billingsley (2012).

4.10   Problems

Problem 4.1 Let X be a $N(0, 1)$ random variable. You will recall that $E(X^2) = 1$ and $E(X^4) = 3$.

(a) Use Chebyshev's inequality to get a bound on $P(|X| > 2)$;
(b) Use the inequality that involves the fourth moment of X to bound $P(|X| > 2)$. Do you get a better bound?
(c) Compare with what you know about the $N(0, 1)$ random variable.

Problem 4.2 Write a Python simulation of Hajek's random multiple access scheme. There are 20 stations. An arrival occurs at each station with probability $\lambda/20$ at each time slot. The stations update their transmission probability as explained in the text. Plot the total backlog in all the stations as a function of time.

Problem 4.3 Consider a multiple access scheme where the N stations independently transmit short reservation packets with duration equal to one time unit with probability p. If the reservation packets collide or no station transmits a reservation packet, the stations try again. Once a reservation is successful, the succeeding station transmits a packet during K time units. After that transmission, the process repeats. Calculate the maximum fraction of time that the channel can be used for transmitting packets. Note: This scheme is called *Reservation Aloha*.

Problem 4.4 Let X be a random variable with mean zero and variance 1. Show that $E(X^4) \geq 1$.

Hint Use the fact that $E((X^2 - 1)^2) \geq 0$.

Problem 4.5 Let X, Y be two random variables. Show that

$$(E(XY))^2 \leq E(X^2)E(Y^2).$$

This is the *Cauchy–Schwarz inequality*.

Hint Use $E((\lambda X - Y)^2) \geq 0$ with $\lambda = E(XY)/E(X^2)$.

Networks: A

5

> **Application:** Social Networks, Communication Networks
> **Topics:** Random Graphs, Queueing Networks

5.1 Spreading Rumors

Picture yourself in a social network. You are connected to a number of "friends" who are also connected to friends. You send a message to some of your friends and they in turn forward it to some of their friends. We are interested in the number of people who eventually get the message.

To explore this question, we model the social network as a random tree of which you are the root. You send a message to a random number of your friends that we model as the children of the root node. Similarly, every node in the graph has a random number of children. Assume that the numbers of children of the different nodes are independent, identically distributed, and have mean μ. The tree is potentially infinite, a clear mathematical idealization. The model ignores cycles in friendships, another simplification.

The model is illustrated in Fig. 5.1. Thus, the graph only models the people who get a copy of the message. For the problem to be non-trivial, we assume that there is a positive probability that some nodes have no children, i.e., that someone does not forward the message. Without this assumption, the message always spreads forever.

We have the following result.

© The Author(s) 2021
J. Walrand, *Probability in Electrical Engineering and Computer Science*,
https://doi.org/10.1007/978-3-030-49995-2_5

Fig. 5.1 The spreading of a
message as a random tree

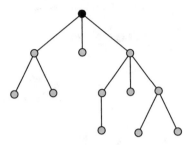

Theorem 5.1 (Spreading of a Message) *Let Z be the number of nodes that eventually receive the message.*

(a) If $\mu < 1$, then $P(Z < \infty) = 1$ and $E(Z) < \infty$;
(b) If $\mu > 1$, then $P(Z = \infty) > 0$.

∎

We prove that result in the next chapter. The result should be intuitive: if $Z < 1$, the spreading dies out, like a population that does not reproduce enough. This model is also relevant for the spread of epidemics or cyber viruses.

5.2   Cascades

If most of your friends prefer Apple over Samsung, you may follow the majority. In turn, your advice will influence other friends. How big is such an influence cascade?

We model that situation with nodes arranged in a line, in the chronological order of their decisions, as shown in Fig. 5.2. Node n listens to the advice of a subset of $\{0, 1, \ldots, n - 1\}$ who have decided before him. Specifically, node n listens to the advice of node $n - k$ independently with probability p_k, for $k = 1, \ldots, n$. If the majority of these friends are blue, node n turns blue; if the majority are red, node n turns red; in case of a tie, node n flips a fair coin and turns red with probability $1/2$ or blue otherwise. Assume that, initially, node 0 is red. Does the fraction of red nodes become larger than 0.5, or does the initial effect of node 0 vanish?

A first observation is that if nodes listen only to their left-neighbor with probability $p \in (0, 1)$, the cascade ends. Indeed, there is a first node that does not listen to its neighbor and then turns red or blue with equal probabilities. Consequently, there will be a string of red nodes followed by a string of blue node, and so on. By symmetry, the lengths of those strings are independent and identically distributed. It is easy to see they have a finite mean. The SLLN then implies that the fraction of red nodes among the first n nodes converges to 0.5. In other words, the influence of the first node vanishes.

The situation is less obvious if $p_k = p < 1$ for all k. Indeed, in this case, as n gets large, node n is more likely to listen to many previous neighbors. The slightly

Fig. 5.2 An influence cascade

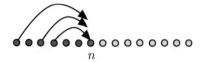

$$n$$

surprising result is that, no matter how small p is, there is a positive probability that all the nodes turn red.

Theorem 5.2 (Cascades) *Assume $p_k = p \in (0, 1]$ for all $k \geq 1$. Then, all nodes turn red with probability at least equal to θ where*

$$\theta = \exp\left\{-\frac{1-p}{p}\right\}.$$

∎

We prove the result in the next chapter. It turns out to be possible that every node listens to at least one previous node. In that case, all the nodes turn red.

5.3 Seeding the Market

Some companies distribute free products to spread their popularity. What is the best fraction of customers who should get free products? To explore this question, let us go back to our model where each node listens only to its left-neighbor with probability p. The system is the same as before, except that each node gets a free product and turns red with probability λ. The fraction of red nodes increases in λ and we write it as $\psi(\lambda)$. If the cost of a product is c and the selling price is s, the company makes a profit $(s - c)\psi(\lambda) - c\lambda$ since it makes a profit $s - c$ from a buyer and loses c for each free product. The company then can select λ to optimize its profit. Next, we calculate $\psi(\lambda)$.

Let $\pi(n - 1)$ be the probability that user $n - 1$ is red. If user n listens to $n - 1$, he turns red unless $n - 1$ is blue and he does not get a free product. If he does not listen to $n - 1$, he turns red with probability 0.5 if he does not get a free product and with probability one otherwise. Thus,

$$\pi(n) = p(1 - (1 - \pi(n-1))(1 - \lambda)) + (1 - p)(0.5(1 - \lambda) + \lambda)$$
$$= p(1 - \lambda)\pi(n - 1) + 0.5\lambda p + 0.5 + 0.5\lambda - 0.5p + 0.5\lambda p.$$

Since $p(1 - \lambda) < 1$, the value of $\pi(n)$ converges to the value $\psi(\lambda)$ that solves the fixed point equation

$$\psi(\lambda) = p(1 - \lambda)\psi(\lambda) + 0.5\lambda p + 0.5 + 0.5\lambda - 0.5p + 0.5\lambda p.$$

Hence,

$$\psi(\lambda) = 0.5\frac{1 + \lambda - p + \lambda p}{1 - p(1 - \lambda)}.$$

To maximize the profit $(s - c)\psi(\lambda) - c\lambda$, we substitute the expression for $\psi(\lambda)$ in the profit and we set the derivative with respect to λ equal to zero. After some algebra, we find that the optimal λ^* is given by

$$\lambda^* = \min\left\{1, \frac{(1 - p)^{1/2} - (1 - p)}{p}\sqrt{\frac{0.5(s - c)}{c}}\right\}.$$

Not surprisingly, λ^* increases with the profit margin $(s - c)/c$ and decreases with p.

5.4   Manufacturing of Consent

Three people walk into a bar. No, this is not a joke. They chat and, eventually, leave with the same majority opinion. As such events repeat, the opinion of the population evolves. We explore a model of this evolution.

Consider a population of $2N \geq 4$ people. Initially, half believe red and the other half believe blue. We choose three people at random. If two are blue and one is red, they all become blue, and they return to the general population. The other cases are similar. The same process then repeats. Let X_n be the number of blue people after n steps, for $n \geq 1$ and let $X_0 = N$. Then X_n is a Markov chain. This Markov chain has two absorbing states: 0 and $2N$. Indeed, if $X_n = k$ for some $k \in \{1, \ldots, 2N - 1\}$, there is a positive probability of choosing three people where two have one opinion and the third has a different one. After their meeting, $X_{n+1} \neq X_n$. The Markov chain is such that $P(1, 0) = 1$ and $P(2N - 1, 2N) = 1$. Moreover, $P(k, k) > 0$, $P(k, k + 1) > 0$, and $P(k, k - 1) > 0$ for all $k \in \{2, \ldots, 2N - 2\}$. Consequently, with probability one,

$$\lim_{n \to \infty} X_n \in \{0, 2N\}.$$

Thus, eventually, everyone is blue or everyone is red. By symmetry, the two limits have probability 0.5.

What is the effect of the media on the limiting consensus? Let us modify our previous model by assuming that when two blue and one red person meet, they all turn blue with probability $1 - p$ and remain as before with probability p. Here p models the power of the media at convincing people to stay red. If two red and one blue meet, they all turn red.

We have, for $k \in \{2, \ldots, 2N - 2\}$,

$$P[X_{n+1} = k + 1 \mid X_n = k] = (1 - p)3 \frac{k(k-1)(2N-k)}{2N(2N-1)(2N-2)} =: p(k).$$

Indeed, X_n increases with probability $1 - p$ from k to $k + 1$ if in the meeting two people are blue and one is red. The probability that the first one is blue is $k/(2N)$ since there are k blue people among $2N$. The probability that the second is also blue is then $(k - 1)/(2N - 1)$. Also, the probability that the third is red is $(2N - k)/(2N - 2)$ since there are $2N - k$ red people among the $2N - 2$ who remain after picking two blue. Finally, there are three orderings in which one could pick one red and two blue.

Similarly, for $k \in \{2, \ldots, 2N - 2\}$,

$$P[X_{n+1} = k - 1 \mid X_n = k] = 3 \frac{(2N-k)(2N-k-1)k}{2N(2N-1)(2N-2)} =: q(k).$$

We want to calculate

$$\alpha(k) = P[T_{2N} < T_0 \mid X_0 = k],$$

where T_0 is the first time that $X_n = 0$ and T_{2N} is the first time that $X_n = 2N$. Then, $\alpha(N)$ is the probability that the population eventually becomes all red.

The first step equations are, for $k \in \{2, \ldots, 2N - 2\}$,

$$\alpha(k) = p(k)\alpha(k+1) + q(k)\alpha(k-1) + (1 - p(k) - q(k))\alpha(k),$$

i.e.,

$$(p(k) + q(k))\alpha(k) = p(k)\alpha(k+1) + q(k)\alpha(k-1).$$

The boundary conditions are $\alpha(1) = 0$, $\alpha(2N - 1) = 1$.

We solve these equations numerically, using Python. Our procedure is as follows. We let $\alpha(1) = 0$ and $\alpha(2) = A$, for some constant A. We then solve recursively

$$\alpha(k+1) = (1 + \frac{q(k)}{p(k)})\alpha(k) - \frac{q(k)}{p(k)}\alpha(k-1), k = 2, 3, \ldots, 2N - 2$$

$$= \left(1 + \frac{2N-k-1}{(1-p)(k-1)}\right)\alpha(k)$$

$$- \frac{2N-k-1)}{(1-p)(k-1)}\alpha(k-1), k = 2, 3, \ldots, 2N - 2.$$

Eventually, we find $\alpha(2N - 1)$. This value is proportional to A. Since $\alpha(2N - 1) = 1$, we then divide all the $\alpha(k)$ by $\alpha(2N - 1)$. Not elegant, but it works. We repeat this process for $p = 0, 0.02, 0.04, \ldots, 0.14$. Figure 5.3 shows the results for $N = 450$, i.e., for a population of 900 people.

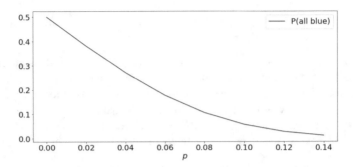

Fig. 5.3 The effect of the media. Here, p is the probability that someone remains red after chatting with two blue people. The graph shows the probability that the whole population turns blue instead of red. A small amount of persuasion goes a long way

5.5   Polarization

In most countries, the population is split among different political and religious persuasions. How is this possible if everyone is faced with the same evidence? One effect is that interactions are not fully mixing. People belong to groups that may converge to a consensus based on the majority opinion of the group.

To model this effect, we consider a population of N people. An adjacency matrix G specifies which people are friends. Here, $G(v, w) = 1$ if v and w are friends and $G(v, w) = 0$ otherwise.

Initially, people are blue or red with equal probabilities. We pick one person at random. If that person has a majority of red friends, she becomes red. If the majority of her friends are blue, she becomes blue. If it is a tie, she does not change. We repeat the process. Note that the graph does not change; it is fixed throughout. We want to explore how the coloring of people evolves over time.

Let $X_n(v) \in \{B, R\}$ be the state of person v at time n, for $n \geq 0$ and $v \in \{1, \ldots, N\}$. We pick v at random. We count the number of red friends and blue friends of v. They are given by

$$\sum_w G(v, w) 1\{X_n(w) = R\} \text{ and } \sum_w G(v, w) 1\{X_n(w) = B\}.$$

Thus,

$$X_{n+1}(v) = \begin{cases} R, & \text{if } \sum_w G(v, w) 1\{X_n(w) = R\} > \sum_w G(v, w) 1\{X_n(w) = B\} \\ B, & \text{if } \sum_w G(v, w) 1\{X_n(w) = R\} < \sum_w G(v, w) 1\{X_n(w) = B\} \\ X_n(v), & \text{otherwise.} \end{cases}$$

We have the following result.

Theorem 5.3 *The state* $X_n = \{X_n(v), v = 1, \ldots, N\}$ *of the system always converges. However, the limit may be random.*

∎

Proof Define the function $V(X_n)$ as follows:

$$V(X_n) = \sum_v \sum_w 1\{X_n(v) \neq X_n(w)\}.$$

That is, $V(X_n)$ is the number of disagreements among friends. The rules of evolution guarantee that $V(X_{n+1}) \leq V(X_n)$ and that $P(V(X_{n+1}) < V(X_n)) > 0$ unless $P(X_{n+1} = X_n) = 1$. Indeed, if the state of v changes, it is to make that person agree with more of her neighbors. Also, if there is no v who can reduce her number of disagreements, then the state can no longer change. These properties imply that the state converges.

A simple example shows that the limit may be random. Consider four people at the vertices of a square that represents G. Assume that two opposite vertices are blue and the other two are red. If the first person v to reconsider her opinion is blue, she turns red, and the limit is all red. If v is red, the limit is all blue. Thus, the limit is equally likely to be all red or all blue. □

In the limit, it may be that a fraction of the nodes are red and the others are blue. For instance, if the nodes are arranged in a line graph, then the limit is alternating sequences of at least two red nodes and sequences of at least two blue nodes.

The properties of the limit depend on the adjacency graph G. One might think that a close group of friends should have the same color, but that is not necessarily the case, as the example of Fig. 5.4 shows.

Fig. 5.4 A close group of friends, the four vertices of the square, do not share the same color

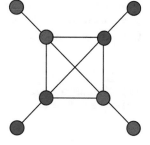

5.6 $M/M/1$ Queue

We discuss a simple model of a queue, called an $M/M/1$ queue. We turn to networks
in the next section. This section uses concepts from continuous-time Markov chains
that we develop in the next chapter. Thus, the discussion here is a bit informal, but
is hopefully clear enough to be read first.

Figure 5.5 illustrates a queue where customers (this is the standard terminology)
arrive and a server serves them one at a time, in a first come, first served order.
The times between arrivals are independent and exponentially distributed with
rate λ. Thus, the average time between two consecutive arrivals is $1/\lambda$, so that λ
customers arrive per unit of time, on average. The service times are independent
and exponentially distributed with rate μ. The durations of the service times and the
arrival times are independent. The expected value of a service time is $1/\mu$. Thus, if
the queue were always full, there would be μ service completions per unit time, on
average. If $\lambda < \mu$, the server can keep up with the arrivals, and the queue should
empty regularly. If $\lambda > \mu$, one can expect the number of customers in the queue to
increase without bound.

In the notation $M/M/1$, the first M indicates that the inter-arrival times are
memoryless, the second M indicates that the service times are memoryless, and the
1 indicates that there is one server. As you may expect, there are related notations
such as $D/M/3$ or $M/G/5$, and so on, where the inter-arrival times and the service
times have other properties and there are multiple servers.

Let X_t be the number of customers in the queue at time t, for $t \geq 0$. We call X_t
the queue length process. The middle part of Fig. 5.5 shows a possible realization
of that process. Observing the queue length process up to some time t provides
information about previous inter-arrival times and service times and also about
when the last arrival occurred and when the last service started. Since the inter-
arrival times and service times are independent and memoryless, this information
is independent of the time until the next arrival or the next service completion. In
particular, given $\{X_s, s \leq t\}$, the likelihood that a new arrival occurs during $(t, t+\epsilon]$
is approximately $\lambda\epsilon$ for $\epsilon \ll 1$; also the likelihood that a service completes during
$(t, t + \epsilon]$ is approximately $\mu\epsilon$ if $X_t > 0$ and zero if $X_t = 0$.

Fig. 5.5 An $M/M/1$ queue,
a possible realization, and its
state transition diagram

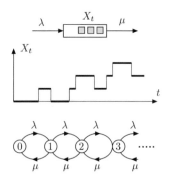

The bottom part of Fig. 5.5 is a *state transition diagram* that indicates the *rates* of transitions. For instance, the arrow from 1 to 2 is marked with λ to indicate that, in $\epsilon \ll 1$ s, the queue length jumps from 1 to 2 with probability $\lambda\epsilon$. The figure shows that arrivals (that increase the queue length) occur at the same rate λ, independently of the queue length. Also, service completions (that reduce the queue length) occur at rate μ as long as the queue is nonempty.

Note that

$$P(X_{t+\epsilon} = 0) = P(X_t = 0, X_{t+\epsilon} = 0) + P(X_t = 1, X_{t+\epsilon} = 0)$$
$$\approx P(X_t = 0)(1 - \lambda\epsilon) + P(X_t = 1)\mu\epsilon.$$

The first identity is the law of total probability: the event $\{X_{t+\epsilon} = 0\}$ is the union of the two disjoint events $\{X_t = 0, X_{t+\epsilon} = 0\}$ and $\{X_t = 1, X_{t+\epsilon} = 0\}$. The second identity uses the fact that $\{X_t = 0, X_{t+\epsilon} = 0\}$ occurs when $X_t = 0$ and there is no arrival during $(t, t+\epsilon]$. This event has probability $P(X_t = 0)$ multiplied by $(1 - \lambda\epsilon)$ since arrivals are independent of the current queue length. The other term is similar.

Now, imagine that π is a pmf on $\mathbb{Z}_{\geq 0} := \{0, 1, \ldots\}$ such that $P(X_t = i) = \pi(i)$ for all time t and $i \in \mathbb{Z}_{\geq 0}$. That is, assume that π is an *invariant distribution* for X_t. In that case, $P(X_{t+\epsilon} = 0) = \pi(0)$, $P(X_t = 0) = \pi(0)$, and $P(X_t = 1) = \pi(1)$. Hence, the previous identity implies that

$$\pi(0) \approx \pi(0)(1 - \lambda\epsilon) + \pi(1)\mu\epsilon.$$

Subtracting $\pi(0)(1 - \lambda\epsilon)$ from both terms gives

$$\pi(0)\lambda\epsilon \approx \pi(1)\mu\epsilon.$$

Dividing by ϵ shows that[1]

$$\pi(0)\lambda = \pi(1)\mu. \tag{5.1}$$

Similarly, for $i \geq 1$, one has

$$P(X_{t+\epsilon} = i) = P(X_t = i - 1, X_{t+\epsilon} = i) + P(X_t = i, X_{t+\epsilon} = i)$$
$$+ P(X_t = i + 1, X_{t+\epsilon} = i)$$
$$\approx P(X_t = i - 1)\lambda\epsilon + P(X_t = i)(1 - \lambda\epsilon - \mu\epsilon) + P(X_t = i + 1)\mu\epsilon.$$

Hence,

$$\pi(i) \approx \pi(i - 1)\lambda\epsilon + \pi(i)(1 - \lambda\epsilon - \mu\epsilon) + \pi(i + 1)\mu\epsilon.$$

---

[1] Technically, the $\approx$ sign is an identity up to a term negligible in ϵ. When we divide by ϵ and let $\epsilon \to 0$, the $\approx$ becomes an identity.

This relation implies that

$$\pi(i)(\lambda + \mu) = \pi(i-1)\lambda + \pi(i+1)\mu, i \geq 1. \tag{5.2}$$

The Eqs. (5.1)–(5.2) are called the *balance equations*. Thus, if π is invariant for X_t, it must satisfy the balance equations. Looking back at our calculations, we also see that if π satisfies the balance equations, and if $P(X_t = i) = \pi(i)$ for all i, then $P(X_{t+\epsilon} = i) = \pi(i)$ for all i. Thus, π is invariant for X_t if and only if it satisfies the balance equations.

One can solve the balance equations (5.1)–(5.2) as follows. Equation (5.1) shows that $\pi(1) = \rho\pi(0)$ with $\rho = \lambda/\mu$. Subtracting (5.1) from (5.2) yields

$$\pi(1)\lambda = \pi(2)\mu.$$

This equation then shows that $\pi(2) = \pi(1)\rho = \pi(0)\rho^2$. Continuing in this way shows that $\pi(n) = \pi(0)\rho^n$ for $n \geq 0$. To find $\pi(0)$, we use the fact that $\sum_n \pi(n) = 1$. That is

$$\sum_{n=0}^{\infty} \pi(0)\rho^n = 1.$$

If $\rho \geq 1$, i.e., if $\lambda \geq \mu$, this is not possible. In that case there is no invariant distribution. If $\rho < 1$, then the previous equation becomes

$$\pi(0)\frac{1}{1-\rho} = 1,$$

so that $\pi(0) = 1 - \rho$ and

$$\pi(n) = (1-\rho)\rho^n, n \geq 0.$$

In particular, when X_t has the invariant distribution π, one has

$$E(X_t) = \sum_{n=0}^{\infty} n(1-\rho)\rho^n = \frac{\rho}{1-\rho} = \frac{\lambda}{\mu - \lambda} =: L.$$

To calculate the average delay W of a customer in the queue, one can use Little's Law $L = \lambda W$. This identity implies that

$$W = \frac{1}{\mu - \lambda}.$$

Another way of deriving this expression is to realize that if a customer finds k other customers in the queue upon his arrival, he has to wait for $k + 1$ service completions before he leaves. Since very service completions lasts $1/\mu$ on average, his average delay is $(k+1)/\mu$. Now, the probability that this customer finds k other customers in the queue is $\pi(k)$. To see this, note that the probability that a customer who enters the queue between time t and $t+\epsilon$ finds k customers in the queue is

$$P[X_t = k \mid X_{t+\epsilon} = X_t + 1].$$

Now, the conditioning event is independent of X_t, because the arrivals occur at rate λ, independently of the queue length. Thus, the expression above is equal to $P(X_t = k) = \pi(k)$. Hence,

$$W = \sum_{k=0}^{\infty} \frac{k+1}{\mu} \pi(k) = \sum_{k=0}^{\infty} \frac{k+1}{\mu} (1-\rho)\rho^k = \frac{1}{\mu - \lambda},$$

as some simple algebra shows.

5.7 Network of Queues

Figure 5.6 shows a representative network of queues. Two types of customers arrive into the network, with respective rates γ_1 and γ_2. The first type goes through queue 1, then queue 3, and should leave the network. However, with probability p_1 these customers must go back to queue 1 and try again. In a communication network, this event models an transmission error where a packet (a group of bits) gets corrupted and has to be retransmitted. The situation is similar for the other type. Thus, in $\epsilon \ll 1$ time unit, a packet of the first type arrives with probability $\gamma_1\epsilon$, independently of what happened previously. This is similar to the arrivals into an $M/M/1$ queue. Also, we assume that the service times are exponentially distributed with rate μ_k in queue k, for $k = 1, 2, 3$.

Let X_t^k be the number of customers in queue k at time t, for $k = 1, 2$ and $t \geq 0$. Let also X_t^3 be the list of customer types in queue 3 at time t. For instance, in Fig. 5.6, one has $X_t^3 = (1, 1, 2, 1)$, from tail to head of the queue to indicate that the customer at the head of the queue is of type 1, that he is followed by a customer of

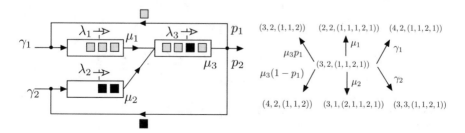

Fig. 5.6 A network of queues

type 2, etc. Because of the memoryless property of the exponential distribution, the process $X_t = (X_t^1, X_t^2, X_t^3)$ is a Markov chain: observing the past up to time t does not help predict the time of the next arrival or service completion.

Figure 5.6 shows the transition rates out of the current state $(3, 2, (1, 1, 2, 1))$. For instance, with rate $\mu_3 p_1$, a service completes in queue 3 and that customer has to go back to queue 1, so that the new state is $(4, 2, (1, 1, 2))$. The other transitions are similar.

One can then, in principle, write down the balance equations and try to solve them. This looks like a very complex task and it seems very unlikely that one could solve these equations analytically. However, a miracle occurs and one has the remarkably simple result stated in the next theorem. Before we state the result, we need to define λ_1, λ_2, and λ_3. As sketched in Fig. 5.6, for $k = 1, 2, 3$, the quantity λ_k is the rate at which customers go through queue k, in the long term. These rates should be such that

$$\lambda_1 = \gamma_1 + \lambda_1 p_1$$

$$\lambda_2 = \gamma_2 + \lambda_2 p_2$$

$$\lambda_3 = \lambda_1 + \lambda_2.$$

For instance, the rate λ_1 at which customers enter queue 1 is the rate γ_1 plus the rate at which customers of type 1 that leave queue 3 are sent back to queue 1. Customers of type 1 go through queue 3 at rate λ_1, since they come out of queue 1 at rate λ_1; also, a fraction p_1 of these customers go back to queue 1. The other expressions can be understood similarly. The equations above are called the *flow conservation equations*.

These equations admit the following solution:

$$\lambda_1 = \frac{\gamma_1}{1 - p_1}, \lambda_2 = \frac{\gamma_2}{1 - p_2}, \lambda_3 = \lambda_1 + \lambda_2.$$

Theorem 5.4 (Invariant Distribution of Network) *Assume $\lambda_k < \mu_k$ and let $\rho_k = \lambda_k/\mu_k$, for $k = 1, 2, 3$. Then the Markov chain X_t has a unique invariant distribution π that is given by*

$$\pi(x_1, x_2, x_3) = \pi_1(x_1)\pi_2(x_2)\pi_3(x_3)$$

$$\pi_k(n) = (1 - \rho_k)\rho_k^n, n \geq 0, k = 1, 2$$

$$\pi_3(a_1, a_2, \ldots, a_n) = p(a_1)p(a_2) \cdots p(a_n)(1 - \rho_3)\rho_3^n,$$

$$n \geq 0, a_k \in \{1, 2\}, k = 1, \ldots, n,$$

where $p(1) = \lambda_1/(\lambda_1 + \lambda_2)$ and $p(2) = \lambda_2/(\lambda_1 + \lambda_2)$.

∎

This result shows that the invariant distribution has a *product form*.

We prove this result in the next chapter. It indicates that under the invariant distribution π, the states of the three queues are independent. Moreover, the state of queue 1 has the same invariant distribution as an $M/M/1$ queue with arrival rate λ_1 and service rate μ_1, and similarly for queue 2. Finally, queue 3 has the same invariant distribution as a single queue with arrival rates λ_1 and λ_2 and service rate μ_3: the length of queue 3 has the same distribution as an $M/M/1$ queue with arrival rate $\lambda_1 + \lambda_3$ and the types of the customers in the queue are independent and of type 1 with probability $p(1)$ and 2 with probability $p(2)$.

This result is remarkable not only for its simplicity but mostly because it is surprising. The independence of the states of the queues is shocking: the arrivals into queue 3 are the departures from the other two queues, so it seems that if customers are delayed in queues 1 and 2, one should have larger values for X_t^1 and X_t^2 and a smaller one for the length of queue 3. Thus, intuition suggests a strong dependency between the queue lengths. Moreover, the fact that the invariant distributions of the queues are the same as for $M/M/1$ queues is also shocking. Indeed, if there are many customers in queue 1, we know that a fraction of them will come back into the queue, so that future arrivals into queue 1 depend on the current queue length, which is not the case for an $M/M/1$ queue. The paradox is explained in a reference.

We use this theorem to calculate the delay of customers in the network.

Theorem 5.5 *For $k = 1, 2$, the average delay W_k of customers of type k is given by*

$$W_k = \frac{1}{1 - p_k}\left(\frac{1}{\mu_k - \lambda_k} + \frac{1}{\mu_3 - \lambda_1 - \lambda_2}\right),$$

where

$$\lambda_1 = \frac{\gamma_1}{1 - p_1} \quad \text{and} \quad \lambda_2 = \frac{\gamma_2}{1 - p_2}.$$

∎

Proof We use Little's Law that says that $L_k = \gamma_k W_k$ where L_k is the average number of customers of type k in the network. Consider the case $k = 1$. The other one is similar. L_1 is the average number of customers in queue 1 plus the average number of customers of type 1 in queue 3.

The average length of queue 1 is $\lambda_1/(\mu_1 - \lambda_1)$ because the invariant distribution of queue 1 is the same as that of an $M/M/1$ queue with arrival rate λ_1 and service rate μ_1.

The average length of queue 3 is $(\lambda_1 + \lambda_2)/(\mu_3 - \lambda_1 - \lambda_2)$ because the invariant distribution of queue 3 is the same as queue with arrival rate λ_1 and λ_2 and service rate μ_3. Also, the probability that any customer in queue 3 is of type 1 is $p(1) = \lambda_1/(\lambda_1 + \lambda_2)$. Thus, the average number of customers of type 1 in queue 3 is

$$p(1)\frac{\lambda_1 + \lambda_2}{\mu_3 - \lambda_1 - \lambda_2} = \frac{\lambda_1}{\mu_3 - \lambda_1 - \lambda_2}.$$

Hence,

$$L_1 = \frac{\lambda_1}{\mu_1 - \lambda_1} + \frac{\lambda_1}{\mu_3 - \lambda_1 - \lambda_2}.$$

Combined with Little's Law, this expression yields W_1.   □

5.8   Optimizing Capacity

We use our network model to optimize the rates of the transmitters. The basic idea is that nodes with more traffic should have faster transmitter. To make this idea precise, we formulate an optimization problem: minimize a delay cost subject to a given budget for buying the transmitters.

We carry out the calculations not because of the importance of the specific example (it is not important!) but because they are representative of problems of this type.

Consider once again the network in Fig. 5.6. Assume that the cost of the transmitters is $c_1\mu_1 + c_2\mu_2 + c_3\mu_3$. The delay cost is $d_1 W_1 + d_2 W_2$ where W_k is the average delay for packets of type k ($k = 1, 2$). The problem is then as follows:

$$\text{Minimize } D(\mu_1, \mu_2, \mu_3) := d_1 W_1 + d_2 W_2$$

$$\text{subject to } C(\mu_1, \mu_2, \mu_3) := c_1\mu_1 + c_2\mu_2 + c_3\mu_3 \leq B.$$

Thus, the objective function is

$$D(\mu_1, \mu_2, \mu_3) = \sum_{k=1,2} \frac{d_k}{1 - p_k}\left(\frac{1}{\mu_k - \lambda_k} + \frac{1}{\mu_3 - \lambda_1 - \lambda_2}\right).$$

We convert the constrained optimization problem into an unconstrained one by replacing the constraint by a penalty. That is, we consider the problem

$$\text{Minimize } D(\mu_1, \mu_2, \mu_3) + \alpha(C(\mu_1, \mu_2, \mu_3) - B),$$

where $\lambda > 0$ is a Lagrange multiplier that penalizes capacities that have a high cost. To solve this problem for a given value of λ, we set to zero the derivative of this expression with respect to each μ_k. For $k = 1, 2$ we find

$$0 = \frac{\partial}{\partial \mu_k} D(\mu_1, \mu_2, \mu_3) + \frac{\partial}{\partial \mu_1} \alpha C(\mu_1, \mu_2, \mu_3)$$

$$= -\frac{d_k}{1 - p_k} \frac{1}{(\mu_k - \lambda_k)^2} + \alpha c_k.$$

For $k = 3$, we find

$$0 = -\frac{d_1/(1 - p_1) + d_2/(1 - p_2)}{(\mu_3 - \lambda_1 - \lambda_2)^2} + \alpha c_3.$$

Hence,

$$\mu_k = \lambda_k + \left(\frac{d_k}{\alpha c_k(1 - p_k)}\right)^{1/2}, \quad \text{for } k = 1, 2$$

$$\mu_3 = \lambda_1 + \lambda_2 + \left(\frac{d_1/(1 - p_1) + d_2/(1 - p_2)}{\alpha c_3}\right)^{1/2}.$$

These identities express μ_1, μ_2, and μ_3 in terms of α. Using these expressions in $C(\mu_1, \mu_2, \mu_3)$, we find that the cost is given by

$$C(\mu_1, \mu_2, \mu_3) = c_1\lambda_1 + c_2\lambda_2 + c_3(\lambda_1 + \lambda_3)$$

$$+ \frac{1}{\sqrt{\alpha}}\left[\sum_{k=1,2}\left(\frac{d_k c_k}{1 - p_k}\right)^{1/2} + c_3^{1/2}\left(\sum_{k=1,2}\frac{d_k}{1 - p_k}\right)^{1/2}\right].$$

Using $C(\mu_1, \mu_2, \mu_3) = B$ then enables to solve for α. As a last step, we substitute that value of α in the expressions for the μ_k. We find,

$$\mu_k = \lambda_k + D\left(\frac{d_k}{c_k(1 - p_k)}\right)^{1/2}, \quad \text{for } k = 1, 2$$

$$\mu_3 = \lambda_1 + \lambda_2 + D\left(\sum_{k=1,2}\frac{d_k}{c_k(1 - p_k)}\right)^{1/2},$$

where

$$D = \frac{B - c_1\lambda_1 - c_2\lambda_2 - c_3(\lambda_1 + \lambda_2)}{\sum_{k=1,2}\left(\frac{d_k c_k}{1 - p_k}\right)^{1/2} + c_3^{1/2}\left(\sum_{k=1,2}\frac{d_k c_k}{1 - p_k}\right)^{1/2}}.$$

These results show that, for $k = 1, 2$, the capacity μ_k increases with d_k, i.e., the cost of delays of packets of type k; it also decreases with c_k, i.e., the cost of providing that capacity.

A numerical solution can be obtained using a scipy optimization tool called *minimize*. Here is the code.

```
import numpy as np
from scipy.optimize import minimize
d = [1, 2] # delay cost coefficients
c = [2, 3, 4] # capacity cost coefficients
l = [3, 2] # rates l[0] = lambda1, etc
p = [0.1, 0.2] # error probabilities
B = 60 # capacity budget
UB = 50 # upper bound on capacity
# x = mu1, mu2, mu3: x[0] = mu1, etc
def objective(x): # objective to minimize
    z = 0
    for k in range(2):
        z = z + (d[k]/(1 - p[k]))*(1/(x[k] - l[k])
                + 1/(x[2] - l[0]-l[1]))
    return z
def constraint(x): # budget constraint >= 0
    z = B
    for k in range(3):
        z = z - c[k]*x[k]
    return z
x0 = [5,5,10] # initial value for optimization
b0 = (l[0], UB)  # lower and upped bound for x[0]
b1 = (l[1], UB)  # lower and upped bound for x[1]
b2 = (l[0]+l[1], UB) # lower and upped bound for x[1]
bnds = (b0,b1,b2) # bounds for the three variables x
con = {'type': 'ineq', 'fun': constraint}
        # specifies constraints
sol = minimize(objective,x0,method='SLSQP',
        bounds = bnds, constraints=con)
# sol will be the solution
print(sol)
```

The code produces an approximate solution. The advantage is that one does not need any analytical skills. The disadvantage is that one does not get any qualitative insight.

5.9   Internet and Network of Queues

Can one model the internet as a network of queues? If so, does the result of the previous section really apply? Well, the mathematical answers are maybe and maybe.

The internet transports packets (groups of bits) from node to node. The nodes are sources and destinations such as computers, webcams, smartphones, etc., and network nodes such as switches or routers. The packets go from buffer to buffer. These buffers look like queues. The service times are the transmission times of packets. The transmission time of a packet (in seconds) is the number of bits in the packet divided by the rate of the transmitter (in bits per second). The packets have random lengths, so the service times are random. So, the internet looks like a network of queues. However, there are some important ways in which our network of queues is not an exact model of the internet. First, the packet lengths are not exponentially distributed. Second, a packet keeps the same number of bits as it moves from one queue to the next. Thus, the service times of a given packet in the different queues are all proportional to each other. Third, the time between the arrival two successive packets from a given node cannot be smaller than the transmission time of the first packet. Thus, the arrival times and the service times in one queue are not independent and the times between arrivals are not exponentially distributed.

The real question is whether the internet can be approximated by a network similar to that of the previous section. For instance, if we use that model, are we very far off when we try to estimate delays of queue lengths? Experiments suggest that the approximation may be reasonable to a first order. One intuitive justification is the diversity of streams of packets. It goes as follows. Consider one specific queue in a large network node of the internet. This node is traversed by packets that come from many different sources and go to many destinations. Thus, successive packets that arrive at the queue may come from different previous nodes, which reduces the dependency of the arrivals and the service times. The service time distribution certainly affects the delays. However, the results obtained assuming an exponential distribution may provide a reasonable estimate.

5.10   Product-Form Networks

The example of the previous sections generalizes as follows. There are $N \geq 1$ queues and $C \geq 1$ classes of customers. At each queue i, customers of class $c \in \{1, \ldots, C\}$ arrive with rate γ_i^c, independently of the past and of other arrivals. Queue i serves customers with rate μ_i. When a customer of class c completes service in queue i, it goes to queue j and becomes a customer of class d with probability $r_{i,j}^{c,d}$, for $i, j \in \{1, \ldots, N\}$ and $c, d \in \{1, \ldots, C\}$. That customer leaves the network with probability $r_{i,0}^c = 1 - \sum_{j=1}^{N} \sum_{d=1}^{C} r_{i,j}^{c,d}$. That is, a customer of class c who completes service in queue i either goes to another queue or leaves the network.

Define λ_i^c as the average rate of customers of class c that go through queue i, for $i \in \{1, \ldots, N\}$ and for $c \in \{1, \ldots, C\}$. Assume that the rate of arrivals of customers of a given class into a queue is equal to the rate of departures of those customers from the queue. Then the rates λ_i^c should satisfy the following *flow conservation equations*:

$$\lambda_i^c = \gamma_i^c + \sum_{j=1}^{N} \sum_{d=1}^{C} r_{j,i}^{d,c}, i \in \{1, \ldots, N\}, c \in \{1, \ldots, C\}.$$

Let also $X(t) = \{X_i(t), i = 1, \ldots, N\}$ where $X_i(t)$ is the configuration of queue i at time $t \geq 0$. That is, $X_i(t)$ is the list of customer classes in queue i, from the tail of the queue to the head of the queue. For instance, $X_i(t) = 132{,}312$ if the customer at the tail of queue i is of class 1, the customer in front of her is of class 3, and so on, and the customer at the head of the queue and being served is of class 2. If the queue is empty, then $X_i(t) = []$, where $[]$ designates the empty string.

One then has the following theorem.

Theorem 5.6 (Product-Form Networks)

(a) *Let* $\{\lambda_i^c, i = 1, \ldots, N; c - 1, \ldots, C\}$ *be a solution of the flow conservation equations. If* $\lambda_i := \sum_{c=1}^{C} \lambda_i^c < \mu_i$ *for* $i = 1, \ldots, N$, *then* X_t *is a Markov chain and its invariant distribution is given by*

$$\pi(x) = A\Pi_{i=1}^{N} g_i(x_i),$$

where

$$g_i(c_1 \cdots c_n) = \frac{\lambda_i^{c_1} \cdots \lambda_i^{c_N}}{\mu_i^n}$$

and A *is a constant such that* π_i *sums to one over all the possible configurations of the queues.*

(b) *If the network is* open *in that every customer can leave the network, then the invariant distribution becomes*

$$\pi(x) = \Pi_{i=1}^{N} \pi_i(x_i),$$

where

$$\pi_i(c_1 \cdots c_n) = \left(1 - \frac{\lambda_i}{\mu_i}\right) \frac{\lambda_i^{c_1} \cdots \lambda_i^{c_n}}{\mu_i^n}.$$

In this case, under the invariant distribution, the queue lengths at time t *are all independent, the length of queue* i *has the same distribution as that of an*

Fig. 5.7 A network of
queues

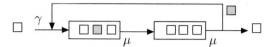

*M/M/1 queue with arrival rate λ_i and service rate μ_i, and the customer classes
are all independent and are equal to c with probability λ_i^c/λ_i.*

The proof of this theorem is the same as that of the particular example given in
the next chapter.

5.10.1 Example

Figure 5.7 shows a network with two types of jobs. There is a single gray job that
visits the two queues as shown. The white jobs go through the two queues once. The
gray job models "hello" messages that the queues keep on exchanging to verify that
the system is alive. For ease of notation, we assume that the service rates in the two
queues are identical.

We want to calculate the average time that the white jobs spend in the system
and compare that value to the case when there is no gray job. That is, we want
to understand the "cost" of using hello messages. The point of the example is to
illustrate the methodology for networks where some customers never leave. The
calculations show the following somewhat surprising result.

Theorem 5.7 *Using a hello message increases the expected delay of the white jobs
by 50%.*

We prove the theorem in the next chapter. In that proof, we use Theorem 5.6 to
calculate the invariant distribution of the system, derive the expected number L of
white jobs in the network, then use Little's Law to calculate the average delay W of
the white jobs as $W = L/\gamma$. We then compare that value to the case where there is
not gray job.

5.11   References

The literature on social networks is vast and growing. The textbook Easley and
Kleinberg (2012) contains many interesting models and result. The text Shah (2009)
studies the propagation of information in networks.

The book Kelly (1979) is the most elegant presentation of the theory of queueing
networks. It is readily available online. The excellent notes Kelly and Yudovina
(2013) discuss recent results. The nice textbook Srikant and Ying (2014) explains
network optimization and other performance evaluation problems. The books

Bremaud (2017) and Lyons and Perez (2017) are excellent sources for deeper studies of networks. The text Walrand (1988) is more clumsy but may be useful.

5.12   Problems

Problem 5.1 There are K users of a social network who collaborate to estimate some quantity by exchanging information. At each step, a pair (i, j) of users is selected uniformly at random and user j sends a message to user i with his estimate. User i then replaces his estimate by the average of his estimate and that of user j. Show that the estimates of all the users converge in probability to the average value of the initial estimates. This is an example of *consensus algorithm*.

Hint Let $X_n(i)$ be the estimate of user i at step n and X_n the vector with components $X_n(i)$. Show that

$$E[X_{n+1}(i) \mid X_n] = (1 - \alpha)X_n(i) + \alpha A,$$

where $\alpha = 1/(2(K - 1))$ and $A = \sum_i X_0(i)/K$. Consequently,

$$E[|X_{n+1}(i) - A| \mid X_n] = (1 - \alpha)|X_n(i) - A|,$$

so that

$$E[|X_{n+1}(i) - A|] = (1 - \alpha)E[|X_n(i) - A|]$$

and

$$E[|X_n(i) - A|] \to 0.$$

Markov's inequality then shows that $P(|X_n(i) - A| > \epsilon) \to 0$ for any $\epsilon > 0$.

Problem 5.2 Jobs arrive at rate γ in the system shown in Fig. 5.8. With probability p, a customer is sent to queue 1, independently of the other jobs; otherwise, the job is sent to queue 2. For $i = 1, 2$, queue i serves the jobs at rate μ_i. Find the value of p that minimizes the average delay of jobs in the system. Compare the resulting average delay to that of the system where the jobs are in one queue and join the available server when they reach the head of the queue, and the fastest server if both are idle, as shown in the bottom part of Fig. 5.8.

Hint The system of the top part of the figure is easy to analyze: with probability p, a job faces the average delay $1/(\mu_1 - \gamma p)$ in the top queue and with probability $1 - p$ the job faces the average delay $1/(\mu_2 - \gamma(1 - p))$, One the finds the value of p that minimizes the expected delay. For the system in the bottom part of the figure, the state is n with $n \geq 2$ when there are at least two jobs and the two servers are

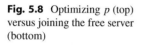

Fig. 5.8 Optimizing p (top) versus joining the free server (bottom)

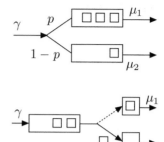

Fig. 5.9 The state transition diagram. Here, $\mu := \mu_1 + \mu_2$

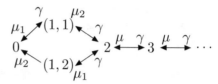

Fig. 5.10 The state transition diagram

busy, or $(1, s)$ where $s \in \{1, 2\}$ indicates which server is busy, or 0 when the system is empty. One then needs to find the invariant distribution of the state, compute the average number of jobs, and use Little's Law to find the average delay. The state transition diagram is shown in Fig. 5.9.

Problem 5.3 This problem compares parallel queues to a single queue. There are N servers. Each server serves customers at rate μ. The customers arrive at rate $N\lambda$. In the first system, the customers are split into N queues, one for each server. Customers arrive at each queue with rate λ. The average delay is that of an $M/M/1$ queue, i.e., $1/(\mu - \lambda)$. In the second system, the customers join a single queue. The customer at the head of the queue then goes to the next available server. Calculate the average delay in this system. Write a Python program to plot the average delays of the two systems as a function $\rho := \lambda/\mu$ for different values of N.

Hint The state diagram is shown in Fig. 5.10.

Problem 5.4 In this problem, we explore a system of parallel queues where the customers join the shortest queue. Customers arrive at rate $N\lambda$ and there are N queues, each with a server who serves customers at rate $\mu > \lambda$. When a customer arrives, she joins the shortest queue. The goal is to analyze the expected delay in the system. Unfortunately, this problem cannot be solved analytically. So, your task is to write a Python program to evaluate the expected delay numerically. The first

Fig. 5.11 The system

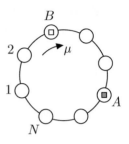

step is to draw the state transition diagram. Approximate the system by discarding customers who arrive when there are already M customers in the system. The second step is to write the balance equations. Finally, one writes a program to solve the equations numerically.

Problem 5.5 Figure 5.11 shows a system of N queues that serve jobs at rate μ. If there is a single job, it takes on average N/μ time units for it to go around the circle. Thus, the average rate at which a job leaves a particular queue is μ/N. Show that when there are two jobs, this rate is $2\mu/(N+1)$.

Networks—B

6

Application: Social Networks, Communication Networks
Topics: Continuous-Time Markov Chains, Product-Form Queueing Networks

6.1   Social Networks

We provide the proofs of the theorems in Sect. 5.1.

Theorem 6.1 (Spreading of a Message) *Let Z be the number of nodes that eventually receive the message.*

(a) If $\mu < 1$, then $P(Z < \infty) = 1$ and $E(Z) < \infty$;
(b) If $\mu > 1$, then $P(Z = \infty) > 0$.

Proof For part (a), let X_n be the number of nodes that are n steps from the root. If $X_n = k$, we can write $X_{n+1} = Y_1 + \cdots + Y_k$ where Y_j is the number of children of node j at level n. By assumption, $E(Y_j) = \mu$ for all j. Hence,

$$E[X_{n+1} \mid X_n = k] = E(Y_1 + \cdots + Y_k) = \mu k.$$

Hence, $E[X_{n+1} \mid X_n] = \mu X_n$. Taking expectations shows that $E(X_{n+1}) = \mu E(X_n), n \geq 0$. Consequently,

$$E(X_n) = \mu^n, n \geq 0.$$

© The Author(s) 2021   93
J. Walrand, *Probability in Electrical Engineering and Computer Science*,
https://doi.org/10.1007/978-3-030-49995-2_6

Now, the sequence $Z_n = X_0 + \cdots + X_n$ is nonnegative and increases to $Z = \sum_{n=0}^{\infty} Z_n$. By MCT, it follows that $E(Z_n) \to Z$. But

$$E(Z_n) = \mu_0 + \cdots + \mu^n = \frac{1 - \mu^{n+1}}{1 - \mu}.$$

Hence, $E(Z) = 1/(1 - \mu) < \infty$. Consequently, $P(Z < \infty) = 1$.

For part (b), one first observes that the theorem does not state that $P(Z = \infty) = 1$. For instance, assume that each node has three children with probability 0.5 and has no child otherwise. Then $\mu = 1.5 > 1$ and $P(Z = 1) = P(X_1 = 0) = 0.5$, so that $P(Z = \infty) \le 0.5 < 1$. We define X_n, Y_j, and Z_n as in the proof of part (a).

Let $\alpha_n = P(X_n > 0)$. Consider the X_1 children of the root. Since α_{n+1} is the probability that there is one survivor after $n + 1$ generations, it is the probability that at least one of the X_1 children of the root has a survivor after n generations. Hence,

$$1 - \alpha_{n+1} = E((1 - \alpha_n)^{X_1}), n \ge 0.$$

Indeed, if $X_1 = k$, the probability that none of the k children of the root has a survivor after n generations is $(1 - \alpha_n)^k$. Hence,

$$\alpha_{n+1} = 1 - E((1 - \alpha_n)^{X_1}) =: g(\alpha_n), n \ge 0.$$

Also, $\alpha_0 = 1$. As $n \to \infty$, one has $\alpha_n \to \alpha^* = P(X_n > 0, \text{ for all } n)$. Figure 6.1 shows that $\alpha^* > 0$. The key observations are that

$$g(0) = 0$$
$$g(1) = P(X_1 > 0) < 1$$
$$g'(0) = E(X_1(1 - \alpha)^{X_1 - 1})\,|_{\alpha=0} = \mu > 1$$
$$g'(1) = E(X_1(1 - \alpha)^{X_1 - 1})\,|_{\alpha=1} = 0,$$

so that the figure is as drawn.   □

Theorem 6.2 (Cascades) *Assume $p_k = p \in (0, 1]$ for all $k \ge 1$. Then, all nodes turn red with probability at least equal to θ where*

$$\theta = \exp\left\{-\frac{1 - p}{p}\right\}.$$

Proof The probability that node n does not listen to anyone is $a_n = (1 - p)^n$. Let X be the index of the first node that does not listen to anyone. Then

Fig. 6.1 The proof that
$\alpha^* > 0$

$$g(\alpha) = 1 - E((1-\alpha)^{X_1})$$

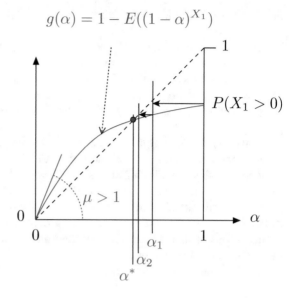

$$P(X > n) = (1-a_1)(1-a_2)\cdots(1-a_n) \le \exp\{-a_1 - \cdots - a_n\}$$

$$= \exp\left\{-\frac{1}{p}((1-p) - (1-p)^{n+1})\right\}.$$

Now,

$$P(X = \infty) = \lim_n P(X > n) \ge \exp\left\{-\frac{1-p}{p}\right\} = \theta.$$

Thus, with probability at least θ, every node listens to at least one previous node. When that is the case, all the nodes turn red. To see this, assume that n is the first blue node. That is not possible since it listened to some previous nodes that are all red. □

6.2   Continuous-Time Markov Chains

Our goal is to understand networks where packets travel from node to node until they reach their destination. In particular, we want to study the delay of packets from source to destination and the backlog in the nodes.

It turns out that the analysis of such systems is much easier in continuous time than in discrete time. To carry out such analysis, we have to introduce continuous-time Markov chains. We do this on a few simple examples.

6.2.1 Two-State Markov Chain

Figure 6.2 illustrates a random process $\{X_t, t \geq 0\}$ that takes values in $\{0, 1\}$. A random process is a collection of random variables indexed by $t \geq 0$. Saying that such a random process is defined means that one can calculate the probability that $\{X_{t_1} = x_1, X_{t_2} = x_2, \ldots, X_{t_n} = x_n\}$ for any value of $n \geq 1$, any $0 \leq t_1 \leq \cdots \leq t_n$, and $x_1, \ldots, x_n \in \{0, 1\}$. We explain below how one could calculate such a probability.

We call X_t the *state* of the process at time t. The possible values $\{0, 1\}$ are also called states. The state X_t evolves according to rules characterized by two positive numbers λ and μ. As Fig. 6.2 shows, if $X_0 = 0$, the state remains equal to zero for a random time T_0 that is exponentially distributed with parameter λ, thus with mean $1/\lambda$. The state X_t then jumps to 1 where it stays for a random time T_1 that is exponentially distributed with rate μ, independent of T_0, and so on. The definition is similar if $X_0 = 1$. In that case, X_t keeps the value 1 for an exponentially distributed time with rate μ, then jumps to 0, etc.

Thus, the pdf of T_0 is

$$f_{T_0}(t) = \lambda \exp\{-\lambda t\} 1\{t \geq 0\}.$$

In particular,

$$P(T_0 \leq \epsilon) \approx f_{T_0}(0)\epsilon = \lambda \epsilon, \text{ for } \epsilon \ll 1.$$

Throughout this chapter, the symbol $\approx$ means "up to a quantity negligible compared to ϵ." It is shown in Theorem 15.3 that exponentially distributed random variable is *memoryless*. That is,

$$P[T_0 > t + s \mid T_0 > t] = P(T_0 > s), s, t \geq 0.$$

The memoryless property and the independence of the exponential times T_k imply that $\{X_t, t \geq 0\}$ *starts afresh* from X_s at time s. Figure 6.3 illustrates that property. Mathematically, it says that given $\{X_t, t \leq s\}$ with $X_s = k$, the process $\{X_{s+t}, t \geq 0\}$ has the same properties as $\{X_t, t \geq 0\}$ given that $X_0 = k$, for $k = 0, 1$ and for any $s \geq 0$. Indeed, if $X_s = 0$, then the residual time that X_t remains in 0 is exponentially distributed with rate λ and is independent of what happened before

Fig. 6.2 A random process on $\{0, 1\}$

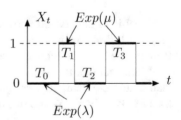

Fig. 6.3 The process X_t
starts afresh from X_s at time s

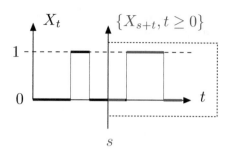

time s, because the time in 0 is memoryless and independent of the previous times
in 0 and 1. This property is written as

$$P[\{X_{s+t}, t \geq 0\} \in A \mid X_s = k; X_t, t \leq s] = P[\{X_t, t \geq 0\} \in A \mid X_0 = k],$$

for $k = 0, 1$, for all $s \geq 0$, and for all sets A of possible trajectories. A generic set
A of trajectories is

$$A = \{(x_t, t \geq 0) \in C_+ \mid x_{t_1} = i_1, \ldots, x_{t_n} = i_n\}$$

for given $0 < t_1 < \cdots < t_n$ and $i_1, \ldots, i_n \in \{0, 1\}$. Here, C_+ is the set of right-
continuous functions of $t \geq 0$ that take values in $\{0, 1\}$.

This property is the continuous-time version of the Markov property for Markov
chains. One says that the process X_t satisfies the *Markov property* and one calls
$\{X_t, t \geq 0\}$ is a *continuous-time Markov chain* (CTMC).

For instance,

$$P[X_{s+2.5} = 1, X_{s+4} = 0, X_{s+5.1} = 0 \mid X_s = 0; X_t, t \leq s]$$

$$= P[X_{2.5} = 1, X_4 = 0, X_{5.1} = 0 \mid X_0 = 0].$$

The Markov property generalizes to situations where s is replaced by a random
τ that is defined by a *causal* rule, i.e., a rule that does not look ahead. For instance,
as in Fig. 6.4, τ can be the second time that X_t visits state 0. Or τ could be the
first time that it visits state 0 after having spent at least 3 time units in state 1. The
property does not extend to non-causal times such as one time unit before X_t visits
state 1. Random times τ defined by causal rules are called *stopping times*. This more
general property is called the *strong Markov property*. To prove this property, one
conditions on the value s of τ and uses the fact that the future evolution does not
depend on this value since the event $\{\tau = s\}$ depends only on $\{X_t, t \leq s\}$.

For $0 < \epsilon \ll 1$ one has

$$P[X_{t+\epsilon} = 1 \mid X_t = 0] \approx \lambda\epsilon.$$

Fig. 6.4 The process X_t
starts afresh from X_τ at the
stopping time τ

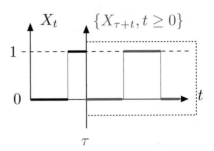

Fig. 6.5 The state transition
diagram

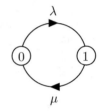

Indeed, the process jumps from 0 to 1 in ϵ time units if the exponential time in 0 is less than ϵ, which has probability approximately $\lambda\epsilon$.

Similarly,

$$P[X_{t+\epsilon} = 0 \mid X_t = 1] \approx \mu\epsilon.$$

We say that the *transition rate* from 0 to 1 is equal to λ and that from 1 to 0 is equal to μ to indicate that the probability of a transition from 0 to 1 in ϵ units of time is approximately $\lambda\epsilon$ and that from 1 to 0 is approximately $\mu\epsilon$.

Figure 6.5 illustrates these transition rates. This figure is called the *state transition diagram*.

The previous two identities imply that

$$P(X_{t+\epsilon} = 1) = P(X_t = 0, X_{t+\epsilon} = 1) + P(X_t = 1, X_{t+\epsilon} = 1)$$
$$= P(X_t{=}0)P[X_{t+\epsilon}{=}1 \mid X_t{=}0] + P(X_t{=}1)P[X_{t+\epsilon}{=}1 \mid X_t{=}1]$$
$$\approx P(X_t = 0)\lambda\epsilon + P(X_t = 1)(1 - P[X_{t+\epsilon} = 0 \mid X_t = 1])$$
$$\approx P(X_t = 0)\lambda\epsilon + P(X_t = 1)(1 - \mu\epsilon).$$

Also, similarly, one finds that

$$P(X_{t+\epsilon} = 0) \approx P(X_t = 0)(1 - \lambda\epsilon) + P(X_t = 1)\mu\epsilon.$$

We can write these identities in a convenient matrix notation as follows. For $t \geq 0$, one defines the row vector π_t as

$$\pi_t = [P(X_t = 0), P(X_t = 1)].$$

One also defines the *transition rate matrix* Q as follows:

$$Q = \begin{bmatrix} -\lambda & \lambda \\ \mu & -\mu \end{bmatrix}.$$

With that notation, the previous identities can be written as

$$\pi_{t+\epsilon} \approx \pi_t(\mathbf{I} + Q\epsilon),$$

where $\mathbf{I}$ is the identity matrix. Subtracting π_t from both sides, dividing by ϵ, and letting $\epsilon \rightarrow 0$, we find

$$\frac{d}{dt}\pi_t = \pi_t Q. \tag{6.1}$$

By analogy with the scalar equation $dx_t/dt = ax_t$ whose solution is $x_t = x_0 \exp\{at\}$, we conclude that

$$\pi_t = \pi_0 \exp\{Qt\}, \tag{6.2}$$

where

$$\exp\{Qt\} := \mathbf{I} + Qt + \frac{1}{2!}Q^2 t^2 + \frac{1}{3!}Q^3 t^3 + \cdots.$$

Note that

$$\frac{d}{dt}\exp\{Qt\} = \mathbf{0} + Q + Q^2 t + \frac{1}{2!}Q^3 t^2 + \cdots = Q\exp\{Qt\}.$$

Observe also that $\pi_t = \pi$ for all $t \geq 0$ if and only if $\pi_0 = \pi$ and

$$\pi Q = 0. \tag{6.3}$$

Indeed, if $\pi_t = \pi$ for all t, then (6.1) implies that $0 = \frac{d}{dt}\pi_t = \pi_t Q = \pi Q$. Conversely, if $\pi_0 = \pi$ with $\pi Q = 0$, then

$$\pi_t = \pi_0 \exp\{Qt\} = \pi \exp\{Qt\} = \pi\left(\mathbf{I} + Qt + \frac{1}{2!}Q^2 t^2 + \frac{1}{3!}Q^3 t^3 + \cdots\right) = \pi.$$

These equations $\pi Q = 0$ are called the *balance equations*. They are

$$[\pi(0), \pi(1)]\begin{bmatrix} -\lambda & \lambda \\ \mu & -\mu \end{bmatrix} = 0,$$

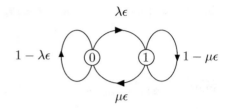

Fig. 6.6 A discrete-time approximation of X_t

i.e.,

$$\pi(0)(-\lambda) + \pi(1)\mu = 0$$
$$\pi(0)\lambda - \pi(1)\mu = 0.$$

These two equations are identical. To determine π, we use the fact that $\pi(0) + \pi(1) = 1$. Combined with the previous identity, we find

$$[\pi(0), \pi(1)] = \left[\frac{\mu}{\lambda + \mu}, \frac{\lambda}{\lambda + \mu} \right].$$

The identity $\pi_{t+\epsilon} \approx \pi_t(\mathbf{I} + Q\epsilon)$ shows that one can view $\{X_{n\epsilon}, n = 0, 1, \ldots\}$ as a discrete-time Markov chain with transition matrix $P = \mathbf{I} + Q\epsilon$. Figure 6.6 shows the transition diagram that corresponds to this transition matrix. The invariant distribution for P is such that $\pi P = \pi$, i.e., $\pi(\mathbf{I} + Q\epsilon) = \pi$, so that $\pi Q = 0$, not surprisingly.

Note that this discrete-time Markov chain is aperiodic because states have self-loops. Thus, we expect that

$$\pi_{n\epsilon} \to \pi, \text{ as } n \to \infty.$$

Consequently, we expect that, in continuous time,

$$\pi_t \to \pi, \text{ as } t \to \infty.$$

6.2.2   Three-State Markov Chain

The previous Markov chain alternates between the states 0 and 1. More general Markov chains visit states in a random order. We explain that feature in our next example with 3 states. Fortunately, this example suffices to illustrate the general case. We do not have to look at Markov chains with 4, 5, ... states to describe the general model.

Fig. 6.7 A three-state
Markov chain

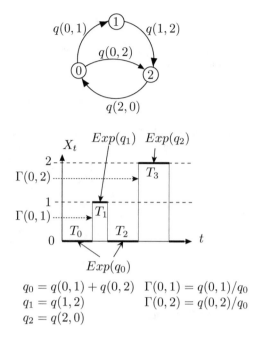

$$q_0 = q(0,1) + q(0,2) \quad \Gamma(0,1) = q(0,1)/q_0$$
$$q_1 = q(1,2) \quad\quad\quad\quad\quad \Gamma(0,2) = q(0,2)/q_0$$
$$q_2 = q(2,0)$$

In the example shown in Fig. 6.7, the rules of evolution are characterized
by positive numbers $q(0,1), q(0,2), q(1,2)$, and $q(2,0)$. One also defines
$q_0, q_1, q_2, \Gamma(0,1)$, and $\Gamma(0,2)$ as in the figure.

If $X_0 = 0$, the state X_t remains equal to 0 for some random time T_0 that
is exponentially distributed with rate q_0. At time T_0, the state jumps to 1 with
probability $\Gamma(0,1)$ or to state 2 otherwise, with probability $\Gamma(0,2)$. If X_t jumps
to 1, it stays there for an exponentially distributed time T_1 with rate q_1 that is
independent of T_0. More generally, when X_t enters state k, it stays there for a
random time that is exponentially distributed with rate q_k that is independent of the
past evolution. From this definition, it should be clear that the process X_t satisfies
the Markov property.

Define $\pi_t = [\pi_t(0), \pi_t(1), \pi_t(2)]$ where $\pi_t(k) = P(X_t = k)$ for $k = 0, 1, 2$.
One has, for $0 < \epsilon \ll 1$,

$$P[X_{t+\epsilon} = 1 \mid X_t = 0] \approx q_0 \epsilon \Gamma(0,1) = q(0,1)\epsilon.$$

Indeed, the process jumps from 0 to 1 in ϵ time units if the exponential time with
rate q_0 is less than ϵ and if the process then jumps to 1 instead of jumping to 2.

Similarly,

$$P[X_{t+\epsilon} = 2 \mid X_t = 0] \approx q_0 \epsilon \Gamma(0,2) = q(0,2)\epsilon.$$

Also,

$$P[X_{t+\epsilon} = 1 \mid X_t = 1] \approx 1 - q_1\epsilon,$$

since this is approximately the probability that the exponential time with rate q_1 is larger than ϵ. Moreover,

$$P[X_{t+\epsilon} = 1 \mid X_t = 2] \approx 0,$$

because the probability that both the exponential time with rate q_2 in state 2 and the exponential time with rate q_0 in state 0 are less than ϵ is roughly $(q_2\epsilon) \times (q_1\epsilon)$, and this is negligible compared to ϵ.

These observations imply that

$$\pi_{t+\epsilon}(1) = P(X_t = 0, X_{t+\epsilon} = 1) + P(X_t = 1, X_{t+\epsilon} = 1) + P(X_t = 2, X_{t+\epsilon} = 1)$$

$$= P(X_t=0)P[X_{t+\epsilon}=1 \mid X_t=0] + P(X_t=1)P[X_{t+\epsilon}=1 \mid X_t=1]$$

$$+ P(X_t = 2)P[X_{t+\epsilon} = 1 \mid X_t = 2]$$

$$\approx \pi_t(0)q(0, 1)\epsilon + \pi_t(1)(1 - q_1\epsilon).$$

Proceeding in a similar way shows that

$$\pi_{t+\epsilon}(0) \approx \pi_t(0)(1 - q_0\epsilon) + \pi_t(2)q(2, 0)\epsilon$$

$$\pi_{t+\epsilon}(2) \approx \pi_t(1)q(1, 2)\epsilon + \pi_t(2)(1 - q_2\epsilon).$$

Similarly to the two-state example, let us define the rate matrix Q as follows:

$$Q = \begin{bmatrix} -q_0 & q(0, 1) & q(0, 2) \\ 0 & -q_1 & q(0, 1) \\ q(2, 0) & 0 & -q_2 \end{bmatrix}.$$

The previous identities can then be written as follows:

$$\pi_{t+\epsilon} \approx \pi_t[\mathbf{I} + Q\epsilon].$$

Subtracting π_t from both sides, dividing by ϵ, and letting $\epsilon \to 0$ then shows that

$$\frac{d}{dt}\pi_t = \pi_t Q.$$

As before, the solution of this equation is

$$\pi_t = \pi_0 \exp\{Qt\}, t \geq 0.$$

The distribution π is invariant if and only if

Fig. 6.8 The transition matrix of the discrete-time approximation

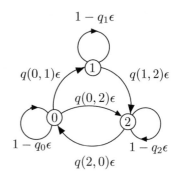

$$\pi Q = 0.$$

Once again, we note that $\{X_{n\epsilon}, n = 0, 1, \ldots\}$ is approximately a discrete-time Markov chain with transition matrix $P = I + Q\epsilon$ shown in Fig. 6.8. This Markov chain is aperiodic, and we conclude that

$$P(X_{n\epsilon} = k) \to \pi(k), \text{ as } n \to \infty.$$

Thus, we can expect that

$$\pi_t \to \pi, \text{ as } t \to \infty.$$

Also, since $X_{n\epsilon}$ is irreducible, the long-term fraction of time that it spends in the different states converge to π, and we can then expect the same for X_t.

6.2.3 General Case

Let $\mathscr{X}$ be a countable or finite set. The process $\{X_t, t \geq 0\}$ is defined as follows. One is given a probability distribution π on $\mathscr{X}$ and a *rate matrix* $Q = \{q(i, j), i, j \in \mathscr{X}\}$.

By definition, Q is such that

$$q(i, j) \geq 0, \forall i \neq j \text{ and } \sum_j q(i, j) = 0, \forall x.$$

Definition 6.1 (Continuous-Time Markov Chain) A continuous-time Markov chain with initial distribution π and rate matrix Q is a process $\{X_t, t \geq 0\}$ such that $P(X_0 = i) = \pi(i)$. Also,

$$P[X_{t+\epsilon} = j | X_t = i, X_u, u < t] = 1\{i = j\} + \epsilon q(i, j) + o(\epsilon).$$

◇

Fig. 6.9 Construction of a
continuous-time Markov
chain

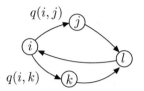

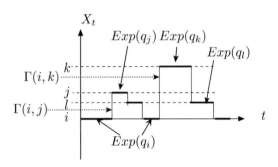

This definition means that the process jumps from i to $j \neq i$ with probability $q(i, j)\epsilon$ in $\epsilon \ll 1$ time units. Thus, $q(i, j)$ is the probability of jumping from i to j, per unit of time. Note that the sum of these expressions over all j gives 1, as should be.

One construction of this process is as follows. Say that $X_t = i$. One then chooses a random time τ that is exponentially distributed with rate $q_i := -q(i, i)$. At time $t + \tau$, the process jumps and goes to state y with probability $\Gamma(i, j) = q(i, j)/q_i$ for $j \neq i$ (Fig. 6.9).

Thus, if $X_t = i$, the probability that $X_{t+\epsilon} = j$ is the probability that the process jumps in $(t, t + \epsilon)$, which is $q_i \epsilon$, times the probability that it then jumps to j, which is $\Gamma(i, j)$. Hence,

$$P[X_{t+\epsilon} = j | X_t = i] = q_i \epsilon \frac{q(i, j)}{q_i} = q(i, j)\epsilon,$$

up to $o(\epsilon)$. Thus, the construction yields the correct transition probabilities.

As we observed in the examples,

$$\frac{d}{dt}\pi_t = \pi_t Q,$$

so that

$$\pi_t = \pi_0 \exp\{Qt\}.$$

Moreover, a distribution π is invariant if and only if it solves the balance equations

$$0 = \pi Q.$$

These equations, state by state, say that

$$\pi(i)q_i = \sum_{j \neq i} \pi(j)q(j, i), \forall i \in \mathcal{X}.$$

These equations express the equality of the rate of leaving a state and the rate of entering that state.

Define

$$P_t(i, j) = P[X_{s+t} = j \mid X_s = i], \text{ for } i, j \in \mathcal{X} \text{ and } s, t \geq 0.$$

The Markov property implies that

$$P(X_{t_1} = i_1, \ldots, X_{t_n} = i_n) = P(X_{t_1} = i_1)P_{t_2-t_1}(i_1, i_2)P_{t_3-t_2}(i_2, i_3) \cdots P_{t_n-t_{n-1}}(i_{n-1}, i_n),$$

for all $i_1, \ldots, i_n \in \mathcal{X}$ and all $0 < t_1 < \cdots < t_n$.

Moreover, this identity implies the Markov property. Indeed, if it holds, one has

$$P[X_{t_{m+1}} = i_{m+1}, \ldots, X_{t_n} = i_n \mid X_{t_1} = i_1, \ldots, X_{t_m} = i_m]$$

$$= \frac{P(X_{t_1} = i_1, \ldots, X_{t_n} = i_n)}{P(X_{t_1} = i_1, \ldots, X_{t_m} = i_m)}$$

$$= \frac{P(X_{t_1} = i_1)P_{t_2-t_1}(i_1, i_2)P_{t_3-t_2}(i_2, i_3) \cdots P_{t_n-t_{n-1}}(i_{n-1}, i_n)}{P(X_{t_1} = i_1)P_{t_2-t_1}(i_1, i_2)P_{t_3-t_2}(i_2, i_3) \cdots P_{t_{m-1}-t_{m-2}}(i_{m-2}, i_{m-1})}$$

$$= P_{t_m-t_{m-1}}(i_{m-1}, i_m) \cdots P_{t_n-t_{n-1}}(i_{n-1}, i_n).$$

Hence,

$$P[X_{t_{m+1}} = i_{m+1}, \ldots, X_{t_n} = i_n \mid X_{t_1} = i_1, \ldots, X_{t_m} = i_m]$$

$$= \frac{P(X_{t_{m-1}} = i_{m-1})P_{t_m-t_{m-1}}(i_{m-1}, i_m) \cdots P_{t_n-t_{n-1}}(i_{n-1}, i_n)}{P(X_{t_{m-1}} = i_{m-1})P_{t_m-t_{m-1}}(i_{m-1}, i_m)}$$

$$= \frac{P(X_{t_{m-1}} = i_{m-1}, \ldots, X_{t_n} = i_n)}{P(X_{t_{m-1}} = i_{m-1})}$$

$$= P[X_{t_m} = i_m, \ldots, X_{t_n} = i_n \mid X_{t_{m-1}} = i_{m-1}].$$

If X_t has the invariant distribution, one has

$$P(X_{t_1} = i_1, \ldots, X_{t_n} = i_n) = \pi(i_1)P_{t_2-t_1}(i_1, i_2)P_{t_3-t_2}(I_2, i_3) \cdots P_{t_n-t_{n-1}}(i_{n-1}, i_n),$$

for all $i_1, \ldots, i_n \in \mathcal{X}$ and all $0 < t_1 < \cdots < t_n$.

Here is the result that corresponds to Theorem 15.1. We define irreducibility, transience, and null and positive recurrence as in discrete time. There is no notion of periodicity in continuous time.

Theorem 6.1 (Big Theorem for Continuous-Time Markov Chains)
Consider a continuous-time Markov chain.

(a) If the Markov chain is irreducible, the states are either all transient, all positive recurrent, or all null recurrent. We then say that the Markov chain is transient, positive recurrent, or null recurrent, respectively.
(b) If the Markov chain is positive recurrent, it has a unique invariant distribution π and $\pi(i)$ is the long-term fraction of time that X_t is equal to i. Moreover, the probability $\pi_t(i)$ that the Markov chain X_t is in state i converges to $\pi(i)$.
(c) If the Markov chain is not positive recurrent, it does not have an invariant distribution and the fraction of time that it spends in any state goes to zero.

∎

6.2.4 Uniformization

We saw earlier that a CTMC can be approximated by a discrete-time Markov chain that has a time step $\epsilon \ll 1$. There are two other DTMCs that have a close relationship with the CTMC: the jump chain and the uniformized chain. We explain these chains for the CTMC X_t in Fig. 6.7.

The *jump chain* is X_t observed when it jumps. As Fig. 6.7 shows, this DTMC has a transition matrix equal to Γ where

$$\Gamma(i, j) = \begin{cases} q(i, j)/q_i, & \text{if } i \neq j \\ 0, & \text{if } i = j. \end{cases}$$

Let v be the invariant distribution of this jump chain. That is, $v = v\Gamma$. Since $v(i)$ is the long-term fraction of time that the jump chain is in state i, and since the CTMC X_t spends an average time $1/q_i$ in state i whenever it visits that state, the fraction of time that X_t spends in state i should be proportional to $v(i)/q_i$. That is, one expects

$$\pi(i) = Av(i)/q_i$$

for some constant A. That is, one should have

$$\sum_j [Av(i)/q_i] q(i, j) = 0.$$

To verify that equality, we observe that

$$\sum_j [v(i)/q_i] q(i, j) = \sum_{j \neq i} v(i) \Gamma(i, j) + v(i) q(i, i)/q_i = v(i) - v(i) = 0.$$

We used the fact that $v\Gamma = v$ and $q(i, i) = -q_i$.

The *uniformized chain* is not the jump chain. It is a discrete-time Markov chain obtained from the CTMC as follows. Let $\lambda \geq q_i$ for all i. The rate at which X_t changes state is q_i when it is in state i. Let us add a dummy jump from i to i with rate $\lambda - q_i$. The rate of jumps, including these dummy jumps, of this new Markov chain Y_t is now constant and equal to λ.

The transition matrix P of Y_t is such that

$$P(i, j) = \begin{cases} (\lambda - q_i)/\lambda, & \text{if } i = j \\ q(i, j)/\lambda, & \text{if } i \neq j. \end{cases}$$

To see this, assume that $Y_t = i$. The next jump will occur with rate λ. With probability $(\lambda - q_i)/\lambda$, it is a dummy jump from i to i. With probability q_i/λ it is an actual jump where Y_t jumps to $j \neq i$ with probability $\Gamma(i, j)$. Hence, Y_t jumps from i to i with probability $(\lambda - q_i)/\lambda$ and from i to $j \neq i$ with probability $(q_i/\lambda)\Gamma(i, j) = q(i, j)/\lambda$.

Note that

$$P = \mathbf{I} + \frac{1}{\lambda} Q,$$

where $\mathbf{I}$ is the identity matrix.

Now, define Z_n to be the jump chain of Y_t, i.e., the Markov chain with transition matrix P. Since the jumps of Y_t occur at rate λ, independently of the value of the state Y_t, we can simulate Y_t as follows. Let N_t be a Poisson process with rate λ. The jump times $\{t_1, t_2, \ldots\}$ of N_t will be the jump times of Y_t. The successive values of Y_t are those of Z_n. Formally,

$$Y_t = Z_{N_t}.$$

That is, if $N_t = n$, then we define $Y_t = Z_n$. Since the CTMC Y_t spends $1/\lambda$ on average between jumps, the invariant distribution of Y_t should be the same as that of X_t, i.e., π. To verify this, we check that $\pi P = \pi$, i.e., that

$$\pi \left(\mathbf{I} + \frac{1}{\lambda} Q \right) = \pi.$$

That identity holds since $\pi Q = 0$. Thus, the DTMC Z_n has the same invariant distribution as X_t. Observe that Z_n is not the same as the jump chain of X_t. Also, it is not a discrete-time approximation of X_t. This DTMC shows that a CTMC can be seen as a DTMC where one replaces the constant time steps by i.i.d. exponentially distributed time steps between the jumps.

6.2.5   Time Reversal

As a preparation for our study of networks of queues, we note the following result.

Theorem 6.2 (Kelly's Lemma) *Let Q be the rate matrix of a Markov chain on $\mathcal{X}$. Let also $\tilde{Q}$ be another rate matrix on $\mathcal{X}$. Assume that π is a distribution on $\mathcal{X}$ and that*

$$q_i = \tilde{q}_i, i \in \mathcal{X} \text{ and}$$

$$\pi(i)q(i, j) = \pi(j)\tilde{q}(j, i), \forall i \neq j.$$

Then $\pi Q = 0$.

■

Proof We have

$$\sum_{j\neq i} \pi(j)q(j, i) = \sum_{j\neq i} p(i)\tilde{q}(i, j) = p(i) \sum_{j\neq i} \tilde{q}(i, j) = p(i)\tilde{q}_i = p(i)q_i,$$

so that $\pi Q = 0$.

□

The following result explains the meaning of $\tilde{Q}$ in the previous theorem. We state it without proof.

Theorem 6.3 *Assume that X_t has the invariant distribution π. Then X_t reversed in time is a Markov chain with rate matrix $\tilde{Q}$ given by*

$$\tilde{q}(i, j) = \frac{\pi(j)q(j, i)}{\pi(i)}.$$

■

6.3   Product-Form Networks

Theorem 6.4 (Invariant Distribution of Network) *Assume $\lambda_k < \mu_k$ and let $\rho_k = \lambda_k/\mu_k$, for $k = 1, 2, 3$. Then the Markov chain X_t has a unique invariant distribution π that is given by*

$$\pi(x_1, x_2, x_3) = \pi_1(x_1)\pi_2(x_2)\pi_3(x_3)$$

$$\pi_k(n) = (1 - \rho_k)\rho_k^n, n \geq 0, k = 1, 2$$

$$\pi_3(a_1, a_2, \ldots, a_n) = p(a_1)p(a_2)\cdots p(a_n)(1 - \rho_3)\rho_3^n,$$

$$n \geq 0, a_k \in \{1, 2\}, k = 1, \ldots, n,$$

Fig. 6.10 The network (top) and a guess for its time-reversal (bottom). The bottom network is obtained from the top one by reversing the flows of customers. It is a bold guess that the arrivals have exponential inter-arrival times and their rates are independent of the current queue lengths

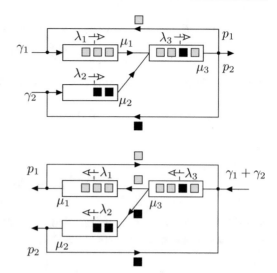

where $p(1) = \lambda_1/(\lambda_1 + \lambda_2)$ and $p(2) = \lambda_2/(\lambda_1 + \lambda_2)$.

Proof Figure 6.10 shows a guess for the time-reversal of the network.

Let Q be the rate matrix of the top network and $\tilde{Q}$ that of the bottom one. Let also π be as stated in the theorem. We show that $\pi, Q, \tilde{Q}$ satisfy the conditions of Kelly's Lemma.

For instance, we verify that

$$\pi([3, 2, [1, 1, 2, 1]])q([3, 2, [1, 1, 2, 1]], [4, 2, [1, 1, 2]])$$
$$= \pi([4, 2, [1, 1, 2]])\tilde{q}([4, 2, [1, 1, 2]], [3, 2, [1, 1, 2, 1]]).$$

Looking at the figure, we can see that

$$q([3, 2, [1, 1, 2, 1]], [4, 2, [1, 1, 2]]) = \mu_3 p_1$$
$$\tilde{q}([4, 2, [1, 1, 2]], [3, 2, [1, 1, 2, 1]]) = \mu_1 p_1.$$

Thus, the previous identity reads

$$\pi([3, 2, [1, 1, 2, 1]])\mu_3 p_1 = \pi([4, 2, [1, 1, 2]])\mu_1 p_1,$$

i.e.,

$$\pi([3, 2, [1, 1, 2, 1]])\mu_3 = \pi([4, 2, [1, 1, 2]])\mu_1.$$

Given the expression for π, this is

$$(1 - \rho_1)\rho_1^3 \times (1 - \rho_2)\rho_2^2 \times p(1)p(1)p(2)p(1)(1 - \rho_3)\rho_3^4\mu_3$$
$$= (1 - \rho_1)\rho_1^4 \times (1 - \rho_2)\rho_2^2 \times p(1)p(1)p(2)(1 - \rho_3)\rho_3^3\mu_1.$$

After simplifications, this identity is seen to be equivalent to

$$p(1)\rho_3\mu_3 = \rho_1\mu_1,$$

i.e.,

$$\frac{\lambda_1}{\lambda_3}\frac{\lambda_3}{\mu_3}\mu_3 = \frac{\lambda_1}{\mu_1}\mu_1$$

and this equation is seen to be satisfied. A similar argument shows that Kelly's lemma is satisfied for all pairs of states.   □

6.4   Proof of Theorem 5.7

The first step in using the theorem is to solve the flow conservation equations. Let us call class 1 that of the white jobs and class 2 that of the gray job. Then we see that

$$\lambda_1^1 = \lambda_2^1 = \gamma, \lambda_1^2 = \lambda_2^2 = \alpha$$

solve the flow conservation equations for any $\alpha > 0$. We have to assume $\gamma < \mu$ for the services to be able to keep up with the white jobs. With this assumption, we can choose α small enough so that $\lambda_1 = \lambda_2 = \lambda := \gamma + \alpha < \min\{\mu_1, \mu_2\}$.

The second step is to use the theorem to obtain the invariant distribution. It is

$$\pi(x_1, x_2) = Ah(x_1)h(x_2)$$

with

$$h(x_i) = \left(\frac{\gamma}{\mu}\right)^{n_1(x_i)}\left(\frac{\alpha}{\mu}\right)^{n_2(x_i)} = \rho_1^{n_1(x_i)}\rho_2^{n_2(x_i)},$$

where $\rho_1 = \gamma/\mu$, $\rho_2 = \alpha/\mu$, and $n_c(x)$ is the number of jobs of class c in x_i, for $c = 1, 2$. To calculate A, we note that there are $n + 1$ states x_i with n class 1 jobs and 1 class 2 job, and 1 state x_i with n classes 1 jobs and no class 2 job. Indeed, the class 2 customer can be in $n + 1$ positions in the queue with the n customers of class 1.

Also, all the possible pairs (x_1, x_2) must have one class 2 customer either in queue 1 or in queue 2. Thus,

$$1 = \sum_{(x_1, x_2)} \pi(x_1, x_2) = A \sum_{m=0}^{\infty} \sum_{n=0}^{\infty} G(m, n),$$

where

$$G(m, n) = (m + 1)\rho_1^{m+n}\rho_2 + (n + 1)\rho_1^{m+n}\rho_2.$$

In this expression, the first term corresponds to the states with m class 1 customers and one class 2 customer in queue 1 and n customers of class 1 in queue 2; the second term corresponds to the states with m customer of class 1 in queue 1, and n customers of class 1 and one customer of class 2 in queue 2. Thus, $AG(m, n)$ is the probability that there are m customers of class 1 in the first queue and n customers of class 1 in the second queue.

Hence,

$$1 = A \sum_{m=0}^{\infty} \sum_{n=0}^{\infty} [(m + 1)\rho_1^{m+n}\rho_2 + (n + 1)\rho_1^{m+n}\rho_2] = 2A \sum_{m=0}^{\infty} \sum_{n=0}^{\infty} (m + 1)\rho_1^{m+n}\rho_2,$$

by symmetry of the two terms. Thus,

$$1 = 2A\rho_2 \left[\sum_{m=0}^{\infty} (m + 1)\rho_1^m \right] \left[\sum_{n=0}^{\infty} \rho_1^n \right].$$

To compute the sum, we use the following identities:

$$\sum_{n=0}^{\infty} \rho^n = (1 - \rho)^{-1}, \text{ for } 0 < \rho < 1$$

and

$$\sum_{n=0}^{\infty} (n + 1)\rho^n = \frac{\partial}{\partial \rho} \sum_{n=0}^{\infty} \rho^{n+1} = \frac{\partial}{\partial \rho} [(1 - \rho)^{-1} - 1] = (1 - \rho)^{-2}.$$

Thus, one has

$$1 = 2A\rho_2(1 - \rho_1)^{-3},$$

so that

$$A = \frac{(1 - \rho_1)^3}{2\rho_2}.$$

Third, we calculate the expected number L of jobs of class 1 in the two queues. One has

$$
L = \sum_{m=0}^{\infty} \sum_{n=0}^{\infty} A(m+n) G(m,n)
$$

$$
= \sum_{m=0}^{\infty} \sum_{n=0}^{\infty} A(m+n)(m+1) \rho_1^{m+n} \rho_2 + \sum_{m=0}^{\infty} \sum_{n=0}^{\infty} A(m+n)(n+1) \rho_1^{m+n} \rho_2
$$

$$
= 2 \sum_{m=0}^{\infty} \sum_{n=0}^{\infty} A(m+n)(m+1) \rho_1^{m+n} \rho_2,
$$

where the last identity follows from the symmetry of the two terms. Thus,

$$
L = 2 \sum_{m=0}^{\infty} \sum_{n=0}^{\infty} A m(m+1) \rho_1^{m+n} \rho_2 + 2 \sum_{m=0}^{\infty} \sum_{n=0}^{\infty} A n(m+1) \rho_1^{m+n} \rho_2
$$

$$
= 2A\rho_2 \left[\sum_{m=0}^{\infty} m(m+1) \rho_1^m \right] \left[\sum_{n=0}^{\infty} \rho_1^n \right] + 2A\rho_2 \left[\sum_{m=0}^{\infty} (m+1) \rho_1^m \right] \left[\sum_{n=0}^{\infty} n \rho_1^n \right]
$$

$$
= 2A\rho_2 \left[\sum_{m=0}^{\infty} m(m+1) \rho_1^m \right] (1-\rho_1)^{-1} + 2A\rho_2 (1-\rho)^{-2} \left[\sum_{n=0}^{\infty} n \rho_1^n \right].
$$

To calculate the sums, we use the fact that

$$
\sum_{m=0}^{\infty} m(m+1) \rho^m = \rho \sum_{m=0}^{\infty} m(m+1) \rho^{m-1}
$$

$$
= \rho \frac{\partial^2}{\partial \rho^2} \sum_{m=0}^{\infty} \rho^{m+1} = \rho \frac{\partial^2}{\partial \rho^2} [(1-\rho)^{-1} - 1]
$$

$$
= 2\rho(1-\rho)^{-3}.
$$

Also,

$$
\sum_{n=0}^{\infty} n \rho_1^n = \rho_1 \sum_{n=0}^{\infty} n \rho_1^{n-1} = \rho_1 \sum_{n=0}^{\infty} (n+1) \rho_1^n = \rho_1 (1-\rho_1)^{-2}.
$$

Hence,

$$
L = 2A\rho_2 \times 2\rho(1-\rho)^{-3} \times (1-\rho_1)^{-1} + 2A\rho_2(1-\rho)^{-2} \times \rho_1(1-\rho_1)^{-2}
$$

$$
= 6A\rho_2 \rho_1 (1-\rho_1)^{-4}.
$$

Substituting the value for A that we derived above, we find

$$L = 3\frac{\rho_1}{1 - \rho_1}.$$

Finally, we get the average time W that jobs of class 1 spend in the network: $W = L/\gamma$.

Without the gray job, the expected delay W' of the white jobs would be the sum of delays in two M/M/1 queues, i.e., $W' = L'/\gamma$ where

$$L' = 2\frac{\rho_1}{1 - \rho_1}.$$

Hence, we find that

$$W = 1.5W',$$

so that using a hello message increases the average delay of the class 1 customers by 50%.

6.5 References

The time-reversal arguments are developed in Kelly (1979). That book also explains many other models that can be analyzed using that approach. See also Bremaud (2008), Lyons and Perez (2017), Neely (2010).

Digital Link—A

7

Application: Transmitting bits across a physical medium
Topics: MAP, MLE, Hypothesis Testing

7.1   Digital Link

A *digital link* consists of a transmitter and a receiver. It transmits bits over some physical medium that can be a cable, a phone line, a laser beam, an optical fiber, an electromagnetic wave, or even a sound wave. This contrasts with an analog system that transmits signals without converting them into bits, as in Fig. 7.1.

An elementary such system[1] consists of a phone line and, to send a bit 0, the transmitter applies a voltage -1 Volt across its end of the line for T seconds; to send a bit 1, it applies the voltage $+1$ Volt for T second. The receiver measures the voltage across its end of the line. If the voltage that the receiver measures is negative, it decides that the transmitter must have sent a 0; if it is positive, it decides that the transmitter sent a 1. This system is not error-free. The receiver gets a noisy and attenuated version of what the transmitter sent. Thus, there is a chance that a 0 is mistaken for a 1, and vice versa. Various coding techniques are used to reduce the chances of such errors Fig. 7.2 shows the general structure of a digital link.

In this chapter, we explore the operating principles of digital links and their characteristics. We start with a discussion of Bayes' rule and of detection theory. We apply these ideas to a simple model of communication link. We then explore a coding scheme that makes the transmissions faster. We conclude the chapter with a

---

[1] We are ignoring many details of synchronization.

© The Author(s) 2021
J. Walrand, *Probability in Electrical Engineering and Computer Science*,
https://doi.org/10.1007/978-3-030-49995-2_7

Fig. 7.1 An analog
communication system

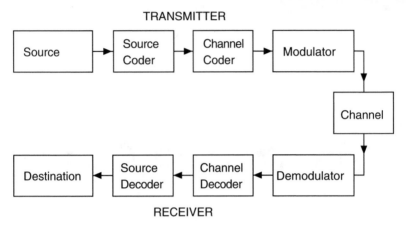

Fig. 7.2 Components of a digital link

discussion of modulation and detection schemes that actual transmission systems, such as ADSL and Cable Modems, use.

7.2   Detection and Bayes' Rule

The receiver gets some signal S and tries to guess what the transmitter sent. We explore a general model of this problem and we then apply it to concrete situations.

7.2.1   Bayes' Rule

The basic formulation is that there are N possible exclusive circumstances $C_1, \ldots, C_N$ under which a particular symptom S can occur. By exclusive, we mean that exactly one circumstance occurs (Fig. 7.3). Each circumstance C_i has some *prior probability* p_i and q_i is the probability that S occurs under circumstance C_i. Thus,

$$p_i = P(C_i) \text{ and } q_i = P[S \mid C_i], \text{ for } i = 1, \ldots, N,$$

Fig. 7.3 The symptom and
its possible circumstances.
Here, $p_i = P(C_i)$ and
$q_i = P[S \mid C_i]$

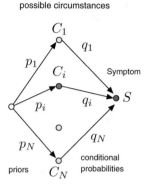

where

$$p_i \geq 0, q_i \in [0, 1] \text{ for } i = 1, \ldots, N \text{ and } \sum_{i=1}^{N} p_i = 1.$$

The *posterior probability* π_i that circumstance C_i is in effect given that S is observed can be computed by using *Bayes' rule* as we explain next. One has

$$\pi(i) = P[C_i|S] = \frac{P(C_i \text{ and } S)}{P(S)}$$

$$= \frac{P(C_i \text{ and } S)}{\sum_{j=1}^{N} P(C_j \text{ and } S)} = \frac{P[S|C_i]P(C_i)}{\sum_{j=1}^{N} P[S|C_j]P(C_j)}$$

$$= \frac{p_i q_i}{\sum_{j=1}^{N} p_j q_j}.$$

Given the importance of this result, we state it as a theorem.

Theorem 7.1 (Bayes' Rule)   *One has*

$$\pi_i = \frac{p_i q_i}{\sum_{j=1}^{N} p_j q_j}, i = 1, \ldots, N. \tag{7.1}$$

∎

This rule is very simple but is a canonical example of how observations affect our beliefs. It is due to Thomas Bayes (Fig. 7.4).

Fig. 7.4 Thomas Bayes,
1701–1761

7.2.2   Circumstances vs. Causes

In the previous section we were careful to qualify the C_i as possible *circumstances*, not as *causes*. The distinction is important. Say that you go to a beach, eat an ice cream, and leave with a sunburn. Later, you meet a friend who did not go to the beach, did not eat an ice cream, and did not get sunburned. More generally, the probability that someone got sunburned is larger if that person ate an ice cream. However, it would be silly to qualify the ice cream as the cause of the sunburn.

Unfortunately, confusing correlation and causation is a prevalent mistake.

7.2.3   MAP and MLE

Given the previous model, we see that the most likely circumstance under which the symptom occurs, which we call the *Maximum A Posteriori (MAP)* estimate of the circumstance given the symptom, is

$$MAP = \arg\max_i \pi_i = \arg\max_i p_i q_i.$$

The notation is that if $h(\cdot)$ is a function, then $\arg\max_x h(x)$ is any value of x that achieves the maximum of $h(\cdot)$. Thus, if $x^* = \arg\max_x h(x)$, then $h(x^*) \geq h(x)$ for all x.

Thus, the MAP is the most likely circumstance, *a posteriori*, that is, after having observed the symptom.

Note that if all the prior probabilities are equal, i.e., if $p_i = 1/N$ for all i, then the MAP maximizes q_i. In general, the estimate that maximizes q_i is called the *Maximum Likelihood Estimate (MLE)* of the circumstance given the symptom. That is,

$$MLE = \arg\max_i q_i.$$

That is, the MLE is the circumstance that makes the symptom most likely.

More generally, one has the following definitions.

Definition 7.1 (MAP and MLE) Let (X, Y) be discrete random variables. Then

$$MAP[X|Y = y] = \arg\max_{x} P(X = x \text{ and } Y = y)$$

and

$$MLE[X|Y = y] = \arg\max_{x} P[Y = y|X = x].$$

◇

These definitions extend in the natural way to the continuous case, as we will get to see later.

Example: Ice Cream and Sunburn

As an example, say that on a particular summer day in Berkeley 500 out of 100,000 people eat ice cream, among which 50 get sunburned and that among the 99,500 who do not eat ice cream, 600 get sunburned. Then, the MAP of *eating ice cream* given *sunburn* is *No* but the MLE is *Yes*. Indeed, we see that

$$P(sunburn\ and\ ice\ cream) = 50 < P(sunburn\ and\ no\ ice\ cream) = 600,$$

so that among those who have a sunburn, a minority eat ice cream, so that it is more likely that a sunburn person did not eat ice cream. Hence, the MAP if No. However, the fraction of people who have a sunburn is larger among those who eat ice cream (10%) than among those who do not (0.6%). Hence, the MLE is Yes.

7.2.4 Binary Symmetric Channel

We apply the concepts of MLE and MAP to a simplified model of a communication link. Figure 7.5 illustrates the model, called a *binary symmetric channel (BSC)*.

In this model, the transmitter sends a 0 or a 1 and the receiver gets the transmitted bit with probability $1 - p$, otherwise it gets the opposite bit. Thus, the channel makes an error with probability p. We assume that if the transmitter sends successive bits, the errors are i.i.d.

Fig. 7.5 The binary symmetric channel

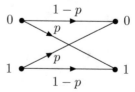

Fig. 7.6 MAP for BSC.
Here, $\alpha = P(X = 1)$ and p is
the probability of a channel
error

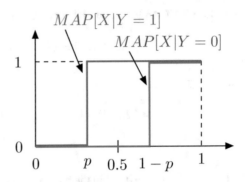

Note that if $p = 0$ or $p = 1$, then one can recover exactly every bit that is sent. Also, if $p = 0.5$, then the output is independent of the input and no useful information goes through the channel. What happens in the other cases?

Call $X \in \{0, 1\}$ the input of the channel and $Y \in \{0, 1\}$ its output. Assume that you observe $Y = 1$ and that $P(X = 1) = \alpha$, so that $P(X = 0) = 1 - \alpha$. We have the following result illustrated in Fig. 7.6.

Theorem 7.2 (MAP and MLE for BSC) *For the BSC with $p < 0.5$,*

$$MAP[X|Y = 0] = 1\{\alpha > 1 - p\}, \, MAP[X|Y = 1] = 1\{\alpha > p\}$$

and

$$MLE[X|Y] = Y.$$

∎

To understand the MAP results, consider the case $Y = 1$. Since $p < 0.5$, we are inclined to think that $X = 1$. However, if α is small, this is unlikely. The result is that $X = 1$ is more likely than $X = 0$ if $\alpha > p$, i.e., if the prior is "stronger" than the noise. The case $Y = 0$ is similar.

Proof In the terminology of Bayes' rule, the event $Y = 1$ is the symptom. Also, the prior probabilities are

$$p_0 = 1 - \alpha \text{ and } p_1 = \alpha,$$

and the conditional probabilities are

$$q_0 = P[Y = 1|X = 0] = p \text{ and } q_1 = P[Y = 1|X = 1] = 1 - p.$$

Hence,

$$MAP[X|Y = 1] = \arg \max_{i \in \{0,1\}} p_i q_i.$$

Thus,

$$MAP[X|Y = 1] = \begin{cases} 1, & \text{if } p_1 q_1 = \alpha(1 - p) > p_0 q_0 = (1 - \alpha)p \\ 0, & \text{otherwise.} \end{cases}$$

Hence, $MAP[X|Y = 1] = 1\{\alpha > p\}$. That is, when $Y = 1$, your guess is that $X = 1$ if the prior that $X = 1$ is larger than the probability that the channel makes an error.

Also,

$$MLE[X|Y = 1] = \arg \max_{i \in \{0,1\}} q_i.$$

In this case, since $p < 0.5$, we see that $MLE[X|Y = 1] = 1$, because $Y = 1$ is more likely when $X = 1$ than when $X = 0$. Thus, the MLE ignores the prior and always guesses that $X = 1$ when $Y = 1$, even though the prior probability $P(X = 1) = \alpha$ may be very small.

Similarly, we see that

$$MAP[X|Y = 0] = \arg \max_{i \in \{0,1\}} p_i(1 - q_i).$$

Thus,

$$MAP[X|Y = 0] = \begin{cases} 1, & \text{if } p_1(1 - q_1) = \alpha p > p_0(1 - q_0)(1 - \alpha)(1 - p) \\ 0, & \text{otherwise.} \end{cases}$$

Hence, $MAP[X|Y = 0] = 1\{\alpha > 1 - p\}$. Thus, when $Y = 0$, you guess that $X = 1$ if $X = 1$ is more likely a priori than the channel being correct.

Also, $MLE[X|Y = 0] = 0$ because $p < 0.5$, irrespectively of α.   □

7.3 Huffman Codes

Coding can improve the characteristics of a digital link. We explore Huffman codes in this section.

Say that you want to transmit strings of symbols A, B, C, D across a digital link. The simplest method is to encode these symbols as $00, 01, 10$, and 11, respectively. In so doing, each symbol requires transmitting two bits. Assuming that there is no error, if the receiver gets the bits 0100110001, it recovers the string $BADAB$.

Fig. 7.7 David Huffman, 1925–1999

Now assume that the strings are such that the symbols occur with the following frequencies: $(A, 55\%)$, $(B, 30\%)$, $(C, 10\%)$, $(D, 5\%)$. Thus, A occurs 55% of the time, and similarly for the other symbols. In this situation, one may design a code where A requires fewer bits than D.

The *Huffman code* (Huffman 1952, Fig. 7.7) for this example is as follows:

$$A = 0, B = 10, C = 110, D = 111.$$

The average number of bits required per symbol is

$$1 \times 55\% + 2 \times 30\% + 3 \times 10\% + 3 \times 5\% = 1.6.$$

Thus, one saves 20% of the transmissions and the resulting system is 25% faster (ah! arithmetics). Note that the code is such that, when there is no error, the receiver can recover the symbols uniquely from the bits it gets. For instance, if the receiver gets 110100111, the symbols are $CBAD$, without ambiguity.

The reason why there is no possible ambiguity is that one can picture the bits as indicating the path in a tree that ends with a leaf of the tree, as shown in Fig. 7.8. Thus, starting with the first bit received, one walks down the tree until one reaches a leaf. One then repeats for the subsequent bits. In our example, when the bits are 110100111, one starts at the top of the tree, then one follows the branches 110 and reaches leaf C, then one restarts from the top and follows the branches 10 and gets to the leaf B, and so on. Codes that have this property of being uniquely decodable in one pass are called *prefix-free codes*.

The construction of the code is simple. As shown in Fig. 7.8, one joins the two symbols with the smallest frequency of occurrence, here C and D, with branches 0 and 1 and assigns the group CD the sum of the symbol frequencies, here 0.15. One then continues in the same way, joining CD and B and assigning the group BCD the frequency $0.3 + 0.15 = 0.45$. Finally, one joins A and BCD. The resulting tree specifies the code.

The following property is worth noting.

Fig. 7.8 Huffman code

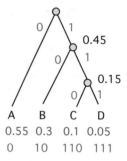

A B C D
0.55 0.3 0.1 0.05
0 10 110 111

Theorem 7.3 (Optimality of Huffman Code) *The Huffman code has the smallest average number of bits per symbol among all prefix-free codes.*

∎

Proof See Chap. 8. □

It should be noted that other codes have a smaller average length, but they are not symbol-by-symbol codes and are more complex. One code is based on the observation that there are only 2^{nH} likely strings of $n \gg 1$ symbols, where

$$H = -\sum_X x \log_2(x).$$

In this expression, x is the frequency of symbol X and the sum is over all the symbols. This expression H is the *entropy* of the distribution of the symbols. Thus, by listing all these strings and assigning nH bits to identify them, one requires only nH bits for n symbols, or H bits per symbol (See Sect. 15.7.).

In our example, one has

$$H = -0.55 \log_2(0.55) - 0.3 \log_2(0.3)$$
$$- 0.1 \log_2(0.1) - 0.05 \log_2(0.05) = 1.54.$$

Thus, for this example, the savings over the Huffman code are not spectacular, but it is easy to find examples for which they are. For instance, assume that there are only two symbols A and B with frequencies p and $1 - p$, for some $p \in (0, 1)$. The Huffman code requires one bit per symbol, but codes based on long strings require only $-p \log_2(p) - (1 - p) \log_2(1 - p)$ bits per symbol. For $p = 0.1$, this is 0.47, which is less than half the number of bits of the Huffman code.

Coding based on long strings of symbols are discussed in Sect. 15.7.

7.4   Gaussian Channel

In the previous sections, we had a simplified model of a channel as a BSC. In this section, we examine a more realistic model of the channel that captures the physical characteristic of the noise. In this model, the transmitter sends a bit $X \in \{0, 1\}$ and the receiver gets Y where

$$Y = X + Z.$$

In this identity, $Z =_D \mathcal{N}(0, \sigma^2)$ and is independent of X. We say that this is an *additive Gaussian noise channel*.

Figure 7.9 shows the densities of Y when $X = 0$ and when $X = 1$. Indeed, when $X = x$, we see that $Y =_D \mathcal{N}(x, \sigma^2)$.

Assume that the receiver observes Y. How should it decide whether $X = 0$ or $X = 1$? Assume again that $P(X = 1) = p_1 = \alpha$ and $P(X = 0) = p_0 = 1 - \alpha$.

In this example, $P[Y = y|X = 0] = 0$ for all values of y. Indeed, Y is a continuous random variable. So, we must change a little our discussion of Bayes' rule. Here is how to do it. Pretend that we do not measure Y with infinite precision but that we instead observe that $Y \in (y, y + \epsilon)$ where $0 < \epsilon \ll 1$. Thus, the symptom is $Y \in (y, y + \epsilon)$ and it now has a positive probability. In fact,

$$q_0 = P[Y \in (y, y + \epsilon)|X = 0] \approx f_0(y)\epsilon,$$

by definition of the density $f_0(y)$ of Y when $X = 0$. Similarly,

$$q_1 = P[Y \in (y, y + \epsilon)|X = 1] \approx f_1(y)\epsilon.$$

Hence,

$$MAP[X|Y \in (y, y + \epsilon)] = \arg \max_{i \in \{0,1\}} p_i f_i(y)\epsilon.$$

Since the result does not depend on ϵ, we write

$$MAP[X|Y = y] = \arg \max_{i \in \{0,1\}} p_i f_i(y).$$

Fig. 7.9 The pdf of Y is f_0 when $X = 0$ and f_1 when $X = 1$

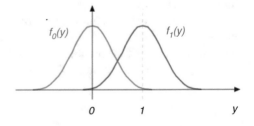

Similarly,

$$MLE[X|Y = y] = \arg \max_{i \in \{0,1\}} f_i(y).$$

We can verify that

$$MAP[X|Y = y] = 1 \left\{ y \geq \frac{1}{2} + \sigma^2 \log \left(\frac{p_0}{p_1} \right) \right\}. \qquad (7.2)$$

Also, the resulting probability of error is

$$P \left(\mathcal{N}(0, \sigma^2) \geq \frac{1}{2} + \sigma^2 \log \left(\frac{p_0}{p_1} \right) \right) p_0$$

$$+ P \left(\mathcal{N}(1, \sigma^2) \leq \frac{1}{2} + \sigma^2 \log \left(\frac{p_0}{p_1} \right) \right) p_1.$$

Also,

$$MLE[X|Y = y] = 1\{y \geq 0.5\}.$$

If we choose the MLE detection rule, the system has the same probability of error as a BSC channel with

$$p = p(\sigma^2) := P(\mathcal{N}(0, \sigma^2) > 0.5) = P \left(\mathcal{N}(0, 1) > \frac{0.5}{\sigma} \right).$$

Simulation
Figure 7.10 shows the simulation results when $\alpha = 0.5$ and $\sigma = 1$. The code is in the Jupyter notebook for this chapter.

7.4.1   BPSK

The system in the previous section was very simple and corresponds to a practical transmission scheme called *Binary Phase Shift Keying (BPSK)*. In this system, instead of sending a constant voltage for T seconds to represent either a bit 0 or a bit 1, the transmitter sends a sine wave for T seconds and the phase of that sine wave depends on whether the transmitter sends a 0 or a 1 (Fig. 7.11).
   Specifically, to send bit 0, the transmitter sends the signal

$$\mathbf{s}_0 = \{s_0(t) = A \sin(2\pi f t), t \in [0, T]\}.$$

Here, T is a multiple of the period, so that $fT = k$ for some integer k. To send a bit 1, the transmitter sends the signal $\mathbf{s}_1 = -\mathbf{s}_0$. Why all this complication? The

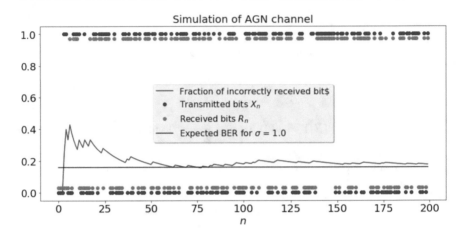

Fig. 7.10 Simulation of the AGN channel with $\alpha = 0.5$ and $\sigma = 1$

Fig. 7.11 The signal that the transmitter sends when using BPSK

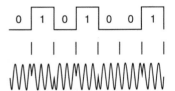

signal is a sine wave around frequency f and the designer can choose a frequency that the transmission medium transports well. For instance, if the transmission is wireless, the frequency f is chosen so that the antennas radiate and receive that frequency well. The *wavelength* of the transmitted electromagnetic wave is the speed of light divided by f and it should be of the same order as the physical length of the antenna. For instance, 1GHz corresponds to a wavelength of one foot and it can be transmitted and received by suitably shaped cell phone antennas.

In any case, the transmitter sends the signal s_i to send a bit i, for $i = 0, 1$. The receiver attempts to detect whether s_0 or $s_1 = -s_0$ was sent. To do this, it multiplies the received signal by a sine wave at the frequency f, then computes the average value of the product. That is, if the receiver gets the signal $\mathbf{r} = \{r_t, 0 \leq t \leq T\}$, it computes

$$\frac{1}{T} \int_0^T r_t \sin(2\pi f t) dt.$$

You can verify that if $\mathbf{r} = s_0$, then the result is $A/2$ and if $\mathbf{r} = s_1$, then the result is $-A/2$. Thus, the receiver guesses that bit 0 was transmitted if this average value is positive and that bit 1 was transmitted otherwise.

The signal that the receiver gets is not s_i when the transmitter sends s_i. Instead, the receiver gets an attenuated and noisy version of that signal. As a result, after doing its calculation, the receiver gets $B + Z$ or $-B + Z$ where B is some constant that depends on the attenuation, Z is a $\mathcal{N}(0, \sigma^2)$ random variable and σ^2 reflects the power of the noise.

Accordingly, the detection problem amounts to detecting the mean value of a Gaussian random variable, which is the problem that we discussed earlier.

7.5   Multidimensional Gaussian Channel

When using $BPSK$, the transmitter has a choice between two signals: s_0 and s_1. Thus, in T seconds, the transmitter sends one bit. To increase the transmission rate, communication engineers devised a more efficient scheme called *Quadrature Amplitude Modulation (QAM)*. When using this scheme, a transmitter can send a number k of bits every T seconds. The scheme can be designed for different values of k. When $k = 1$, the scheme is identical to BPSK. For $k > 1$, there are 2^k different signals and each one is of the form

$$a\cos(2\pi f t) + b\sin(2\pi f t),$$

where the coefficients (a, b) characterize the signal and correspond to a given string of k-bits. These coefficients form a constellation as shown in Fig. 7.12 in the case of QAM-16, which corresponds to $k = 4$.

When the receiver gets the signal, it multiplies it by $2\cos(2\pi f t)$ and computes the average over T seconds. This average value should be the coefficient a if there was not attenuation and no noise. The receiver also multiplies the signal by $2\sin(2\pi f t)$ and computes the average over T seconds. The result should be the coefficient b. From the value of (a, b), the receiver can tell the four bits that the transmitter sent.

Because of the noise (we can correct for the attenuation), the receiver gets a pair of values $\mathbf{Y} = (Y_1, Y_2)$, as shown in the figure. The receiver essentially finds the constellation point closest to the measured point $\mathbf{Y}$ and reads off the corresponding bits.

Fig. 7.12 A QAM-16 constellation

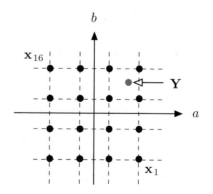

The values of $|a|$ and $|b|$ are bounded, because of a power constraint on the transmitter. Accordingly, a constellation with more points (i.e., a larger value of k) has points that are closer together. This proximity increases the likelihood that the noise misleads the receiver. Thus, the size of the constellation should be adapted to the power of the noise. This is in fact what actual systems do. For instance, a cable modem and an ADSL modem divide the frequency band into small channels and they measure the noise power in each channel and choose the appropriate constellation for each. WiFi, LTE, and 5G systems use a similar scheme.

7.5.1   MLE in Multidimensional Case

We can summarize the effect of modulation, demodulation, amplification to compensate for the attenuation and the noise as follows. The transmitter sends one of the sixteen vectors $\mathbf{x}_k = (a_k, b_k)$ shown in Fig. 7.12. Let us call the transmitted vector $\mathbf{X}$. The vector that the receiver computes is $\mathbf{Y}$.

Assume first that

$$\mathbf{Y} = \mathbf{X} + \mathbf{Z}$$

where $\mathbf{Z} = (Z_1, Z_2)$ and Z_1, Z_2 are i.i.d. $N(0, \sigma^2)$ random variables. That is, we assume that the errors in Y_1 and Y_2 are independent and Gaussian. In this case, we can calculate the conditional density $f_{\mathbf{Y}|\mathbf{X}}[\mathbf{y}|\mathbf{x}]$ as follows. Given $\mathbf{X} = \mathbf{x}$, we see that $Y_1 = x_1 + Z_1$ and $Y_2 = x_2 + Z_2$. Since Z_1 and Z_2 are independent, it follows that Y_1 and Y_2 are independent as well. Moreover, $Y_1 = N(x_1, \sigma^2)$ and $Y_2 = N(x_2, \sigma^2)$. Hence,

$$f_{\mathbf{Y}|\mathbf{X}}[\mathbf{y}|\mathbf{x}] = \frac{1}{\sqrt{2\pi\sigma^2}} \exp\left\{-\frac{(y_1 - x_1)^2}{2\sigma^2}\right\} \frac{1}{\sqrt{2\pi\sigma^2}} \exp\left\{-\frac{(y_2 - x_2)^2}{2\sigma^2}\right\}.$$

Recall that $MLE[\mathbf{X}|\mathbf{Y} = \mathbf{y}]$ is the value of $\mathbf{x} \in \{\mathbf{x}_1, \ldots, \mathbf{x}_{16}\}$ that maximizes this expression. Accordingly, it is the value $\mathbf{x}_k$ that minimizes

$$||\mathbf{x}_k - \mathbf{y}||^2 = (x_1 - y_1)^2 + (x_2 - y_2)^2.$$

Thus, $MLE[\mathbf{X}|\mathbf{Y}]$ is indeed the constellation point that is the closest to the measured value $\mathbf{Y}$.

7.6   Hypothesis Testing

There are many situations where the MAP and MLE are not satisfactory guesses. This is the case for designing alarms, medical tests, failure detection algorithms, and many other applications. We describe an important formulation, called the *hypothesis testing problem*.

7.6.1   Formulation

We consider the case where $X \in \{0, 1\}$ and where one assumes a distribution of Y given X. The goal will be to solve the following problem:

$$\text{Maximize } PCD := P[\hat{X} = 1 | X = 1]$$

$$\text{subject to } PFA := P[\hat{X} = 1 | X = 0] \leq \beta.$$

Here, PCD is the *probability of correct detection*, i.e., of detecting that $X = 1$ when it is actually equal to 1. Also, PFA is the *probability of false alarm*, i.e., of declaring that $X = 1$ when it is in fact equal to zero. The constant β is a given bound on the probability of false alarm.[2]

For making sense of the terminology, think that $X = 1$ means that your house is on fire. It is not reasonable to assume a prior probability that $X = 1$, so that the MAP formulation is not appropriate. Also, the MLE amounts to assuming that $P(X = 1) = 1/2$, which is not suitable here. In the hypothesis testing formulation, the goal is to detect a fire with the largest possible probability, subject to a bound on the probability of false alarm. That is, one wishes to make the fire detector as sensitive as possible, but not so sensitive that it produces frequent false alarms.

One has the following useful concept.

Definition 7.2 (Receiver Operating Characteristic (ROC)) If the solution of the problem is $PCD = R(\beta)$, the function $R(\beta)$ is called the *Receiver Operating Characteristic (ROC)*.

$\diamond$

A typical ROC is shown in Fig. 7.13. The terminology comes from the fact that this function depends on the conditional distributions of Y given $X = 0$ and given $X = 1$, i.e., of the signal that is received about X.

Note the following features of that curve. First, $R(1) = 1$ because if one is allowed to have $PFA = 1$, then one can choose $\hat{X} = 1$ for all observations; in that case $PCD = 1$.

Second, the function $R(\beta)$ is concave. To see this, let $0 \leq \beta_1 < \beta_2 \leq 1$ and assume that $g_i(Y)$ achieves $P[g_i(Y) = 1 | X = 1] = R(\beta_i)$ and $P[g_i(Y) = 1 | X = 0] = \beta_i$ for $i = 1, 2$. Choose $\epsilon \in (0, 1)$ and define $X' = g_1(Y)$ with probability ϵ and $X' = g_2(Y)$ otherwise. Then,

$$P[X' = 1 | X = 0] = \epsilon P[g_1(Y) = 1 | X = 0] + (1 - \epsilon) P[g_2(Y) = 1 | X = 0]$$

$$= \epsilon \beta_1 + (1 - \epsilon) \beta_2.$$

---

[2]If H_0 means that you are healthy and H_1 means that you have a disease, PFA is the probability of a *false positive* test and $1 - PCD$ is the probability of a *false negative* test. These are also called type I and type II errors in the literature. PFA is also called the p-value of the test.

Fig. 7.13 The Receiver
Operating Characteristic is
the maximum probability of
correct detection $R(\beta)$ as a
function of the bound β on
the probability of false alarm

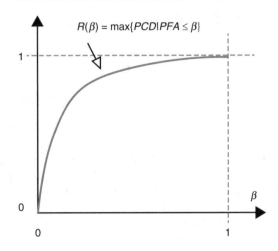

Also,

$$P[X' = 1|X = 1] = \epsilon P[g_1(Y) = 1|X = 1] + (1 - \epsilon)P[g_2(Y) = 1|X = 1]$$
$$= \epsilon R(\beta_1) + (1 - \epsilon)R(\beta_2).$$

Now, the decision rule $\hat{X}$ that maximizes $P[\hat{X} = 1|X = 1]$ subject to $P[\hat{X} = 1|X = 0] = \epsilon\beta_1 + (1 - \epsilon)\beta_2$ must be at least as good as X'. Hence,

$$R(\epsilon\beta_1 + (1 - \epsilon)\beta_2) \geq \epsilon\beta_1 + (1 - \epsilon)\beta_2.$$

This inequality proves the concavity of $R(\beta)$.

Third, the function $R(\beta)$ is nondecreasing. Intuitively, if one can make a larger PFA, one can decide $\hat{X} = 1$ with a larger probability, which increases PCD. To show this formally, let $\beta_2 = 1$ in the previous derivation.

Fourth, note that it may not be the case that $R(0) = 0$. For instance, assume that $Y = X$. In this case, one chooses $\hat{X} = Y = X$, so that $PCD = 1$ and $PFA = 0$.

7.6.2   Solution

The solution of the hypothesis testing problem is stated in the following theorem.

Theorem 7.4 (Neyman–Pearson (1933)) *The decision $\hat{X}$ that maximizes PCD subject to $PFA \leq \beta$ is given by*

$$\hat{X} = \begin{cases} 1, & \textit{if } L(Y) > \lambda \\ 1 \textit{ w.p. } \gamma, & \textit{if } L(Y) = \lambda \\ 0, & \textit{if } L(Y) < \lambda. \end{cases} \tag{7.3}$$

Fig. 7.14 Jerzy Neyman, 1894–1981

In these expressions,

$$L(y) = \frac{f_{Y|X}[y|1]}{f_{Y|X}[y|0]}$$

is the likelihood ratio, *i.e., the ratio of the likelihood of y when $X = 1$ divided by its likelihood when $X = 0$. Also, $\lambda > 0$ and $\gamma \in [0, 1]$ are chosen so that the resulting $\hat{X}$ satisfies*

$$P[\hat{X} = 1 | X = 0] = \beta.$$

∎

Thus, if $L(Y)$ is large, $\hat{X} = 1$. The fact that $L(Y)$ is large means that the observed value Y is much more likely when $X = 1$ than when $X = 0$. One is then inclined to decide that $X = 1$, i.e. to guess $\hat{X} = 1$. The situation is similar when $L(Y)$ is small. By adjusting λ, one controls the sensitivity of the detector. If λ is small, one tends to choose $\hat{X} = 1$ more frequently, which increases PCD but also PFA. One then chooses λ so that the detector is just sensitive enough so that $PFA = \beta$. In some problems, one may have to hedge the guess for the critical value λ as we will explain in examples (Fig. 7.14).

We prove this theorem in the next chapter. Let us consider a number of examples.

7.6.3 Examples

Gaussian Channel

Recall our model of the scalar Gaussian channel:

$$Y = X + Z,$$

where $Z = N(0, \sigma^2)$ and is independent of X. In this model, $X \in \{0, 1\}$ and the receiver tries to guess X from the received signal Y.

We looked at two formulations: MLE and MAP. In the MLE, we want to find the value of X that makes Y most likely. That is,

$$MLE[X|Y = y] = \arg\max_x f_{Y|X}[y|x].$$

The answer is $MLE[X|Y] = 0$ if $Y < 0.5$ and $MLE[X|Y] = 1$, otherwise.

The MAP is the most likely value of X in $\{a, b\}$ given Y. That is,

$$MAP[X|Y = y] = \arg\max_x P[X = x|Y = y].$$

To calculate the MAP, one needs to know the prior probability p_0 that $X = 0$. We found out that $MAP[X|Y = y] = 1$ if $y \geq 0.5 + \sigma^2 \log(p_0/p_1)$ and $MAP[X|Y = y] = 0$ otherwise.

In the hypothesis testing formulation, we choose a bound β on $PFA = P[\hat{X} = 1|X = 0]$. According to Theorem 7.4, we should calculate the likelihood ratio $L(Y)$. We find that

$$L(y) = \frac{\exp\left\{-\frac{(y-1)^2}{2\sigma^2}\right\}}{\exp\left\{-\frac{y^2}{2\sigma^2}\right\}} = \exp\left\{\frac{2y-1}{2\sigma^2}\right\}.$$

Note that, for any given λ, $P(L(Y) = \lambda) = 0$. Moreover, $L(y)$ is strictly increasing in y. Hence, (7.3) simplifies to

$$\hat{X} = \begin{cases} 1, & \text{if } y \geq y_0 \\ 0, & \text{otherwise.} \end{cases}$$

We choose y_0 so that $PFA = \beta$, i.e., so that

$$P[\hat{X} = 1|X = 0] = P[Y \geq y_0|X = 0] = \beta.$$

Now, given $X = 0$, $Y = N(0, \sigma^2)$. Hence, y_0 is such that

$$P(N(0, \sigma^2) \geq y_0) = \beta,$$

i.e., such that

$$P\left(N(0, 1) \geq \frac{y_0}{\sigma}\right) = \beta.$$

For instance, Fig. 3.7 shows that if $\beta = 5\%$, then $y_0/\sigma = 1.65$. Figure 7.15 illustrates the solution.

Fig. 7.15 The solution of the hypothesis testing problem for a Gaussian channel

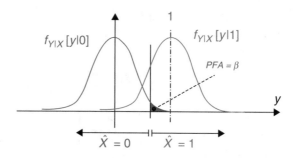

Fig. 7.16 The ROC a Gaussian channel $Y = X + Z$ where $X \in \{0, 1\}$ and $Z = N(0, \sigma^2)$

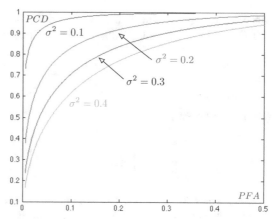

Let us calculate the ROC for the Gaussian channel. Let $y(\beta)$ be such that $P(N(0, 1) \geq y(\beta)) = \beta$, so that $y_0 = y(\beta)\sigma$. The probability of correct detection is then

$$PCD = P[\hat{X} = 1 | X = 1] = P[Y \geq y_0 | X = 1] = P(N(1, \sigma^2) \geq y_0)$$

$$= P(N(0, \sigma^2) \geq y_0 - 1) = P(N(0, 1) \geq \sigma^{-1} y_0 - \sigma^{-1})$$

$$= P(N(0, 1) \geq y(\beta) - \sigma^{-1}).$$

Figure 7.16 shows the ROC for different values of σ, obtained using Python. Not surprisingly, the performance of the system degrades when the channel is noisier.

Mean of Exponential RVs

In this second example, we are testing the mean of exponential random variables. The story is that a machine produces lightbulbs that have an exponentially distributed lifespan with mean $1/\lambda_x$ when $X = x \in \{0, 1\}$. Assume that $\lambda_0 < \lambda_1$. The interpretation is that the machine is defective when $X = 1$ and produces lightbulbs that have a shorter lifespan.

Let $Y = (Y_1, \ldots, Y_n)$ be the observed lifespans of n bulbs. We want to detect that $X = 1$ with $PFA \leq \beta = 5\%$.

We find

$$L(y) = \frac{f_{Y|X}[y|1]}{f_{Y|X}[y|0]} = \frac{\Pi_{i=1}^n \lambda_1 \exp\{-\lambda_1 y_i\}}{\Pi_{i=1}^n \lambda_0 \exp\{-\lambda_0 y_i\}}$$

$$= \left(\frac{\lambda_1}{\lambda_0}\right)^n \exp\left\{-(\lambda_1 - \lambda_0) \sum_{i=1}^n y_i\right\}.$$

Since $\lambda_1 > \lambda_0$, we find that $L(y)$ is strictly decreasing in $\sum_i y_i$ and also that $P(L(Y) = \lambda) = 0$ for all λ. Thus, (7.3) simplifies to

$$\hat{X} = \begin{cases} 1, & \text{if } \sum_{i=1}^n Y_i \leq a \\ 0, & \text{otherwise}, \end{cases}$$

where a is chosen so that

$$P\left[\sum_{i=1}^n Y_i \leq a | X = 0\right] = \beta = 5\%.$$

Now, when $X = 0$, the Y_i are i.i.d. random variables that are exponentially distributed with mean $1/\lambda_0$. The distribution of their sum is rather complicated. We approximate it using the Central Limit Theorem.

We have[3]

$$\frac{Y_1 + \cdots + Y_n - n\lambda_0^{-1}}{\sqrt{n}} \approx N(0, \lambda_0^{-2}).$$

Now,

$$\sum_{i=1}^n Y_i \leq a \Leftrightarrow \frac{Y_1 + \cdots + Y_n - n\lambda_0^{-1}}{\sqrt{n}} \leq \frac{a - n\lambda_0^{-1}}{\sqrt{n}}.$$

Hence,

$$P\left[\sum_{i=1}^n Y_i \leq a | X = 0\right] \approx P\left(N(0, \lambda_0^{-2}) \leq \frac{a - n\lambda_0^{-1}}{\sqrt{n}}\right)$$

$$= P\left(N(0, 1) \leq \lambda_0 \frac{a - n\lambda_0^{-1}}{\sqrt{n}}\right).$$

---

[3]Recall that $\text{var}(Y_i) = \lambda_0^{-2}$.

Hence, if we want this probability to be equal to 5%, by (3.2), we must choose a so that

$$\lambda_0 \frac{a - n\lambda_0^{-1}}{\sqrt{n}} = 1.65,$$

i.e.,

$$a = (n + 1.65\sqrt{n})\lambda_0^{-1}.$$

One point is worth noting for this example. We see that the calculation of $\hat{X}$ is based on $Y_1 + \cdots + Y_n$. Thus, although one has measured the individual lifespans of the n bulbs, the decision is based only on their sum, or equivalently on their average.

Bias of a Coin

In this example, we observe n coin flips. Given $X = x \in \{0, 1\}$, the coins are i.i.d. $B(p_x)$. That is, given $X = x$, the outcomes $Y_1, \ldots, Y_n$ of the coin flips are i.i.d. and equal to 1 with probability p_x and to zero otherwise. We assume that $p_1 > p_0 = 0.5$. That is, we want to test whether the coin is fair or biased.

Here, the random variables Y_i are discrete. We see that

$$P[Y_i = y_i, i = 1, \ldots, n | X = x] = \Pi_{i=1}^n p_x^{Y_i}(1 - p_x)^{1-Y_i}$$
$$= p_x^S(1 - p_x)^{n-S} \text{ where } S = Y_1 + \cdots + Y_n.$$

Hence,

$$L(Y_1, \ldots, Y_n) = \frac{P[Y_i = y_i, i = 1, \ldots, n | X = 1]}{P[Y_i = y_i, i = 1, \ldots, n | X = 0]}$$
$$= \left(\frac{p_1}{p_0}\right)^S \left(\frac{1 - p_1}{1 - p_0}\right)^{n-S} = \left(\frac{1 - p_1}{1 - p_0}\right)^n \left(\frac{p_1(1 - p_0)}{p_0(1 - p_1)}\right)^S.$$

Since $p_1 > p_0$, we see that the likelihood ratio is increasing in S. Thus, the solution of the hypothesis testing problem is

$$\hat{X} = 1\{S \geq n_0\},$$

where n_0 is such that $P[S \geq n_0 | X = 0] \approx \beta$. To calculate n_0, we approximate S, when $X = 0$, by using the Central Limit Theorem. We have

$$P[S \geq n_0 | X = 0] = P\left[\frac{S - np_0}{\sqrt{n}} \geq \frac{n_0 - np_0}{\sqrt{n}} \Big| X = 0\right]$$
$$\approx P\left(N(0, p_0(1 - p_0)) \geq \frac{n_0 - np_0}{\sqrt{n}}\right)$$
$$= P\left(N(0, 0.25) \geq \frac{n_0 - np_0}{\sqrt{n}}\right) = P\left(N(0, 1) \geq \frac{2n_0 - n}{\sqrt{n}}\right).$$

Say that $\beta = 5\%$, then we need

$$\frac{2n_0 - n}{\sqrt{n}} = 1.65,$$

by (3.2). Hence,

$$n_0 = 0.5n + 0.83\sqrt{n}.$$

Discrete Observations

In the examples that we considered so far, the random variable $L(Y)$ is continuous. In such cases, the probability that $L(Y) = \lambda$ is always zero, and there is no need to randomize the choice of $\hat{X}$ for specific values of Y. In our next examples, that need arises.

First consider, as usual, the problem of choosing $\hat{X} \in \{0, 1\}$ to maximize the probability of correct detection $P[\hat{X} = 1|X = 1]$ subject to a bound $P[\hat{X} = 1|X = 0] \leq \beta$ on the probability of false alarm. However, assume that we make no observation. In this case, the solution is to choose $\hat{X} = 1$ with probability β. This choice meets the bound on the probability of false alarm and achieves a probability of correct detection equal to β. This randomized choice is better than always deciding $\hat{X} = 0$.

Now consider a more complex example where $Y \in \{A, B, C\}$ and

$$P[Y = A|X = 1] = 0.2, P[Y = B|X = 1] = 0.2, P[Y = C|X = 1] = 0.6$$

$$P[Y = A|X = 0] = 0.2, P[Y = B|X = 0] = 0.5, P[Y = C|X = 0] = 0.3.$$

Accordingly, the values of the likelihood ratio $L(y) = P[Y = y|X = 1]/P[Y = y|X = 0]$ are as follows:

$$L(A) = 1, L(B) = 0.4 \text{ and } L(C) = 2.$$

We rank the observations in increasing order of the values of L, as shown in Fig. 7.17.

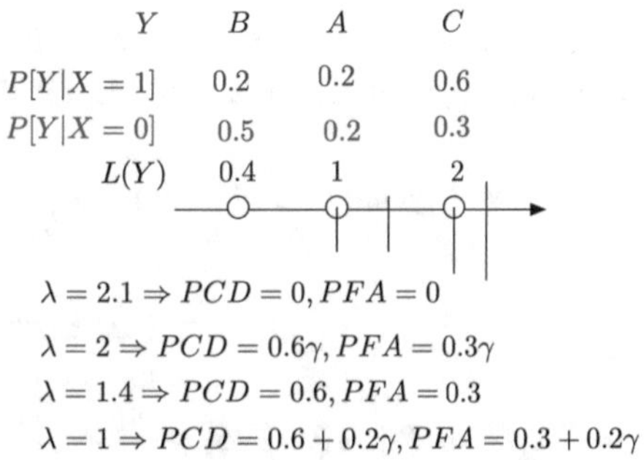

Fig. 7.17 The three possible observations

Y	B	A	C
$P[Y|X = 1]$	0.2	0.2	0.6
$P[Y|X = 0]$	0.5	0.2	0.3
$L(Y)$	0.4	1	2

$\lambda = 2.1 \Rightarrow PCD = 0, PFA = 0$

$\lambda = 2 \Rightarrow PCD = 0.6\gamma, PFA = 0.3\gamma$

$\lambda = 1.4 \Rightarrow PCD = 0.6, PFA = 0.3$

$\lambda = 1 \Rightarrow PCD = 0.6 + 0.2\gamma, PFA = 0.3 + 0.2\gamma$

Fig. 7.18 The ROC for the discrete observation example

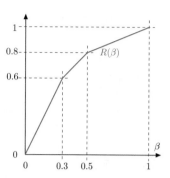

The solution of the hypothesis testing problem amounts to choosing a threshold λ and a randomization γ so that

$$P[\hat{X} = 1|Y] = 1\{L(Y) > \lambda\} + \gamma 1\{L(Y) = \lambda\}.$$

Also, we choose λ and γ so that $P[\hat{X} = 1|X = 0] = \beta$.

Figure 7.17 shows that if we choose $\lambda = 2.1$, then $L(Y) < \lambda$, for all values of Y, so that we always decide $\hat{X} = 0$. Accordingly, $PCD = 0$ and $PFA = 0$.

The figure also shows that if we choose $\lambda = 2$ and a parameter γ, then we decide $\hat{X} = 1$ when $L(Y) = 2$ with probability γ. Thus, if $X = 0$, we decide $\hat{X} = 1$ with probability 0.3γ, because $Y = C$ with probability 0.3 when $X = 0$ and this is precisely when $L(Y) = 2$ and we randomize with probability γ. The figure shows other examples.

It should be clear that as we reduce λ from 2.1 to 0.39, the probability that we decide $\hat{X} = 1$ when $X = 0$ increases from 0 to 1. Also, by choosing the parameter γ suitably when λ is set to a possible value of $L(Y)$, we can adjust PFA to any value in $[0, 1]$.

For instance, we can have $PFA = 0.05$ if we choose $\lambda = 2$ and $\gamma = 0.05/0.3$. Similarly, we can have $PFA = 0.4$ by choosing $\lambda = 1$ and $\gamma = 0.5$. Indeed, in this case, we decide $\hat{X} = 1$ when $Y = C$ and also with probability 0.5 when $Y = A$, so that this occurs with probability $0.3 + 0.2 \times 0.5 = 0.4$ when $X = 0$. The corresponding PCD is then $0.6 + 0.2 \times 0.5 = 0.7$.

Figure 7.18 shows PCD as a function of the bound on PFA.

7.7 Summary

- MAP and MLE;
- BPSK;
- Huffman Codes;
- Independent Gaussian Errors;
- Hypothesis Testing: Neyman–Pearson Theorem.

7.7.1   Key Equations and Formulas

Bayes' Rule	$\pi_i = p_i q_i / (\sum_j p_j q_j)$	Theorem 7.1
$MAP[X\|Y = y]$	$\arg\max_x P[X = x\|Y = y]$	Definition 7.1
$MLE[X\|Y = y]$	$\arg\max_x P[Y = y\|X = x]$	Definition 7.1
Likelihood Ratio	$L(y) = f_{Y\|X}[y\|1]/f_{Y\|X}[y\|0]$	Theorem 7.4
Gaussian Channel	$MAP[X\|Y = y] = 1\{y \geq \frac{1}{2} + \sigma^2 \log(\frac{p_0}{p_1})\}$	(7.2)
Neyman–Pearson Theorem	$P[\hat{X} = 1\|Y] = 1\{L(Y) > \lambda\} + \gamma 1\{L(Y) = \lambda\}$	Theorem 7.4
ROC	$ROC(\beta) = \max. PCD$ s.t. $PFA \leq \beta$	Definition 7.2

7.8   References

Detection theory is obviously a classical topic. It is at the core of digital communication (see e.g., Proakis (2000)). The Neyman–Pearson Theorem is introduced in Neyman and Pearson (1933). For a discussion of hypothesis testing, see Lehmann (2010). For more details on digital communication and, in particular, on wireless communication, see the excellent presentation in Tse and Viswanath (2005).

7.9   Problems

Problem 7.1 Assume that when $X = 0, Y = \mathcal{N}(0, 1)$ and when $X = 1, Y = \mathcal{N}(0, \sigma^2)$ with $\sigma^2 > 1$. Calculate $MLE[X|Y]$.

Problem 7.2 Let X, Y be i.i.d. $U[0, 1]$ random variables. Define $V = X + Y$ and $W = X - Y$.

(a) Show that V and W are uncorrelated;
(b) Are V and W independent? Prove or disprove.

Problem 7.3 A digital link uses the QAM-16 constellation shown in Fig. 7.12 with $\mathbf{x}_1 = (1, -1)$. The received signal is $\mathbf{Y} = \mathbf{X} + \mathbf{Z}$ where $\mathbf{Z} =_D \mathcal{N}(\mathbf{0}, \sigma^2\mathbf{I})$. The receiver uses the MAP. Simulate the system using Python to estimate the fraction of errors for $\sigma = 0.2, 0.3$.

Problem 7.4 Use Python to verify the CLT with i.i.d. $U[0, 1]$ random variables X_n. That is, generate the random variables $\{X_1, \ldots, X_N\}$ for $N = 10000$. Calculate

$$Y_n = \frac{X_{100n+1} + \cdots + X_{(n+1)100} - 50}{10}, n = 0, 1, \ldots, 99.$$

Plot the empirical cdf of $\{Y_0, \ldots, Y_{99}\}$ and compare with the cdf of a $\mathcal{N}(0, 1/12)$ random variable.

Problem 7.5 You are testing a digital link that corresponds to a BSC with some error probability $\epsilon \in [0, 0.5)$.

(a) Assume you observe the input and the output of the link. How do you find the MLE of ϵ.
(b) You are told that the inputs are i.i.d. bits that are equal to 1 with probability 0.6 and to 0 with probability 0.4. You observe n outputs. How do you calculate the MLE of ϵ.
(c) The situation is as in the previous case, but you are told that ϵ has pdf $4 - 8x$ on $[0, 0.5)$. How do you calculate the MAP of ϵ given n outputs.

Problem 7.6 The situation is the same as in the previous problem. You observe n inputs and outputs of the BSC. You want to solve a hypothesis problem to detect that $\epsilon > 0.1$ with a probability of false alarm at most equal to 5%. Assume that n is very large and use the CLT.

Problem 7.7 The random variable X is such that $P(X = 1) = 2/3$ and $P(X = 0) = 1/3$. When $X = 1$, the random variable Y is exponentially distributed with rate 1. When $X = 0$, the random variable Y is uniformly distributed in $[0, 2]$. (*Hint:* Be careful about the case $Y > 2$.)

(a) Find $MLE[X|Y]$;
(b) Find $MAP[X|Y]$;
(c) Solve the following hypothesis testing problem:

$$\text{Maximize } P[\hat{X} = 1|X = 1]$$

$$\text{subject to } P[\hat{X} = 1|X = 0] \le 5\%.$$

Problem 7.8 Simulate the following communication channel. There is an i.i.d. source that generates symbols $\{1, 2, 3, 4\}$ according to a prior distribution $\pi = [p_1, p_2, p_3, p_4]$. The symbols are modulated by QPSK scheme, i.e. they are mapped to constellation points $(\pm 1, \pm 1)$. The communication is on a baseband Gaussian channel, i.e. if the sent signal is (x_1, x_2), the received signal is

$$y_1 = x_1 + Z_1,$$

$$y_2 = x_2 + Z_2,$$

where Z_1 and Z_2 are independent $N(0, \sigma^2)$ random variables. Find the MAP detector and ML detector analytically.

Simulate the channel using Python for $\pi = [0.1, 0.2, 0.3, 0.4]$, and $\sigma = 0.1$ and $\sigma = 0.5$. Evaluate the probability of correct detection.

Problem 7.9 Let X be equally likely to take any of the values $\{1, 2, 3\}$. Given X, the random variable Y is $\mathcal{N}(X, 1)$.

(a) Find $MAP[X|Y]$;
(b) Calculate $MLE[X|Y]$;
(c) Calculate $E((X - Y)^2)$.

Problem 7.10 The random variable X is such that $P(X = 0) = P(X = 1) = 0.5$. Given X, the random variables Y_n are i.i.d. $U[0, 1.1 - 0.1X]$. The goal is to guess $\hat{X}$ from the observations Y_n. Each observation has a cost $\beta > 0$. To get nice numerical solutions, we assume that

$$\beta = 0.018 \approx 0.5(1.1)^{-10} \log(1.1).$$

(a) Assume that you have observed $Y^n = (Y_1, \dots, Y_n)$. What is the guess $\hat{X}_n$ based on these observations that maximizes the probability that $\hat{X}_n = X$?
(b) What is the corresponding value of $P(\hat{X}_n = X)$?
(c) Choose n to maximize $P(X = \hat{X}_n) - \beta n$ where $\hat{X}_n$ is chosen on the basis of $Y_1, \dots, Y_n$). *Hint:* You will recall that

$$\frac{d}{dx}(a^x) = a^x \log(a).$$

Problem 7.11 The random variable X is exponentially distributed with mean 1. Given X, the random variable Y is exponentially distributed with rate X.

(a) Find $MLE[X|Y]$;
(b) Find $MAP[X|Y]$;
(c) Solve the following hypothesis testing problem:

$$\text{Maximize } P[\hat{X} = 1|X = a]$$

$$\text{subject to } P[\hat{X} = 1|X = 1] \le 5\%,$$

where $a > 1$ is given.

Problem 7.12 Consider a random variable Y that is exponentially distributed with parameter θ. You observe n i.i.d. samples $Y_1, \dots, Y_n$ of this random variable. Calculate $\hat{\theta} = MLE[\theta|Y_1, \dots, Y_n]$. What is the bias of this estimator, i.e., $E[\hat{\theta} - \theta|\theta]$? Does the bias converge to 0 as n goes to infinity?

Problem 7.13 Assume that $Y =_D U[a, b]$. You observe n i.i.d. samples $Y_1, \ldots, Y_n$ of this random variable. Calculate the maximum likelihood estimator $\hat{a}$ of a and $\hat{b}$ of b. What is the bias of $\hat{a}$ and $\hat{b}$?

Problem 7.14 We are looking at an hypothesis testing problem where $X, \hat{X}$ take values in $\{0, 1\}$. The value of $\hat{X}$ is decided based on the observed value of the random vector $\mathbf{Y}$. We assume that $\mathbf{Y}$ has a density $f_i(\mathbf{y})$ given that $X = i$, for $i = 0, 1$, and we define $L(\mathbf{y}) := f_1(\mathbf{y})/f_0(\mathbf{y})$.

Define $g(\beta)$ to be the maximum value of $P[\hat{X} = 1|X = 1]$ subject to $P[\hat{X} = 1|X = 0] \leq \beta$ for $\beta \in [0, 1]$. Then (choose the correct answers, if any)

☐ $g(\beta) \geq 1 - \beta$;
☐ $g(\beta) \geq \beta$;
☐ The optimal decision is described by a function $h(\mathbf{y}) = P[\hat{X} = 1|\mathbf{Y} = \mathbf{y}]$ and this function is nondecreasing in $f_1(\mathbf{y})/f_0(\mathbf{y})$.

Problem 7.15 Given $\theta \in \{0, 1\}$, $\mathbf{X} = \theta(1, 1)' + \mathbf{V}$ where V_1 and V_2 are independent and uniformly distributed in $[-2, 2]$. Solve the hypothesis testing problem:

$$\text{Maximize } P[\hat{\theta} = 1|\theta = 1]$$

$$\text{s.t. } P[\hat{\theta} = 1|\theta = 0] \leq 5\%.$$

Problem 7.16 Given $\theta = 1$, $X =_D Exp(1)$ and, given $\theta = 0$, $X =_D U[0, 2]$.

(a) Find $\hat{\theta} = HT[\theta|X, \beta]$, defined as the random variable $\hat{\theta}$ determined from X that maximizes $P[\hat{\theta} = 1|\theta = 1]$ subject to $P[\hat{\theta} = 1|\theta = 0] \leq \beta$;
(b) Compute the resulting value of $\alpha(\beta) = P[\hat{\theta} = 1|\theta = 1]$;
(c) Sketch the ROC curve $\alpha(\beta)$ for $\beta \in [0, 1]$.

Problem 7.17 You observe a random sequence $\{X_n, n = 0, 1, 2, \ldots\}$. With probability p, $\theta = 0$ and this sequence is i.i.d. Bernoulli with $P(X_n = 0) = P(X_n = 1) = 0.5$. With probability $1 - p$, $\theta = 1$ and the sequence is a stationary Markov chain on $\{0, 1\}$ with transition probabilities $P(0, 1) = P(1, 0) = \alpha$. The parameter α is given in $(0, 1)$.

(1) Find $MAP[\theta|X_0, \ldots, X_n]$;
(2) Discuss the convergence of $\hat{\theta}_n$;
(3) Discuss the composite hypothesis testing problem where $\alpha < 0.5$ when $\theta = 1$ and $\alpha = 0.5$ when $\theta = 0$.

Problem 7.18 If $\theta = 0$, the sequence $\{X_n, n \geq 0\}$ is a Markov chain on a finite set $\mathcal{X}$ with transition matrix P_0. If $\theta = 1$, the transition matrix is P_1. In both cases, $X_0 = x_0$ is known. Find $MLE[\theta|X_0, \ldots, X_n]$.

Digital Link—B

8

Topics: Optimality of Huffman Codes, LDPC Codes, Proof of Neyman–Pearson Theorem, Jointly Gaussian RVs, Statistical Tests, ANOVA

8.1   Proof of Optimality of the Huffman Code

We stated the following result in Chap. 7. Here, we provide a proof.

Theorem 8.1 (Optimality of Huffman Code) *The Huffman code has the smallest average number of bits per symbol among all prefix-free codes (Fig. 8.1).*

∎

Proof The argument in Huffman (1952) is by induction on the number of symbols. Assume that the Huffman code has an average path length $L(n)$ that is minimum for n symbols and that there is some other tree T with a smaller average path length $A(n + 1)$ than the Huffman code for $n + 1$ symbols. Let X and Y be the two least frequent symbols and $x \geq y$ their frequencies. We can pick these symbols in T so that their path lengths are maximum and such that Y has the largest path length in T. Otherwise, we could swap Y in T with a more frequent symbol and reduce the average path length. Accept for now the claim that we can also pick X and Y so that they are siblings in T. By merging X and Y into their parent Z with frequency $z = x + y$, we have constructed a code for n symbols with average path length $A(n+1)-z$. Hence, $L(n) \leq A(n+1)-z$. Now, the Huffman code for $n+1$ symbol would merge X and Y also, so that its average path length is $L(n + 1) := L(n) + z$.

© The Author(s) 2021
J. Walrand, *Probability in Electrical Engineering and Computer Science*,
https://doi.org/10.1007/978-3-030-49995-2_8

Fig. 8.1 Huffman code

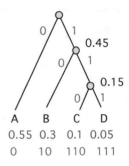

0.45

0.15

A   B   C   D

0.55 0.3   0.1   0.05

0   10   110   111

Thus, $L(n + 1) \leq A(n + 1)$, which contradicts the assumption that the Huffman code is not optimal for $n + 1$ symbols. It remains to prove the claim about X and Y being siblings. First note that Y having the maximum path length, it cannot be an only child, for otherwise, we would replace its parent by Y and reduce the path length. Say that Y has a sibling V other than X. By swapping V and X, one does not increase the average path length, since the frequency of V is not smaller than that of X. This concludes the proof. $\square$

8.2   Proof of Neyman–Pearson Theorem 7.4

The idea of the proof is to consider any other decision rule that produces an estimate $\tilde{X}$ with $P[\tilde{X} = 1 | X = 0] \leq \beta$ and to show that

$$P[\tilde{X} = 1 | X = 1] \leq P[\hat{X} | X = 1], \tag{8.1}$$

where $\hat{X}$ is specified by the theorem. To show this, we note that

$$(\hat{X} - \tilde{X})(L(Y) - \lambda) \geq 0.$$

Indeed, when $L(Y) - \lambda > 0$, one has $\hat{X} = 1 \geq \tilde{X}$, so that the expression above is indeed nonnegative. Similarly, when $L(Y) - \lambda < 0$, one has $\hat{X} = 0 \leq \tilde{X}$, so that the expression is again nonnegative.

Taking the expected value of this expression given $X = 0$, we find

$$E[\hat{X}L(Y)|X = 0] - E[\tilde{X}L(Y)|X = 0]$$

$$\geq \lambda(E[\hat{X}|X = 0] - E[\tilde{X}|X = 0]). \tag{8.2}$$

Now,

$$E[\hat{X}|X = 0] = P[\hat{X} = 1|X = 0] = \beta \geq P[\tilde{X} = 1|X = 0] = E[\tilde{X}|X = 0].$$

Hence, (8.2) implies that

$$E[\hat{X}L(Y)|X = 0] \geq E[\tilde{X}L(Y)|X = 0]. \tag{8.3}$$

Observe that, for any function $g(Y)$, one has

$$E[g(Y)L(Y)|X = 0] = \int g(y)L(y)f_{Y|X}[y|0]dy$$

$$= \int g(y)\frac{f_{Y|X}[y|1]}{f_{Y|X}[y|0]}f_{Y|X}[y|0]dy$$

$$= \int g(y)f_{Y|X}[y|1]dy$$

$$= E[g(Y)|X = 1].$$

Note that this result continues to hold even for a function $g(Y, Z)$ where Z is a random variable that is independent of X and Y. In particular,

$$E[\hat{X}L(Y)|X = 0] = E[\hat{X}|X = 1] = P[\hat{X} = 1|X = 1].$$

Similarly,

$$E[\tilde{X}L(Y)|X = 0] = P[\tilde{X} = 1|X = 1].$$

Combining these results with (8.3) gives (8.1).

□

8.3   Jointly Gaussian Random Variables

In many systems, the errors in the different components of the measured vector $\mathbf{Y}$ are not independent. A suitable model for this situation is that

$$\mathbf{Y} = \mathbf{X} + A\mathbf{Z},$$

where $\mathbf{Z} = (Z_1, Z_2)$ is a pair of i.i.d. $N(0, 1)$ random variables and A is some 2×2 matrix. The key idea here is that the components of the noise vector $A\mathbf{Z}$ will not be independent in general. For instance, if the two rows of A are identical, so are the two components of $A\mathbf{Z}$. Thus, this model allows to capture a dependency between the errors in the two components. The model also suggests that the dependency comes from the fact that the errors are different linear combinations of the same fundamental sources of noise.

For such a model, how does one compute $MLE[\mathbf{X}|\mathbf{Y}]$? We explain in the next section that

$$f_{\mathbf{Y}|\mathbf{X}}[\mathbf{y}|\mathbf{x}] = \frac{1}{2\pi|A|} \exp\left\{-\frac{1}{2}(\mathbf{y} - \mathbf{x})'(AA')^{-1}(\mathbf{y} - \mathbf{x})\right\}, \tag{8.4}$$

where A' is the transposed of matrix A, i.e., $A'(i, j) = A(j, i)$ for $i, j \in \{1, 2\}$.

Consequently, the MLE is the value $\mathbf{x}_k$ of $\mathbf{x}$ that minimizes

$$(\mathbf{y} - \mathbf{x})'(AA')^{-1}(\mathbf{y} - \mathbf{x}) = ||A^{-1}\mathbf{y} - A^{-1}\mathbf{x}||^2.$$

(For simplicity, we assume that A is invertible.)

That is, we want to find the vector $\mathbf{x}_k$ such that $A^{-1}\mathbf{x}_k$ is the closest to $A^{-1}\mathbf{y}$. One way to understand this result is to note that

$$\mathbf{W} := A^{-1}\mathbf{Y} = A^{-1}\mathbf{X} + \mathbf{Z} =: \mathbf{V} + \mathbf{Z}.$$

Thus, if we calculate $A^{-1}\mathbf{Y}$ from the measured vector $\mathbf{Y}$, we find that its components are i.i.d. $N(0, 1)$ for a given value of $\mathbf{X}$. Hence, it is easy to calculate $MLE[\mathbf{V}|\mathbf{W} = \mathbf{w}]$: it is the closest value to $\mathbf{w}$ in the set $\{A^{-1}\mathbf{x}_1, \ldots, A^{-1}\mathbf{x}_{16}\}$ of possible values of $\mathbf{V}$. It is then reasonable to expect that we can recover the MLE of $\mathbf{X}$ by multiplying the MLE of $\mathbf{V} = A^{-1}\mathbf{X}$ by A, i.e., that

$$MLE[\mathbf{X}|\mathbf{Y} = \mathbf{y}] = A \times MLE[\mathbf{V}|\mathbf{W} = A^{-1}\mathbf{y}].$$

8.3.1 Density of Jointly Gaussian Random Variables

Our goal in this section is to explain (8.4) and more general versions of this result.

We start by stating the main definition and a result that we prove later.

Definition 8.1 (Jointly Gaussian $N(\mu_{\mathbf{Y}}, \Sigma_{\mathbf{Y}})$ Random Variables) The random variables $\mathbf{Y} = (Y_1, \ldots, Y_n)'$ are jointly Gaussian with mean $\mu_{\mathbf{Y}}$ and covariance $\Sigma_{\mathbf{Y}}$, which we write as $\mathbf{Y} =_D N(\mu_{\mathbf{Y}}, \Sigma_{\mathbf{Y}})$, if

$$\mathbf{Y} = A\mathbf{X} + \mu_{\mathbf{Y}} \text{ with } \Sigma_{\mathbf{Y}} = AA',$$

where $\mathbf{X}$ is a vector of independent $N(0, 1)$ random variables.

$\diamond$

Here is the main result.

Theorem 8.2 (Density of $N(\mu_{\mathbf{Y}}, \Sigma_{\mathbf{Y}})$ Random Variables) *Let $\mathbf{Y} =_D N(\mu_{\mathbf{Y}}, \Sigma_{\mathbf{Y}})$. Then*

$$f_{\mathbf{Y}}(\mathbf{y}) = \frac{1}{\sqrt{|\Sigma_{\mathbf{Y}}|}(2\pi)^{n/2}} \exp\left\{-\frac{1}{2}(\mathbf{y} - \mu_{\mathbf{Y}})'\Sigma_{\mathbf{Y}}^{-1}(\mathbf{y} - \mu_{\mathbf{Y}})\right\}. \tag{8.5}$$

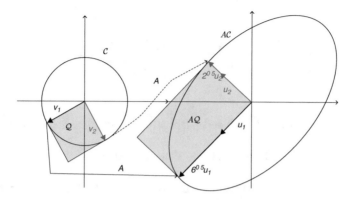

Fig. 8.2 The level curves of f_Y

■

The level curves of this jpdf are ellipses, as sketched in Fig. 8.2.

Note that this joint distribution is determined by the mean and the covariance matrix. In particular, if $\mathbf{Y}' = (\mathbf{V}', \mathbf{W}')$ are jointly Gaussian, then the joint distribution is characterized by the mean and $\Sigma_{\mathbf{V}}$, $\Sigma_{\mathbf{W}}$ and $\text{cov}(\mathbf{V}, \mathbf{W})$. We know that if $\mathbf{V}$ and $\mathbf{W}$ are independent, then they are uncorrelated, i.e., $\text{cov}(\mathbf{V}, \mathbf{W}) = 0$. Since the joint distribution is characterized by the mean and covariance, we conclude that if they are uncorrelated, they are independent. We note this fact as a theorem.

Theorem 8.3 (Jointly Gaussian RVs Are Independent Iff Uncorrelated) *Let* $\mathbf{V}$ *and* $\mathbf{W}$ *be jointly Gaussian random variables. Then, there are independent if and only if they are uncorrelated.*

■

We will use the following result.

Theorem 8.4 (Linear Combinations of JG Are JG) *Let* $\mathbf{V}$ *and* $\mathbf{W}$ *be jointly Gaussian. Then* $A\mathbf{V} + \mathbf{a}$ *and* $B\mathbf{W} + \mathbf{b}$ *are jointly Gaussian.*

■

Proof By definition, $\mathbf{V}$ and $\mathbf{W}$ are jointly Gaussian if they are linear functions of i.i.d. $N(0, 1)$ random variables. But then $A\mathbf{V} + \mathbf{a}$ and $B\mathbf{W} + \mathbf{b}$ are linear functions of the same i.i.d. $N(0, 1)$ random variables, so that they are jointly Gaussian. More explicitly, there are some i.i.d. $\mathcal{N}(0, 1)$ random variables $\mathbf{X}$ so that

$$\begin{bmatrix} \mathbf{V} \\ \mathbf{W} \end{bmatrix} = \begin{bmatrix} \mathbf{c} \\ \mathbf{d} \end{bmatrix} + \begin{bmatrix} C \\ D \end{bmatrix} \mathbf{X},$$

so that

$$\begin{bmatrix} A\mathbf{V} + \mathbf{a} \\ B\mathbf{W} + \mathbf{b} \end{bmatrix} = \begin{bmatrix} \mathbf{a} + A\mathbf{c} \\ \mathbf{b} + B\mathbf{d} \end{bmatrix} + \begin{bmatrix} AC \\ BD \end{bmatrix} \mathbf{X}.$$

□

As an example, let X, Y be independent $N(0, 1)$ random variables. Then,

$$X + Y \text{ and } X - Y \text{ are independent.}$$

Indeed, these random variables are jointly Gaussian by Theorem 8.4. Also, they are uncorrelated since

$$E((X + Y)(X - Y)) - E(X + Y)E(X - Y) = E(X^2 - Y^2) = 0.$$

Hence, they are independent by Theorem 8.3.

We devote the remainder of this section to the derivation of (8.5). We explain in Theorem B.13 how to calculate the p.d.f. of $A\mathbf{X} + \mathbf{b}$ from the density of $\mathbf{X}$. We recall the result here for convenience:

$$f_{\mathbf{Y}}(\mathbf{y}) = \frac{1}{|A|} f_{\mathbf{X}}(\mathbf{x}) \text{ where } A\mathbf{x} + \mathbf{b} = \mathbf{y}. \tag{8.6}$$

Let us apply (8.6) to the case where $\mathbf{X}$ is a vector of n i.i.d. $N(0, 1)$ random variables. In this case,

$$f_{\mathbf{X}}(\mathbf{x}) = \Pi_{i=1}^{n} f_{X_i}(x_i) = \Pi_{i=1}^{n} \frac{1}{\sqrt{2\pi}} \exp\left\{ -\frac{x_i^2}{2} \right\}$$

$$= \frac{1}{(2\pi)^{n/2}} \exp\left\{ -\frac{||\mathbf{x}||^2}{2} \right\}.$$

Then, (8.6) gives

$$f_{\mathbf{Y}}(\mathbf{y}) = \frac{1}{|A|} \frac{1}{(2\pi)^{n/2}} \exp\left\{ -\frac{||\mathbf{x}||^2}{2} \right\},$$

where $A\mathbf{x} + \mu_{\mathbf{Y}} = \mathbf{y}$. Thus,

$$\mathbf{x} = A^{-1}(\mathbf{y} - \mu_{\mathbf{Y}})$$

and

$$||\mathbf{x}||^2 = ||A^{-1}(\mathbf{y} - \mu_{\mathbf{Y}})||^2 = (\mathbf{y} - \mu_{\mathbf{Y}})'(A^{-1})'A^{-1}(\mathbf{y} - \mu_{\mathbf{Y}}),$$

where we used the facts that $||\mathbf{z}||^2 = \mathbf{z}'\mathbf{z}$ and $(M\mathbf{v})' = \mathbf{v}'M'$.

Recall the definition of the covariance matrix:

$$\Sigma_{\mathbf{Y}} = E((\mathbf{Y} - E(\mathbf{Y}))(\mathbf{Y} - E(\mathbf{Y}))').$$

Since $\mathbf{Y} = A\mathbf{X} + \mu_{\mathbf{Y}}$ and $\Sigma_{\mathbf{X}} = \mathbf{I}$, the identity matrix, we see that

$$\Sigma_{\mathbf{Y}} = A\Sigma_{\mathbf{X}}A' = AA'.$$

In particular,

$$|\Sigma_{\mathbf{Y}}| = |A|^2.$$

Hence, we find that

$$f_{\mathbf{Y}}(\mathbf{y}) = \frac{1}{\sqrt{|\Sigma_{\mathbf{Y}}|}(2\pi)^{n/2}} \exp\left\{ -\frac{1}{2}(\mathbf{y} - \mu_{\mathbf{Y}})'\Sigma_{\mathbf{Y}}^{-1}(\mathbf{y} - \mu_{\mathbf{Y}}) \right\}.$$

This is precisely (8.5).

8.4 Elementary Statistics

This section explains some basic statistical tests that are at the core of "data science."

8.4.1 Zero-Mean?

Consider the following hypothesis testing problem. The random variable Y is $\mathcal{N}(\mu, 1)$. We want to decide between two hypotheses:

$$H_0 : \mu = 0. \tag{8.7}$$

$$H_1 : \mu \neq 0. \tag{8.8}$$

We know that $P[|Y| > 2 \mid H_0] \approx 5\%$. That is, if we reject H_0 when $|Y| > 2$, the probability of "false alarm," i.e., of rejecting the hypothesis when it is correct is 5%. This is what all the tests that we will discuss in this chapter do. However, there are many tests that achieve the same false alarm probability. For instance, we could reject H_0 when $Y > 1.64$ and the probability of false alarm would also be 5%. Or, we could reject H_0 when Y is in the interval $[1, 1.23]$. The probability of that event under H_0 is also about 5%.

Thus, there are many tests that reject H_0 with a probability of false alarm equal to 5%. Intuitively, we feel that the first one—rejecting H_0 when $|Y| > 2$—is more sensible than the others. This intuition probably comes from the idea that the alternative hypothesis $H_1 : \mu \neq 0$ appears to be a symmetric assumption about the likely values of μ. That is, we do not have a reason to believe that under H_1 the mean μ is more likely to be positive than negative. We just know that it is nonzero. Given this symmetry, it is intuitively reasonable that the test should be symmetric. However, there are many symmetric tests! So, we need a more careful justification.

To justify the test $|Y| > 2$, we note the following simple result.

Theorem 8.5 *Consider the following hypothesis testing problem: Y is $\mathcal{N}(\mu, 1)$ and*

$$H_0 : \mu = 0$$

$$H_1 : \mu \text{ has a symmetric distribution about } 0.$$

Then, the Neyman–Pearson test with probability of false alarm 5% is to reject H_0 when $|Y| > 2$.

∎

Proof We know that the Neyman–Pearson test is a likelihood ratio test. Thus, it suffices to show that the likelihood ratio is increasing in $|Y|$. Assume that the density of μ under H_1 is $h(x)$. (The same argument goes through it μ is a mixed random variable.) Then the pdf $f_1(y)$ of Y under H_1 is as follows:

$$f_1(y) = \int h(x)f(y - x)dx,$$

where $f(x) = (1/\sqrt{2\pi})\exp\{-0.5y^2\}$ is the pdf of a $\mathcal{N}(0, 1)$ random variable. Consequently, the likelihood ratio $L(y)$ of Y is given by

$$L(y) = \frac{f_1(y)}{f(y)} = \int h(x)\frac{f(y - x)}{f(y)}dx = \int h(x)\exp\{-xy\}\exp\left\{-\frac{x^2}{2}\right\}dx$$

$$= 0.5\int [h(x) + h(-x)]\exp\{-xy\}\exp\left\{-\frac{x^2}{2}\right\}dx$$

$$= 0.5\int h(x)[\exp\{xy\} + \exp\{-xy\}]dx,$$

where the fourth identity comes from $h(x) = 0.5h(x) + 0.5h(-x)$, since $h(x) = h(-x)$. This expression shows that $L(y) = L(-y)$. Also,

$$L'(y) = 0.5 \int h(x)x[\exp\{xy\} - \exp\{-xy\}]dx = \int_0^\infty h(x)x[\exp\{xy\} - \exp\{-xy\}]dx,$$

by symmetry of the integrand. For $y > 0$ and $x > 0$, we see that the last integrand is positive, so that $L'(y) > 0$ for $y > 0$.

Hence, $L(y)$ is symmetric and increasing in $y > 0$, so that it is an increasing function of $|y|$, which completes the proof. □

As a simple application, say that you buy 100 light bulbs from brand A and 100 from brand B. You want to test whether that have the same mean lifetime. You measure the lifetimes $\{X_1^A, \ldots, X_{100}^A\}$ and $\{X_1^B, \ldots, X_{100}^B\}$ of the bulbs of the two batches and you calculate

$$Y = \frac{(X_1^A + \cdots X_{100}^A) - (X_1^B + \cdots X_{100}^B)}{\sigma \sqrt{N}},$$

where σ is the standard deviation of $X_n^A + X_n^B$ that we assume to be known.

By the CLT, it is reasonable to approximate Y by a $\mathcal{N}(0, 1)$ random variable. Thus, we reject the hypothesis that the bulbs of the two brands have the same average lifetime if $|Y| > 2$.

Of course, assuming that σ is known is not realistic. The next test is then more practical.

8.4.2 Unknown Variance

A practically important variation of the previous example is when the variance σ^2 is not known. In that case, the Neyman–Pearson test is to decide H_1 when

$$\frac{|\hat{\mu}|}{\hat{\sigma}} > \lambda,$$

where $\hat{\mu}$ is the sample mean of the Y_m, as before,

$$\hat{\sigma}^2 = \frac{1}{n-1} \sum_{m=1}^{n} (Y_m - \hat{\mu})^2$$

is the sample variance, and λ is such that $P(\frac{|t_{n-1}|}{\sqrt{n-1}} > t_{n-1}) = \beta$.

Here, t_{n-1} is a random variable with a t distribution with $n - 1$ degrees of freedom. By definition, this means that

$$t_{n-1} = \frac{\mathcal{N}(0, 1)}{\sqrt{\chi_{n-1}^2/(n-1)}},$$

Fig. 8.3 The projection error

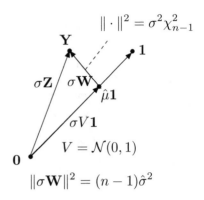

$$\| \cdot \|^2 = \sigma^2 \chi^2_{n-1}$$

$$\| \sigma \mathbf{W} \|^2 = (n-1)\hat{\sigma}^2$$

where χ^2_{n-1} is the sum of the squares of $n-1$ i.i.d. $\mathcal{N}(0,1)$ random variables.

Thus, this *chi-squared* test is very similar to the previous one, except that one replaces the standard deviation σ by it estimate $\hat{\sigma}$ and the threshold λ is adjusted (increased) to reflect the uncertainty in σ. Statistical packages provide routines to calculate the appropriate value of λ. (See scipy.stats.chisquare for Python.)

Figure 8.3 explains the result. The rotation symmetry of $\mathbf{Z}$ implies that we can assume that $V = Z_1$ and that $\mathbf{W} = (0, Z_2, \ldots, Z_n)$. As in the previous examples, one uses the symmetry assumption under H_1 to prove that the likelihood ratio is monotone in $\hat{\mu}/\hat{\sigma}$.

Coming back to our lightbulbs example, what should we do if we have different number of bulbs of the two brands? The next test covers that situation.

8.4.3   Difference of Means

You observe $\{X_n, n = 1, \ldots, n_1\}$ and $\{Y_n, n = 1, \ldots, n_2\}$. Assume that these random variables are all independent and that the X_n are $\mathcal{N}(\mu_1, 1)$ and the Y_n are $\mathcal{N}(\mu_2, 1)$. We want to test whether $\mu_1 = \mu_2$.

Define

$$Z = \frac{1}{n_1^{-1} + n_2^{-1}} \left(\frac{X_1 + \cdots + X_{n_1}}{n_1} - \frac{Y_1 + \cdots + Y_{n_2}}{n_2} \right).$$

Then $Z = \mathcal{N}(\mu, 1)$ where $\mu = \mu_1 - \mu_2$. Testing $\mu_1 = \mu_2$ is then equivalent to testing $\mu = 0$. A sensible decision is then to reject the hypothesis that $\mu_1 = \mu_2$ if $|Z| > 2$.

In practice, if n_1 and n_2 are not too small, one can invoke the Central Limit Theorem to justify the same test even when the random variables are not Gaussian. That is typically how this test is used. Also, when the random variables have nonzero means and unknown variances, one then renormalizes them by subtracting their sample mean and dividing by the sample standard deviation.

Needless to say, some care must be taken. It is not difficult to find distributions for which this test does not perform well. This fact helps explain why many poorly conducted statistical studies regularly contradict one another. Many publications decry this *fallacy of the p-value*. The p-value is the name given to the probability of false alarm.

8.4.4 Mean in Hyperplane?

A generalization of the previous example is as follows:

$$H_0 : \mathbf{Y} = \mathcal{N}(\mu, \sigma^2 \mathbf{I}), \mu \in \mathcal{L}$$

$$H_1 : \mathbf{Y} = \mathcal{N}(\mu, \sigma^2 \mathbf{I}), \mu \in \Re^n.$$

Here, $\mathcal{L}$ is an m-dimensional subspace in $\Re^n$.

Here is the test that has a probability of false alarm (deciding H_1 when H_0 is true) less than β: Decide

$$H = H_1 \text{ if and only if } \frac{1}{\sigma^2} \|\mathbf{Y} - \hat{\mu}\|^2 > \beta_{n-m},$$

where

$$\hat{\mu} = \arg\min\{\|\mathbf{Y} - \mathbf{x}\|^2 : \mathbf{x} \in \mathcal{L}\}$$

$$P(\chi_{n-m}^2 > \beta_{n-m}) = \beta.$$

In this expression, χ_{n-m}^2 represents a random variable that has a *chi-square distribution* with $n - m$ degrees of freedom. This means that it is distributed like the sum of $n - m$ random variables that are i.i.d. $\mathcal{N}(0, 1)$.

Figure 8.4 shows that

$$\mathbf{Y} - \hat{\mu} = \sigma \mathbf{Z}.$$

Now, the distribution of $\mathbf{Z}$ is invariant under rotation. Consequently, we can rotate the axes around μ so that $\mathbf{Z} = \sigma(0, \ldots, 0, Z_{m+1}, \ldots, Z_n)$. Thus,

Fig. 8.4 The projection error

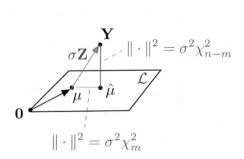

$$\mathbf{Y} - \hat{\mu} = \sigma(0, \ldots, 0, Z_{m+1}, \ldots, Z_n),$$

so that $\|\mathbf{Y} - \hat{\mu}\|^2 = \sigma^2(Z_{m+1}^2 + \cdots + Z_n^2)$, which proves the result.

As in our simple example, this test has a probability of false alarm equal to β. Here also, one can show that the test maximizes the probability of correct detection subject to that bound on the probability of false alarm if under H_1 one knows that μ has a symmetric pmf around $\mathcal{L}$. This means that $\mu = \gamma_i + v_i$ with probability $p_i/2$ and $\gamma_i - v_i$ with probability $p_i/2$ where $\gamma_i \in \mathcal{L}$ and v_i is orthogonal to $\mathcal{L}$, for $i = 1, \ldots, K$. The continuous version of this symmetry should be clear. The verification of this fact is similar to the simple case we discussed above.

8.4.5   ANOVA

Our next model is more general and is widely used. In this model, $\mathbf{Y} = \mathcal{N}(A\gamma, \sigma^2 \mathbf{I})$. We would like to test whether $M\gamma = 0$, which is the H_0 hypothesis. Here, A is a $n \times k$ matrix, with $k < n$. Also, M is a $q \times k$ matrix with $q < k$.

The decision is to reject H_0 if $F > F_0$ where

$$F = \frac{\|\mathbf{Y} - \mu_0\|^2 - \|\mathbf{Y} - \mu_1\|^2}{\|\mathbf{Y} - \mu_1\|^2} \times \frac{n - k}{q}$$

$$\mu_0 = \arg \min_{\mu} \{\|\mathbf{Y} - \mu\|^2 : \mu = A\gamma, M\gamma = 0\}$$

$$\mu_1 = \arg \min_{\mu} \{\|\mathbf{Y} - \mu\|^2 : \mu = A\gamma\}$$

$$\beta = P\left(\frac{\chi_q^2/q}{\chi_{n-k}^2/(n - k)} > F_0\right).$$

In the last expression, the ratio of two χ^2 random variables is said to be an F distribution, in the honor of Sir Ronald A. Fisher who introduced this *F-test* in 1920.

This test has a probability of false alarm equal to β, as Fig. 8.5 shows. This figure represents the situation under H_0, when $\mathbf{Y} = \mu_0 + \sigma \mathbf{Z}$ and shows that F is the ratio of two χ^2 random variables, so that it has an F distribution.

As in the previous examples, the optimality of the test in terms of probability of correct detection requires some symmetry assumptions of μ under H_1.

8.5   LDPC Codes

Low Density Parity Check (LDPC) codes are among the most efficient codes used in practice. Gallager invented these codes in his 1960 thesis (Gallager 1963, Fig. 8.6).

Fig. 8.5 The F-test. The figure shows that F is the ratio of two independent chi-square random variables

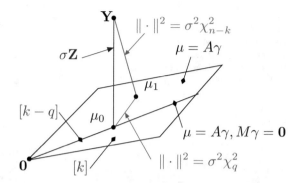

Fig. 8.6 Robert G. Gallager, b. 1931

These codes are used extensively today, for instance, in satellite video transmissions. They are almost optimal for BSC channels and also for many other channels.

The *LDPC codes* are as follows. Let $\mathbf{x} \in \{0, 1\}^n$ be an n-bit string to be transmitted. One augments this string with the m-bit string $\mathbf{y}$ where

$$\mathbf{y} = H\mathbf{x}. \tag{8.9}$$

Here, H is an $m \times n$ matrix with entries in $\{0, 1\}$, one views $\mathbf{x}$ and $\mathbf{y}$ as column vectors and the operations are addition modulo 2. For instance, if

$$H = \begin{bmatrix} 1 & 0 & 1 & 1 & 1 & 0 & 0 & 0 \\ 0 & 1 & 0 & 1 & 1 & 0 & 1 & 0 \\ 1 & 1 & 0 & 0 & 0 & 1 & 0 & 1 \\ 0 & 0 & 1 & 0 & 1 & 1 & 1 & 1 \end{bmatrix}$$

and $\mathbf{x} = [01001010]$, then $\mathbf{y} = [1110]$. This calculation of the parity check bits $\mathbf{y}$ from $\mathbf{x}$ is illustrated by the graph, called Tanner graph, shown in Fig. 8.7.

Thus, instead of simply sending the bit string $\mathbf{x}$, one sends both $\mathbf{x}$ and $\mathbf{y}$. The bits in $\mathbf{y}$ are *parity check* bits. Because of possible transmission errors, the receiver may get $\tilde{\mathbf{x}}$ and $\tilde{\mathbf{y}}$ instead of $\mathbf{x}$ and $\mathbf{y}$. The receiver computes $H\tilde{\mathbf{x}}$ and compares the result with $\tilde{\mathbf{y}}$. The idea is that if $\tilde{\mathbf{y}} = H\tilde{\mathbf{x}}$, then it is likely that $\tilde{\mathbf{x}} = \mathbf{x}$ and $\tilde{\mathbf{y}} = \mathbf{y}$. In other words, it is unlikely that errors would have corrupted $\mathbf{x}$ and $\mathbf{y}$ in a way that these

Fig. 8.7 Tanner graph
representation of the LDPC
code. The graph shows the
nonzero entries of H, so that
$\mathbf{y} = H\mathbf{x}$. The receiver gets $\tilde{\mathbf{x}}$
and $\tilde{\mathbf{y}}$ instead of $\mathbf{x}$ and $\mathbf{y}$. The
nodes x_j are called *message
nodes* and the nodes y_i are
called *check nodes*

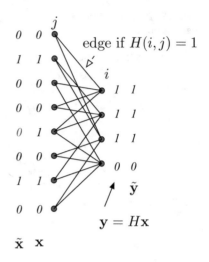

vectors would still satisfy the relation $\tilde{\mathbf{y}} = H\tilde{\mathbf{x}}$. Thus, one expects the scheme to be
good at detecting errors, at least if the matrix H is well chosen.

In addition to detecting errors, the LDPC code is used for *error correction*. If
$\tilde{\mathbf{y}} \neq H\tilde{\mathbf{x}}$, one tries to find the least number of components of $\tilde{\mathbf{x}}$ and $\tilde{\mathbf{y}}$ that can
be changed to satisfy the equations. These would be the most likely transmission
errors, if we assume that bit errors are i.i.d. have a very small probability. However,
searching for the possible combinations of components to change is exponentially
hard. Instead, one uses iterative algorithms that approximate the solution.

We illustrate a commonly used decoding algorithm, called *belief propagation
(BP)*. We assume that each received bit is erroneous with probability $\epsilon \ll 1$ and
correct with probability $\bar{\epsilon} = 1 - \epsilon$, independently of the other bits. We also assume
that the transmitted bits x_j are equally likely to be 0 or 1. This implies that the parity
check bits y_i are also equally likely to be 0 or 1, by symmetry. In this algorithm, the
message nodes x_j and the check nodes y_i exchange beliefs along the links of the
graph of Fig. 8.7 about the probability that the x_j are equal to 1.

In steps $1, 3, 5, \ldots$ of the algorithm, each node x_j sends to each node y_i to which
it is attached an estimate of $P(x_j = 1)$. Each node y_i then combines these estimates
to send back new estimates to each x_j about $P(x_j = 1)$. Here is the calculation
that the y nodes perform. Consider a situation shown in Fig. 8.8 where node y_1 gets
the estimates $a = P(x_1 = 1), b = P(x_2 = 1), c = P(x_3 = 1)$. Assume also that
$\tilde{y}_1 = 1$, from which node y_1 calculates $P[y_1 = 1|\tilde{y}_1] = 1 - \epsilon = \bar{\epsilon}$, by Bayes' rule.
Since the graph shows that $x_1 + x_2 + x_3 = y_1$, node y_1 estimates the probability that
$x_1 = 1$ as the probability that an odd number of bits among $\{x_2, x_3, y_1\}$ are equal to
one (Fig. 8.9).

To see how to do the calculation, assume that $x_1, \ldots, x_n$ are independent $\{0, 1\}$-
random variables with $p_i = P(x_i = 1)$. Note that

$$1 - (1 - 2x_1) \times \cdots \times (1 - 2x_n)$$

Fig. 8.8 Node y_1 gets
estimates from x nodes and
calculates new estimates

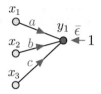

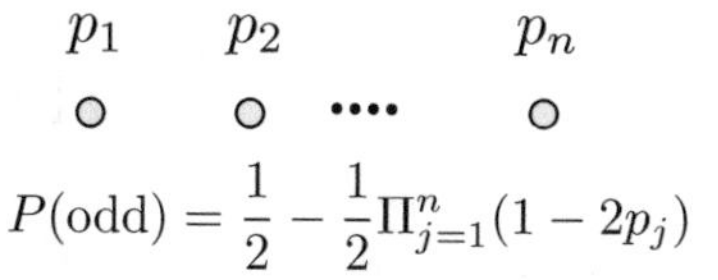

$$
\begin{array}{cccc}
p_1 & p_2 & & p_n \\
\circ & \circ & \bullet\bullet\bullet\bullet & \circ
\end{array}
$$

$$P(\text{odd}) = \frac{1}{2} - \frac{1}{2}\Pi_{j=1}^{n}(1 - 2p_j)$$

Fig. 8.9 Each node j is equal to one w.p. p_j and to zero otherwise, independently of the other nodes. The probability that an odd number of nodes are one is given in the figure

is equal to zero if the number of variables that are equal to one among $\{x_1, \ldots, x_n\}$ is even and is equal to two if it is odd. Thus, taking expectation,

$$2P(\text{odd}) = 1 - \Pi_{i=1}^{n}(1 - 2p_i),$$

so that

$$P(\text{odd}) = \frac{1}{2} - \frac{1}{2}\Pi_{i=1}^{n}(1 - 2p_i). \tag{8.10}$$

Thus, in Fig. 8.8, one finds that

$$P(x_1 = 1) = P(\text{odd among } x_2, x_3, y_1)$$

$$= \frac{1}{2} - \frac{1}{2}(1 - 2b)(1 - 2c)(1 - 2\bar{\epsilon}). \tag{8.11}$$

The y-nodes in Fig. 8.7 use that procedure to calculate new estimates and send them to the x-nodes.

In steps 2, 4, 6, $\ldots$ of the algorithm, each x_j nodes combines the estimates of $P(x_j = 1)$ it gets from $\tilde{x}_j$ and from the y-nodes in the previous steps to calculate new estimates. Each node x_j assumes that the different estimates it got are derived from independent observations. That is, node x_j gets opinions about $P(x_j = 1)$ from independent experts, namely $\tilde{x}_j$ and the y_i to which it is attached in the graph. Node x_j will merge the opinion of these experts to calculate new estimates.

How should one merge the opinions of independent experts? Say that N experts make independent observations $Y_1, \ldots, Y_N$ and provide estimates $p_i = P[X = 1|Y_i]$. Assume that the prior probability is that $P(X = 1) = P(X = 0) = 1/2$. How should one estimate $P[X = 1|p_1, \ldots, p_N]$? Here is the calculation.

Fig. 8.10 Merging the
opinion of independent
experts about $P(X = 1)$
when the prior is $1/2$

$$P(X = 1) =?$$

$$d = \frac{abc}{abc + (1 - a)(1 - b)(1 - c)}$$

One has

$$
\begin{aligned}
P[X = 1 | Y_1, \ldots, Y_N] &= \frac{P(X = 1, Y_1, \ldots, Y_N)}{P(Y_1, \ldots, Y_N)} \\
&= \frac{P[Y_1, \ldots, Y_N | X = 1] P(X = 1)}{\sum_{x=0,1} P[Y_1, \ldots, Y_N | X = x] P(X = x)} \\
&= \frac{P[Y_1 | X = 1] \times \cdots \times P[Y_N | X = 1]}{\sum_{x=0,1} P[Y_1 | X = x] \times \cdots \times P[Y_N | X = x]}.
\end{aligned}
$$

$$(8.12)$$

Now,

$$
P[Y_n | X = x] = \frac{P(X = x, Y_n)}{P(X = x)} = \frac{P[X = x | Y_n] P(Y_n)}{1/2}.
$$

Thus,

$$
P[Y_n | X = 0] = 2(1 - p_n) P(Y_n) \text{ and } P[Y_n | X = 1] = 2 p_n P(Y_n).
$$

Substituting these expressions in (8.12), one finds that

$$
P[X = 1 | Y_1, \ldots, Y_N] = \frac{p_1 \cdots p_N}{p_1 p_2 \cdots p_N + (1 - p_1) \cdots (1 - p_N)},
$$

$$(8.13)$$

as shown in Fig. 8.10.

Let us apply this rule to the situation shown in Fig. 8.11. In the figure, node x_1 gets an estimate ϵ of $P(x_1 = 1)$ from observing $\tilde{x}_1 = 0$. It also gets estimates a, b, c

Fig. 8.11 Node x_1 gets estimates of $P(x_1 = 1)$ from y nodes and calculates new estimates

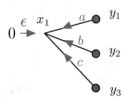

from the nodes y_1, y_2, y_3 and node x_1 assumes that these estimates were based on independent observations.

To calculate a new estimate that it will send to node y_1, node x_1 combines the estimates from $\tilde{x}_1$, y_2 and y_3. This estimate is

$$\frac{\epsilon bc}{\epsilon bc + \bar{\epsilon}\bar{b}\bar{c}}, \tag{8.14}$$

where $\bar{b} = 1 - b$ and $\bar{c} = 1 - c$. In the next step, node x_1 will send that estimate to node y_1. It also calculates estimates for nodes y_2 and y_3.

Summing up, the algorithm is as follows. At each odd step, node x_j sends $X(i, j)$ to each node y_i. At each even step, node y_i sends $Y(i, j)$ to each node x_j. One has

$$Y(i, j) = \frac{1}{2} - \frac{1}{2}(1 - 2\epsilon)(1 - 2\tilde{y}_i)\Pi_{s \in A(i,j)}(1 - 2X(i, s)), \tag{8.15}$$

where $A(i, j) = \{s \neq j \mid H(i, s) = 1\}$ and

$$X(i, j) = \frac{N(i, j)}{N(i, j) + D(i, j)}, \tag{8.16}$$

where

$$N(i, j) = P[x_j = 1|\tilde{x}_j]\Pi_{\{v \neq i|H(v,j)=1\}}Y(v, j)$$

and

$$D(i, j) = P[x_j = 0|\tilde{x}_j]\Pi_{\{v \neq i|H(v,j)=1\}}(1 - Y(v, j))$$

with

$$P[x_j = 1|\tilde{x}_j] = \epsilon + (1 - 2\epsilon)\tilde{x}_j.$$

Also, node x_j can update its probability of being 1 by merging the opinions of the experts as

$$X(j) = \frac{N(j)}{N(j) + D(j)}, \tag{8.17}$$

Fig. 8.12 Belief propagation applied to the example of Fig. 8.7. The horizontal axis is the step of the algorithm. The vertical axis is the best guess for each $x(i)$ at that step. For clarity, we separated the guesses by 0.1. The final detection is [0, 1, 0, 0, 1, 0, 1, 0], which is intuitively the best guess

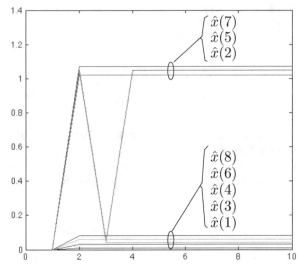

where

$$N(j) = P[x_j = 1|\tilde{x}_j]\Pi_{\{v|H(v,j)=1\}}Y(v, j)$$

and

$$D(j) = P[x_j = 0|\tilde{x}_j]\Pi_{\{v|H(v,j)=1\}}(1 - Y(v, j)).$$

After enough iterations, one makes the detection decisions $x_j = 1\{X(j) \geq 0.5\}$.

Figure 8.12 shows the evolution over time of the estimated probabilities that the x_j are equal to one. Our code is a direct implementations of the formulas in this section. More sophisticated implementations use sums of logarithms instead of products.

Simulations, and a deep theory, show that this algorithm performs well if the graph does not have small cycles. In such a case, the assumption that the estimates are obtained from independent observations is almost correct.

8.6   Summary

- LDPC Codes;
- Jointly Gaussian Random Variables, independent if uncorrelated;
- Proof of Neyman–Pearson Theorem;
- Testing properties of the mean.

8.6.1 Key Equations and Formulas

LDPC	$\mathbf{y} = H\mathbf{x}$	(8.9)
P(odd)	$P(\sum_j X_j = 1) = 0.5 - 0.5\Pi_j(1 - 2p_j)$	(8.10)
Fusion of Experts	$P[X = 1 | Y_1, \ldots, Y_n] = \Pi_j p_j / (\Pi_j p_j + \Pi_j \bar{p}_j)$	(8.13)
Jointly Gaussian	$N(\mu, \Sigma) \Leftrightarrow f_\mathbf{X} = \ldots$	(8.4)
If $\mathbf{X}, \mathbf{Y}$ are J.G., then	$\mathbf{X} \perp \mathbf{Y} \Rightarrow \mathbf{X}, \mathbf{Y}$ are independent	Theorem 8.3

8.7 References

The book (Richardson and Urbanke 2008) is a comprehensive reference on LDPC codes and iterative decoding techniques.

8.8 Problems

Problem 8.1 Construct two Gaussian random variables that are not jointly Gaussian. *Hint:* Let $X =_D \mathcal{N}(0, 1)$ and Z be independent random variables with $P(Z = 1) = P(Z = -1) = 1/2$. Define $Y = XZ$. Show that X and Y meet the requirements of the problem.

Problem 8.2 Assume that $X =_D (Y + Z)/\sqrt{2}$ where Y and Z are independent and distributed like X. Show that $X = \mathcal{N}(0, \sigma^2)$ for some $\sigma^2 \geq 0$. *Hint:* First show that $E(X) = 0$. Second, show by induction that $X =_D (V_1 + \cdots + V_m)/\sqrt{m}$ for $m = 2^n$. where the V_i are i.i.d. and distributed like X. Conclude using the CLT.

Problem 8.3 Consider Problem 7.8 but assume now that $\mathbf{Z} =_D \mathcal{N}(\mathbf{0}, \Sigma)$ where

$$\Sigma = \begin{bmatrix} 0.2 & 0.1 \\ 0.1 & 0.3 \end{bmatrix}.$$

The symbols are equally likely and the receiver uses the MLE. Simulate the system using Python to estimate the fraction of errors.

Tracking—A

9

> **Application:** Estimation, Tracking
> **Topics:** LLSE, MMSE, Kalman Filter

9.1 Examples

A GPS receiver uses the signals it gets from satellites to estimate its location (Fig. 9.1). Temperature and pressure sensors provide signals that a computer uses to estimate the state of a chemical reactor.

A radar measures electromagnetic waves that an object reflects and uses the measurements to estimate the position of that object (Fig. 9.2).

Similarly, your car's control computer estimates the state of the car from measurements it gets from various sensors (Fig. 9.3).

9.2 Estimation Problem

The basic *estimation problem* can be formulated as follows. There is a pair of continuous random variables (X, Y). The problem is to estimate X from the observed value of Y.

This problem admits a few different formulations:

- **Known Distribution**: We know the joint distribution of (X, Y);
- **Off-Line**: We observe a set of sample values of (X, Y);
- **On-Line**: We observe successive values of samples of (X, Y);

© The Author(s) 2021
J. Walrand, *Probability in Electrical Engineering and Computer Science*,
https://doi.org/10.1007/978-3-030-49995-2_9

Fig. 9.1 Estimating the
location of a device from
satellite signals

Fig. 9.2 Estimating the
position of an object from
radar signals

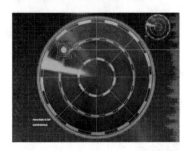

Fig. 9.3 Estimating the state
of a vehicle from sensor
signals

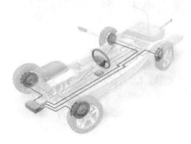

The objective is to choose the inference function $g(\cdot)$ to minimize the expected error $C(g)$ where

$$C(g) = E(c(X, g(Y))).$$

In this expression, $c(X, \hat{X})$ is the cost of guessing $\hat{X}$ when the actual value is X. A standard example is

$$c(X, \hat{X}) = |X - \hat{X}|^2.$$

We will also study the case when $X \in \Re^d$ for $d > 1$. In such a situation, one uses $c(X, \hat{X}) = ||X - \hat{X}||^2$. If the function $g(\cdot)$ can be arbitrary, the function that minimizes $C(g)$ is the *Minimum Mean Squares Estimate (MMSE)* of X given Y. If the function $g(\cdot)$ is restricted to be linear, i.e., of the form $a + BY$, the linear function that minimizes $C(g)$ is the *Linear Least Squares Estimate (LLSE)* of X given Y. One may also restrict $g(\cdot)$ to be a polynomial of a given degree. For instance, one may define the Quadratic Least Squares Estimate $QLSE$ of X given Y. See Fig. 9.4.

Fig. 9.4 Least squares
estimates of X given Y:
$LLSE$ is linear, $QLSE$ is
quadratic, and $MMSE$ can be
an arbitrary function

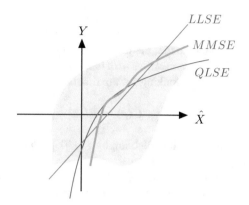

As we will see, a general method for the off-line inference problem is to choose
a parametric class of functions $\{g_w, w \in \Re^d\}$ and to then minimize the empirical
error

$$\sum_{k=1}^{K} c(X_k, g_w(Y_k))$$

over the parameters w. Here, the (X_k, Y_k) are the observed samples. The parametric
function could be linear, polynomial, or a *neural network*.

For the on-line problem, one also chooses a similar parametric family of
functions and one uses a stochastic gradient descent algorithm of the form

$$w(k + 1) = w(k) - \gamma \nabla_w c(X_{k+1}, g_w(Y_{k+1})),$$

where ∇ is the gradient with respect to w and $\gamma > 0$ is a small step size. The
justification for this approach is that, since γ is small, by the SLLN, the update
tends to be in the direction of

$$-\sum_{i=k}^{k+K-1} \nabla_w c(X_{i+1}, g_w(Y_{i+1})) \approx -K \nabla E(c(X_k, g_w(Y_k))) = -K \nabla C(g_w),$$

which would correspond to a gradient algorithm to minimize $C(g_w)$.

9.3   Linear Least Squares Estimates

In this section, we study the linear least squares estimates. Recall the setup that we
explained in the previous section. There is a pair (X, Y) of random variables with
some joint distribution and the problem is to find the function $g(Y) = a + bY$ that
minimizes

$$C(g) = E(|X - g(Y)|^2).$$

One consider the cases where the distribution is known, or a set of samples has been observed, or one observes one sample at a time.

Assume that the joint distribution of (X, Y) is known. This means that we know the *joint cumulative distribution function (j.c.d.f.)* $F_{X,Y}(x, y)$.[1]

We are looking for the function $g(Y) = a + bY$ that minimizes

$$C(g) = E(|X - g(Y)|^2) = E(|X - a - bY|^2).$$

We denote this function by $L[X|Y]$. Thus, we have the following definition.

Definition 9.1 (Linear Least Squares Estimate (LLSE)) The LLSE of X given Y, denoted by $L[X|Y]$, is the linear function $a + bY$ that minimizes

$$E(|X - a - bY|^2).$$

◇

Note that

$$C(g) = E(X^2 + a^2 + b^2Y^2 - 2aX - 2bXY + 2abY)$$
$$= E(X^2) + a^2 + b^2E(Y^2) - 2aE(X) - 2bE(XY) + 2abE(Y).$$

To find the values of a and b that minimize that expression, we set to zero the partial derivatives with respect to a and b. This gives the following two equations:

$$0 = 2a - 2E(X) + 2bE(Y) \tag{9.1}$$

$$0 = 2bE(Y^2) - 2E(XY) + 2aE(Y). \tag{9.2}$$

Solving these equations for a and b, we find that

$$L[X|Y] = a + bY = E(X) + \frac{\text{cov}(X, Y)}{\text{var}(Y)}(Y - E(Y)),$$

where we used the identities

$$\text{cov}(X, Y) = E(XY) - E(X)E(Y) \text{ and } \text{var}(Y) = E(Y^2) - E(Y)^2.$$

We summarize this result as a theorem.

---

[1] See Appendix B.

Theorem 9.1 (Linear Least Squares Estimate) *One has*

$$L[X|Y] = E(X) + \frac{cov(X, Y)}{var(Y)}(Y - E(Y)). \qquad (9.3)$$

■

As a first example, assume that

$$Y = \alpha X + Z, \qquad (9.4)$$

where X and Z are zero-mean and independent. In this case, we find [2]

$$cov(X, Y) = E(XY) - E(X)E(Y)$$

$$= E(X(\alpha X + Z)) = \alpha E(X^2)$$

$$var(Y) = \alpha^2 var(X) + var(Z) = \alpha^2 E(X^2) + E(Z^2).$$

Hence,

$$L[X|Y] = \frac{\alpha E(X^2)}{\alpha^2 E(X^2) + E(Z^2)} Y = \frac{\alpha^{-1} Y}{1 + SNR^{-1}},$$

where

$$SNR := \frac{\alpha^2 E(X^2)}{\sigma^2}$$

is the *signal-to-noise ratio*, i.e., the ratio of the power $E(\alpha^2 X^2)$ of the signal in Y divided by the power $E(Z^2)$ of the noise. Note that if SNR is small, then $L[X|Y]$ is close to zero, which is the best guess about X if one does not make any observation. Also, if SNR is very large, then $L[X|Y] \approx \alpha^{-1} Y$, which is the correct guess if $Z = 0$.

As a second example, assume that

$$X = \alpha Y + \beta Y^2, \qquad (9.5)$$

where[3] $Y =_D U[0, 1]$. Then,

---

[2] Indeed, $E(XZ) = E(X)E(Z) = 0$, by independence.

[3] Thus,

$$E(Y^k) = (1 + k)^{-1}.$$

Fig. 9.5 The figure shows $L[\alpha Y + \beta Y^2|Y]$ when $Y =_D U[0, 1]$

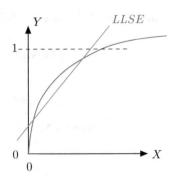

$$E(X) = \alpha E(Y) + \beta E(Y^2) = \alpha/2 + \beta/3;$$

$$\text{cov}(X, Y) = E(XY) - E(X)E(Y)$$

$$= E(\alpha Y^2 + \beta Y^3) - (\alpha/2 + \beta/3)(1/2)$$

$$= \alpha/3 + \beta/4 - \alpha/4 - \beta/6$$

$$= (\alpha + \beta)/12$$

$$\text{var}(Y) = E(Y^2) - E(Y)^2 = 1/3 - (1/2)^2 = 1/12.$$

Hence,

$$L[X|Y] = \alpha/2 + \beta/3 + (\alpha + \beta)(Y - 1/2) = -\beta/6 + (\alpha + \beta)Y.$$

This estimate is sketched in Fig. 9.5. Obviously, if one observes Y, one can compute X. However, recall that $L[X|Y]$ is restricted to being a linear function of Y.

9.3.1 Projection

There is an insightful interpretation of $L[X|Y]$ as a projection that also helps understand more complex estimates. This interpretation is that $L[X|Y]$ is the *projection* of X onto the set $\mathscr{L}(Y)$ of linear functions of Y.

This interpretation is sketched in Fig. 9.6. In that figure, random variables are represented by points and $\mathscr{L}(Y)$ is shown as a plane since the linear combination of points in that set is again in the set. In the figure, the square of the length of a vector from a random variable V to another random variable W is $E(|V - W|^2)$. Also, we say that two vectors V and W are orthogonal if $E(VW) = 0$. Thus, $L[X|Y] = a + bY$ is the projection of X onto $\mathscr{L}(Y)$ if $X - L[X|Y]$ is orthogonal to every linear function of Y, i.e., if

$$E((X - a - bY)(c + dY)) = 0, \forall c, d \in \mathfrak{R}.$$

Fig. 9.6 $L[X|Y]$ is the projection of X onto $\mathscr{L}(Y)$

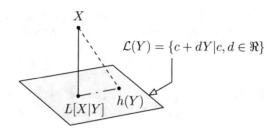

Fig. 9.7 Example of projection

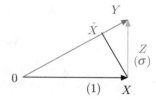

Equivalently,

$$E(X) = a + bE(Y) \text{ and } E((X - a - bY)Y) = 0. \tag{9.6}$$

These two equations are the same as (9.1)–(9.2). We call the identities (9.6) the *projection property*.

Figure 9.7 illustrates the projection when

$$X = \mathcal{N}(0, 1) \text{ and } Y = X + Z \text{ where } Z = \mathcal{N}(0, \sigma^2).$$

In this figure, the length of Z is equal to $\sqrt{E(Z^2)} = \sigma$, the length of X is $\sqrt{E(X^2)} = 1$ and the vectors X and Z are orthogonal because $E(XZ) = 0$.

We see that the triangles $0\hat{X}X$ and $0XY$ are similar. Hence,

$$\frac{||\hat{X}||}{||X||} = \frac{||X||}{||Y||},$$

so that

$$\frac{||\hat{X}||}{1} = \frac{1}{\sqrt{1 + \sigma^2}} = \frac{||Y||}{1 + \sigma^2},$$

since $||Y|| = \sqrt{1 + \sigma^2}$. This shows that

$$\hat{X} = \frac{1}{1 + \sigma^2}Y.$$

To see why the projection property implies that $L[X|Y]$ is the closest point to X in $\mathscr{L}(Y)$, as suggested by Fig. 9.6, we verify that

$$E(|X - L[X|Y]|^2) \leq E(|X - h(Y)|^2),$$

for any given $h(Y) = c + dY$. The idea of the proof is to verify Pythagoras' identity on the right triangle with vertices X, $L[X|Y]$ and $h(Y)$. We have

$$E(|X - h(Y)|^2) = E(|X - L[X|Y] + L[X|Y] - h(Y)|^2)$$
$$= E(|X - L[X|Y]|^2) + E(|L[X|Y] - h(Y)|^2)$$
$$+ 2E((X - L[X|Y])(L[X|Y] - h(Y))).$$

Now, the projection property (9.6) implies that the last term in the above expression is equal to zero. Indeed, $L[X|Y] - h(Y)$ is a linear function of Y. It follows that

$$E(|X - h(Y)|^2) = E(|X - L[X|Y]|^2) + E(|L[X|Y] - h(Y)|^2)$$
$$\geq E(|X - L[X|Y]|^2),$$

as was to be proved.

9.4   Linear Regression

Assume now that, instead of knowing the joint distribution of (X, Y), we observe K i.i.d. samples $(X_1, Y_1), \ldots, (X_K, Y_K)$ of these random variables. Our goal is still to construct a function $g(Y) = a + bY$ so that

$$E(|X - a - bY|^2)$$

is minimized. We do this by choosing a and b to minimize the sum of the squares of the errors based on the samples. That is, we choose a and b to minimize

$$\sum_{k=1}^{K} |X_k - a - bY_k|^2.$$

To do this, we set to zero the derivatives of this sum with respect to a and b. Algebra shows that the resulting values of a and b are such that

$$a + bY = E_K(X) + \frac{\text{cov}_K(X, Y)}{\text{var}_K(Y)}(Y - E_K(Y)), \tag{9.7}$$

where we defined

Fig. 9.8 The linear regression of X over Y

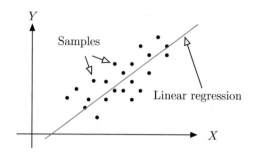

$$E_K(X) = \frac{1}{K} \sum_{k=1}^{K} X_k, \ E_K(Y) = \frac{1}{K} \sum_{k=1}^{K} Y_k,$$

$$\text{cov}_K(X, Y) = \frac{1}{K} \sum_{k=1}^{K} X_k Y_k - E_K(X) E_K(Y),$$

$$\text{var}_K(Y) = \frac{1}{K} \sum_{k=1}^{K} Y_k^2 - E_K(Y)^2.$$

That is, the expression (9.7) is the same as (9.3), except that the expectation is replaced by the sample mean. The expression (9.7) is called the *linear regression* of X over Y. It is shown in Fig. 9.8.

One has the following result.

Theorem 9.2 (Linear Regression Converges to LLSE) *As the number of samples increases, the linear regression approaches the LLSE.*

∎

Proof As $K \to \infty$, one has, by the Strong Law of Large Numbers,

$$E_K(X) \to E(X), \ E_K(Y) \to E(Y),$$

$$\text{cov}_K(X, Y) \to \text{cov}(X, Y), \ \text{var}_K(Y) \to \text{var}(Y).$$

Combined with the expressions for the linear regression and the LLSE, these properties imply the result. □

Formula (9.3) and the linear regression provide an intuitive meaning of the covariance $\text{cov}(X, Y)$. If this covariance is zero, then $L[X|Y]$ does not depend on Y. If it is positive (negative), it increases (decreases, respectively) with Y. Thus, $\text{cov}(X, Y)$ measures a form of dependency in terms of linear regression. For

Fig. 9.9 The random
variables X and Y are
uncorrelated. Note that they
are not independent

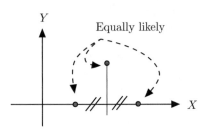

instance, the random variables in Fig. 9.9 are uncorrelated since $L[X|Y]$ does not
depend on Y.

9.5   A Note on Overfitting

In the previous section, we examined the problem of finding the linear function $a +
bY$ that best approximates X, in the mean squared error sense. We could develop the
corresponding theory for quadratic approximations $a + bY + cY^2$, or for polynomial
approximations of a given degree. The ideas would be the same and one would have
a similar projection interpretation.

In principle, a higher degree polynomial approximates X better than a lower
degree one since there are more such polynomials. The question of fitting the
parameters with a given number of observations is more complex.

Assume you observe N data points $\{(X_n, Y_n), n = 1, \ldots, N\}$. If the values Y_n
are different, one can define the function $g(\cdot)$ by $g(Y_n) = X_n$ for $n = 1, \ldots, N$.
This function achieves a zero-mean squared error. What is then the point of looking
for a linear function, or a quadratic, or some polynomial of a given degree? Why not
simply define $g(Y_n) = X_n$?

Remember that the goal of the estimation is to discover a function $g(\cdot)$ that is
likely to work well for data points we have not yet observed. For instance, we hope
that $E(C(X_{N+1}, g(Y_{N+1})))$ is small, where (X_{N+1}, Y_{N+1}) has the same distribution
as the samples (X_n, Y_n) we have observed for $n = 1, \ldots, N$.

If we define $g(Y_n) = X_n$, this does not tell us how to calculate $g(Y_{N+1})$ for a
value Y_{N+1} we have not observed. However, if we construct a polynomial $g(\cdot)$ of
a given degree based on the N samples, then we can calculate $g(Y_{n+1})$. The key
observation is that a higher degree polynomial may not be a better estimate because
it tends to fit noise instead of important statistics.

As a simple illustration of overfitting, say that we observe (X_1, Y_1) and Y_2.
We want to guess X_2. Assume that the samples X_n, Y_n are all independent and
$U[-1, 1]$. If we guess $\hat{X}_2 = 0$, the mean squared error is $E((X_2 - \hat{X}_2)^2) =
E(X_2^2) = 1/3$. If we use the guess $\hat{X}_2 = X_1$ based on the observations, then
$E((X_2 - \hat{X}_2)^2) = E((X_2 - X_1)^2) = 2/3$. Hence, ignoring the observation is better
than taking it into account.

The practical question is how to detect overfitting. For instance, how does one
determine whether a linear regression is better than a quadratic regression? A simple

test is as follows. Say you observed N samples $\{(X_n, Y_n), n = 1, \ldots, N\}$. You remove sample n and compute a linear regression using the $N - 1$ other samples. You use that regression to calculate the estimate $\hat{X}_n$ of X_n based on Y_n. You then compute the squared error $(X_n - \hat{X}_n)^2$. You repeat that procedure for $n = 1, \ldots, N$ and add up the squared errors. You then use the same procedure for a quadratic regression and you compare.

9.6 MMSE

For now, assume that we know the joint distribution of (X, Y) and consider the problem of finding the function $g(Y)$ that minimizes

$$E(|X - g(Y)|^2),$$

per all the possible functions $g(\cdot)$. The best function is called the MMSE of X given Y. We have the following theorem:

Theorem 9.3 (The MMSE Is the Conditional Expectation) *The MMSE of X given Y is given by*

$$g(Y) = E[X|Y],$$

where $E[X|Y]$ is the conditional expectation *of X given Y.*

∎

Before proving this result, we need to define the conditional expectation.

Definition 9.2 (Conditional Expectation) The conditional expectation of X given Y is defined by

$$E[X|Y = y] = \int_{-\infty}^{\infty} x f_{X|Y}[x|y] dx,$$

where

$$f_{X|Y}[x|y] := \frac{f_{X,Y}(x, y)}{f_Y(y)}$$

is the conditional density of X given Y.

◇

Figure 9.10 illustrates the conditional expectation. That figure assumes that the pair (X, Y) is picked uniformly in the shaded area. Thus, if one observes that $Y \in$

Fig. 9.10 The conditional expectation $E[X|Y]$ when the pair (X, Y) is picked uniformly in the shaded area

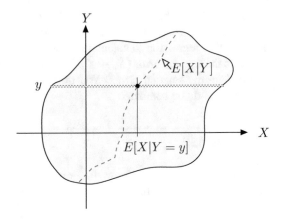

$(y, y + dy)$, the point X is uniformly distributed along the segment that cuts the shaded area at $Y = y$. Accordingly, the average value of X is the mid-point of that segment, as indicated in the figure. The dashed red line shows how that mean value depends on Y and it defines $E[X|Y]$.

The following result is a direct consequence of the definition.

Lemma 9.4 (Orthogonality Property of MMSE)

(a) For any function $\phi(\cdot)$, one has

$$E((X - E[X|Y])\phi(Y)) = 0. \tag{9.8}$$

(b) Moreover, if the function $g(Y)$ is such that

$$E((X - g(Y))\phi(Y)) = 0, \forall \phi(\cdot), \tag{9.9}$$

then $g(Y) = E[X|Y]$.

Proof

(a) To verify (9.8) note that

$$E(E[X|Y]\phi(Y)) = \int_{-\infty}^{\infty} E[X|Y = y]\phi(y) f_Y(y) dy$$

$$= \int_{-\infty}^{\infty} \int_{-\infty}^{\infty} x \frac{f_{X,Y}(x, y)}{f_Y(y)} dx \phi(y) f_Y(y) dy$$

$$= \int_{-\infty}^{\infty} \int_{-\infty}^{\infty} x \phi(y) f_{X,Y}(x, y) dx dy$$

$$= E(X\phi(Y)),$$

which proves (9.8).

Fig. 9.11 The conditional expectation $E[X|Y]$ as the projection of X on the set $\mathscr{G}(Y)$ of functions of Y

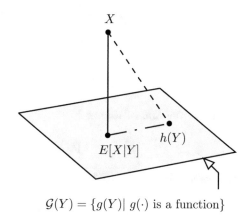

$$\mathcal{G}(Y) = \{g(Y)|\ g(\cdot) \text{ is a function}\}$$

(b) To prove the second part of the lemma, note that

$$E(|g(Y) - E[X|Y]|^2)$$
$$= E((g(Y) - E[X|Y])\{(g(Y) - X) - (E[X|Y] - X)\}) = 0,$$

because of (9.8) and (9.9) with $\phi(Y) = g(Y) - E[X|Y]$.

Note that the second part of the lemma simply says that the projection property characterizes uniquely the conditional expectation. In other words, there is only one projection of X onto $\mathscr{G}(Y)$.

□

We can now prove the theorem.

Proof of Theorem 9.3 The identity (9.8) is the projection property. It states that $X - E[X|Y]$ is orthogonal to the set $\mathscr{G}(Y)$ of functions of Y, as shown in Fig. 9.11.

In particular, it is orthogonal to $h(Y) - E[X|Y]$. As in the case of the LLSE, this projection property implies that

$$E(|X - h(Y)|^2) \geq E(|X - E[X|Y]|^2),$$

for any function $h(\cdot)$. This implies that $E[X|Y]$ is indeed the MMSE of X given Y.

□

From the definition, we see how to calculate $E[X|Y]$ from the conditional density of X given Y. However, in many cases one can calculate $E[X|Y]$ more simply. One approach is to use the following properties of conditional expectation.

Theorem 9.5 (Properties of Conditional Expectation)

(a) Linearity:

$$E[a_1 X_1 + a_2 X_2 | Y] = a_1 E[X_1 | Y] + a_2 E[X_2 | Y];$$

(b) Factoring Known Values:

$$E[h(Y)X|Y] = h(Y)E[X|Y];$$

(c) Independence: If X and Y are independent, then

$$E[X|Y] = E(X).$$

(d) Smoothing:

$$E(E[X|Y]) = E(X);$$

(e) Tower:

$$E[E[X|Y, Z]|Y] = E[X|Y].$$

■

Proof

(a) By Lemma 9.4(b), it suffices to show that

$$a_1 X_1 + a_2 X_2 - (a_1 E[X_1|Y] + a_2 E[X_2|Y])$$

is orthogonal to $\mathscr{G}(Y)$. But this is immediate since it is the sum of two terms

$$a_i(X_i - E[X_i|Y])$$

for $i = 1, 2$ that are orthogonal to $\mathscr{G}(Y)$.
(b) By Lemma 9.4(b), it suffices to show that

$$h(Y)X - h(Y)E[X|Y]$$

is orthogonal to $\mathscr{G}(Y)$, i.e., that

$$E((h(Y)X - h(Y)E[X|Y])\phi(Y)) = 0, \forall \phi(\cdot).$$

Now,

$$E((h(Y)X - h(Y)E[X|Y])\phi(Y)) = E((X - E[X|Y])h(Y)\phi(Y)) = 0,$$

because $X - E[X|Y]$ is orthogonal to $\mathcal{G}(Y)$ and therefore to $h(Y)\phi(Y)$.

(c) By Lemma 9.4(b), it suffices to show that

$$X - E(X)$$

is orthogonal to $\mathcal{G}(Y)$. Now,

$$E((X - E(X))\phi(Y)) = E(X - E(X))E(\phi(Y)) = 0.$$

The first equality follows from the fact that $X - E(X)$ and $\phi(Y)$ are independent since they are functions of independent random variables.[4]

(d) Letting $\phi(Y) = 1$ in (9.8), we find

$$E(X - E[X|Y]) = 0,$$

which is the identity we wanted to prove.

(e) The projection property states that $E[W|Y] = V$ if V is a function of Y and if $W - V$ is orthogonal to $\mathcal{G}(Y)$. Applying this characterization to $W = E[X|Y, Z]$ and $V = E[X|Y]$, we find that to show that $E[E[X|Y, Z]|Y] = E[X|Y]$, it suffices to show that $E[X|Y, Z] - E[X|Y]$ is orthogonal to $\mathcal{G}(Y)$. That is, we should show that

$$E(h(Y)(E[X|Y, Z] - E[X|Y])) = 0$$

for any function $h(Y)$. But $E(h(Y)(X - E[X|Y, Z])) = 0$ by the projection property, because $h(Y)$ is some function of (Y, Z). Also, $E(h(Y)(X - E[X|Y])) = 0$, also by the projection property. Hence,

$$E(h(Y)(E[X|Y, Z] - E[X|Y])) = E(h(Y)(X - E[X|Y]))$$
$$- E(h(Y)(X - E[X|Y, Z])) = 0.$$

□

As an example, assume that X, Y, Z are i.i.d. $U[0, 1]$. We want to calculate

$$E[(X + 2Y)^2|Y].$$

---

[4] See Appendix B.

We find

$$E[(X + 2Y)^2|Y] = E[X^2 + 4Y^2 + 4XY|Y]$$

$$= E[X^2|Y] + 4E[Y^2|Y] + 4E[XY|Y], \text{ by linearity}$$

$$= E(X^2) + 4E[Y^2|Y] + 4E[XY|Y], \text{ by independence}$$

$$= E(X^2) + 4Y^2 + 4Y E[X|Y], \text{ by factoring known values}$$

$$= E(X^2) + 4Y^2 + 4Y E(X), \text{ by independence}$$

$$= \frac{1}{3} + 4Y^2 + 2Y, \text{ since } X =_D U[0, 1].$$

Note that calculating the conditional density of $(X + 2Y)^2$ given Y would have been quite a bit more tedious.

In some situations, one may be able to exploit symmetry to evaluate the conditional expectation. Here is one representative example. Assume that X, Y, Z are i.i.d. Then, we claim that

$$E[X|X + Y + Z] = \frac{1}{3}(X + Y + Z). \tag{9.10}$$

To see this, note that, by symmetry,

$$E[X|X + Y + Z] = E[Y|X + Y + Z] = E[Z|X + Y + Z].$$

Denote by V the common value of these random variables. Note that their sum is

$$3V = E[X + Y + Z|X + Y + Z],$$

by linearity. Thus, $3V = X + Y + Z$, which proves our claim.

$\square$

9.6.1 MMSE for Jointly Gaussian

In general $L[X|Y] \neq E[X|Y]$. As a trivial example, Let $Y =_D U[-1, 1]$ and $X = Y^2$. Then $E[X|Y] = Y^2$ and $L[X|Y] = E(X) = 1/3$ since $\text{cov}(X, Y) = E(XY) - E(X)E(Y) = 0$.

Figure 9.12 recalls that $E[X|Y]$ is the projection of X onto $\mathcal{G}(Y)$, whereas $L[X|Y]$ is the projection of X onto $\mathcal{L}(Y)$. Since $\mathcal{L}(Y)$ is a subspace of $\mathcal{G}(Y)$, one expects the two projections to be different, in general.

However, there are examples where $E[X|Y]$ happens to be linear. We saw one such example in (9.10) and it is not difficult to construct many other examples.

Fig. 9.12 The MMSE and
LLSE are generally different

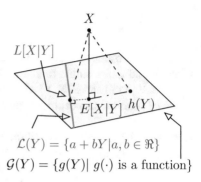

$$\mathcal{L}(Y) = \{a + bY | a, b \in \Re\}$$
$$\mathcal{G}(Y) = \{g(Y)| \ g(\cdot) \text{ is a function}\}$$

There is an important class of problems where this occurs. It is when X and Y are jointly Gaussian. We state that result as a theorem.

Theorem 9.6 (MMSE for Jointly Gaussian RVs)
Let X, Y be jointly Gaussian random variables. Then

$$E[X|Y] = L[X|Y] = E(X) + \frac{cov(X, Y)}{var(Y)}(Y - E(Y)).$$

∎

Proof Note that

$$X - L[X|Y] \text{ and } Y \text{ are uncorrelated.}$$

Also, $X - L[X|Y]$ and Y are two linear functions of the jointly Gaussian random variables X and Y. Consequently, they are jointly Gaussian by Theorem 8.4 and they are independent by Theorem 8.3.

Consequently,

$$X - L[X|Y] \text{ and } \phi(Y) \text{ are independent,}$$

for any $\phi(\cdot)$, because functions of independent random variables are independent by Theorem B.11 in Appendix B. Hence,

$$X - L[X|Y] \text{ and } \phi(Y) \text{ are uncorrelated,}$$

for any $\phi(\cdot)$ by Theorem B.4 of Appendix B.

This shows that

$$X - L[X|Y] \text{ is orthogonal to } \mathcal{G}(Y),$$

and, consequently, that $L[X|Y] = E[X|Y]$. □

9.7 Vector Case

So far, to keep notation at a minimum, we have considered $L[X|Y]$ and $E[X|Y]$ when X and Y are single random variables. In this section, we discuss the vector case, i.e., $L[\mathbf{X}|\mathbf{Y}]$ and $E[\mathbf{X}|\mathbf{Y}]$ when $\mathbf{X}$ and $\mathbf{Y}$ are random vectors. The only difficulty is one of notation. Conceptually, there is nothing new.

Definition 9.3 (LLSE of Random Vectors) Let $\mathbf{X}$ and $\mathbf{Y}$ be random vectors of dimensions m and n, respectively. Then

$$L[\mathbf{X}|\mathbf{Y}] = A\mathbf{y} + \mathbf{b}$$

where A is the $m \times n$ matrix and $\mathbf{b}$ the vector in $\Re^m$ that minimize

$$E(||\mathbf{X} - A\mathbf{Y} - \mathbf{b}||^2).$$

$\diamond$

Thus, as in the scalar case, the LLSE is the linear function of the observations that best approximates $\mathbf{X}$, in the mean squared error sense.

Before proceeding, review the notation of Sect. B.6 for $\Sigma_\mathbf{Y}$ and $cov(\mathbf{X}, \mathbf{Y})$.

Theorem 9.7 (LLSE of Vectors) *Let $\mathbf{X}$ and $\mathbf{Y}$ be random vectors such that $\Sigma_\mathbf{Y}$ is nonsingular.*

(a) Then

$$L[\mathbf{X}|\mathbf{Y}] = E(\mathbf{X}) + cov(\mathbf{X}, \mathbf{Y}) \Sigma_\mathbf{Y}^{-1} (\mathbf{Y} - E(\mathbf{Y})). \tag{9.11}$$

(b) Moreover,

$$E(||\mathbf{X} - L[\mathbf{X}|\mathbf{Y}]||^2) = tr(\Sigma_\mathbf{X} - cov(\mathbf{X}, \mathbf{Y}) \Sigma_\mathbf{Y}^{-1} cov(\mathbf{Y}, \mathbf{X})). \tag{9.12}$$

In this expression, for a square matrix M, $tr(M) := \sum_i M_{i,i}$ is the trace of the matrix.

∎

Proof

(a) The proof is similar to the scalar case. Let $\mathbf{Z}$ be the right-hand side of (9.11). One shows that the error $\mathbf{X} - \mathbf{Z}$ is orthogonal to all the linear functions of $\mathbf{Y}$. One then uses that fact to show that $\mathbf{X}$ is closer to $\mathbf{Z}$ than to any other linear function $h(\mathbf{Y})$ of $\mathbf{Y}$.

First we show the orthogonality. Since $E(\mathbf{X} - \mathbf{Z}) = 0$, we have

$$E((\mathbf{X} - \mathbf{Z})(B\mathbf{Y} + \mathbf{b})') = E((\mathbf{X} - \mathbf{Z})(B\mathbf{Y})') = E((\mathbf{X} - \mathbf{Z})\mathbf{Y}')B'.$$

Next, we show that $E((\mathbf{X} - \mathbf{Z})\mathbf{Y}') = 0$. To see this, note that

$$\begin{aligned}
E((\mathbf{X} - \mathbf{Z})\mathbf{Y}') &= E((\mathbf{X} - \mathbf{Z})(\mathbf{Y} - E(\mathbf{Y}))') \\
&= E((\mathbf{X} - E(\mathbf{X}))(\mathbf{Y} - E(\mathbf{Y}))') \\
&\quad - \mathrm{cov}(\mathbf{X}, \mathbf{Y})\Sigma_{\mathbf{Y}}^{-1} E((\mathbf{Y} - E(\mathbf{Y}))(\mathbf{Y} - E(\mathbf{Y}))') \\
&= \mathrm{cov}(\mathbf{X}, \mathbf{Y}) - \mathrm{cov}(\mathbf{X}, \mathbf{Y})\Sigma_{\mathbf{Y}}^{-1}\Sigma_{\mathbf{Y}} = 0.
\end{aligned}$$

Second, we show that $\mathbf{Z}$ is closer to $\mathbf{X}$ than any linear $h(\mathbf{Y})$. We have

$$\begin{aligned}
E(\|\mathbf{X} - h(\mathbf{Y})\|^2) &= E((\mathbf{X} - h(\mathbf{Y}))'(\mathbf{X} - h(\mathbf{Y}))) \\
&= E((\mathbf{X} - \mathbf{Z} + \mathbf{Z} - h(\mathbf{Y}))'(\mathbf{X} - \mathbf{Z} + \mathbf{Z} - h(\mathbf{Y}))) \\
&= E(\|\mathbf{X} - \mathbf{Z}\|^2) + E(\|\mathbf{Z} - h(\mathbf{Y})\|^2) + 2E((\mathbf{X} - \mathbf{Z})'(\mathbf{Z} - h(\mathbf{Y}))).
\end{aligned}$$

We claim that the last term is equal to zero. To see this, note that

$$E((\mathbf{X} - \mathbf{Z})'(\mathbf{Z} - h(\mathbf{Y}))) = \sum_{i=1}^{n} E((X_i - Z_i)(Z_i - h_i(\mathbf{Y}))).$$

Also,

$$E((X_i - Z_i)(Z_i - h_i(\mathbf{Y}))) = E((\mathbf{X} - \mathbf{Z})(\mathbf{Z} - h(\mathbf{Y}))')_{i,i}$$

and the matrix $E((\mathbf{X} - \mathbf{Z})(\mathbf{Z} - h(\mathbf{Y}))')$ is equal to zero since $\mathbf{X} - \mathbf{Y}$ is orthogonal to any linear function of $\mathbf{Y}$ and, in particular, to $\mathbf{Z} - h(\mathbf{Y})$.

(Note: an alternative way of showing that the last term is equal to zero is to write

$$E((\mathbf{X} - \mathbf{Z})'(\mathbf{Z} - h(\mathbf{Y})) = \mathrm{tr}E((\mathbf{X} - \mathbf{Z})(\mathbf{Z} - h(\mathbf{Y}))') = 0,$$

where the first equality comes from the fact that $\mathrm{tr}(AB) = \mathrm{tr}(BA)$ for matrices of compatible dimensions.)

(b) Let $\tilde{\mathbf{X}} := \mathbf{X} - E[\mathbf{X}|\mathbf{Y}]$ be the estimation error. Thus,

$$\tilde{\mathbf{X}} = \mathbf{X} - E(\mathbf{X}) - \mathrm{cov}(\mathbf{X}, \mathbf{Y})\Sigma_{\mathbf{Y}}^{-1}(\mathbf{Y} - E(\mathbf{Y})).$$

Now, if $\mathbf{V}$ and $\mathbf{W}$ are two zero-mean random vectors and M a matrix,

$$\text{cov}(\mathbf{V} - M\mathbf{W}) = E((\mathbf{V} - M\mathbf{W})(\mathbf{V} - M\mathbf{W})')$$

$$= E(\mathbf{V}\mathbf{V}' - 2M\mathbf{W}\mathbf{V}' + M\mathbf{W}\mathbf{W}'M')$$

$$= \text{cov}(\mathbf{V}) - 2M\text{cov}(\mathbf{W}, \mathbf{V}) + M\text{cov}(\mathbf{W})M'.$$

Hence,

$$\text{cov}(\tilde{\mathbf{X}}) = \Sigma_{\mathbf{X}} - 2\text{cov}(\mathbf{X}, \mathbf{Y})\Sigma_{\mathbf{Y}}^{-1}\text{cov}(\mathbf{Y}, \mathbf{X})$$

$$+ \text{cov}(\mathbf{X}, \mathbf{Y})\Sigma_{\mathbf{Y}}^{-1}\Sigma_{\mathbf{Y}}\Sigma_{\mathbf{Y}}^{-1}\text{cov}(\mathbf{Y}, \mathbf{X})$$

$$= \Sigma_{\mathbf{X}} - \text{cov}(\mathbf{X}, \mathbf{Y})\Sigma_{\mathbf{Y}}^{-1}\text{cov}(\mathbf{Y}, \mathbf{X}).$$

To conclude the proof, note that, for a zero-mean random vector $\mathbf{V}$,

$$E(||\mathbf{V}||^2) = E(\text{tr}(\mathbf{V}\mathbf{V}')) = \text{tr}(E(\mathbf{V}\mathbf{V}')) = \text{tr}(\Sigma_{\mathbf{V}}).$$

$\square$

9.8   Kalman Filter

The Kalman Filter is an algorithm to update the estimate of the state of a system using its output, as sketched in Fig. 9.13. The system has a *state* $X(n)$ and an *output* $Y(n)$ at time $n = 0, 1, \ldots$. These variables are defined through a system of linear equations:

$$X(n + 1) = AX(n) + V(n), n \geq 0; \tag{9.13}$$

$$Y(n) = CX(n) + W(n), n \geq 0. \tag{9.14}$$

In these equations, the random variables $\{X(0), V(n), W(n), n \geq 0\}$ are all orthogonal and zero-mean. The covariance of $V(n)$ is Σ_V and that of $W(n)$ is Σ_W. The filter is developed when the variables are random vectors and A, C are matrices of compatible dimensions.

The objective is to derive recursive equations to calculate

$$\hat{X}(n) = L[X(n)|Y(0), \ldots, Y(n)], n \geq 0.$$

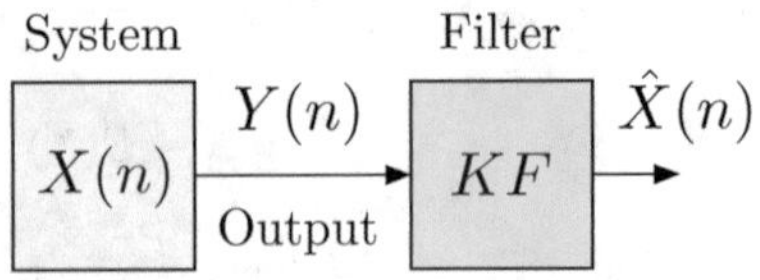

Fig. 9.13 The Kalman Filter computes the LLSE of the state of a system given the past of its output

9.8.1 The Filter

Here is the result, due to Rudolf Kalman (Fig. 9.14), which we prove in the next chapter. Do not panic when you see the equations!

Theorem 9.8 (Kalman Filter) *One has*

$$\hat{X}(n) = A\hat{X}(n-1) + K_n[Y(n) - CA\hat{X}(n-1)] \tag{9.15}$$

$$K_n = S_n C'[CS_n C' + \Sigma_W]^{-1} \tag{9.16}$$

$$S_n = A\Sigma_{n-1}A' + \Sigma_V \tag{9.17}$$

$$\Sigma_n = (I - K_n C)S_n. \tag{9.18}$$

Moreover,

$$S_n = cov(X(n) - A\hat{X}(n-1)) \text{ and } \Sigma_n = cov(X(n) - \hat{X}(n)). \tag{9.19}$$

∎

We will give a number of examples of this result. But first, let us make a few comments.

- Equations (9.15)–(9.18) are recursive: the estimate at time n is a simple linear function of the estimate at time $n-1$ and of the new observation $Y(n)$.
- The matrix K_n is the filter gain. It can be precomputed at time 0.
- The covariance of the error $X(n) - \hat{X}(n)$, Σ_n, can also be precomputed at time 0: it does not depend on the observations $\{Y(0), \ldots, Y(n)\}$. The estimate $\hat{X}(n)$ depends on these observations but the mean squared error does not.
- If $X(0)$ and the noise random variables are Gaussian, then the Kalman filter computes the MMSE.
- Finally, observe that these equations, even though they look a bit complicated, can be programmed in a few lines. This filter is elementary to implement and this explains its popularity.

Fig. 9.14 Rudolf Kalman, 1930–2016

9.8.2 Examples

In this section, we examine a few examples of the Kalman filter.

Random Walk

The first example is a filter to track a "random walk" by making noisy observations.
    Let

$$X(n+1) = X(n) + V(n) \tag{9.20}$$

$$Y(n) = X(n) + W(n) \tag{9.21}$$

$$\text{var}(V(n)) = 0.04, \text{var}(W(n)) = 0.09. \tag{9.22}$$

That is, $X(n)$ has orthogonal increments and it is observed with orthogonal noise.
Figure 9.15 shows a simulation of the filter. The left-hand part of the figure shows
that the estimate tracks the state with a bounded error. The middle part of the figure
shows the variance of the error, which can be precomputed. The right-hand part of
the figure shows the filter with the time-varying gain (in blue) and the filter with the
limiting gain (in green). The filter with the constant gain performs as well as the one
with the time-varying gain, in the limit, as justified by part (c) of the theorem.

Random Walk with Unknown Drift

In the second example, one tracks a random walk that has an unknown drift. This
system is modeled by the following equations:

$$X_1(n+1) = X_1(n) + X_2(n) + V(n) \tag{9.23}$$

$$X_2(n+1) = X_2(n) \tag{9.24}$$

$$Y(n) = X_1(n) + W(n) \tag{9.25}$$

$$\text{var}(V(n)) = 1, \text{var}(W(n)) = 0.25. \tag{9.26}$$

In this model, $X_2(n)$ is the constant but unknown drift and $X_1(n)$ is the value of
the "random walk." Figure 9.16 shows a simulation of the filter. It shows that the

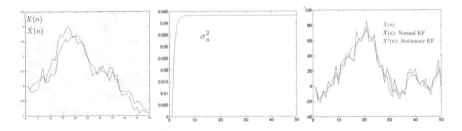

Fig. 9.15 The Kalman Filter for (9.20)–(9.22)

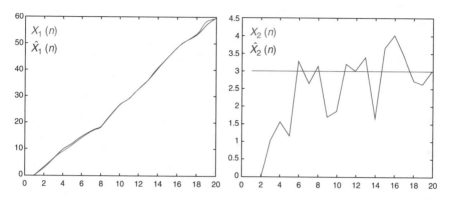

Fig. 9.16 The Kalman Filter for (9.23)–(9.26)

filter eventually estimates the drift and that the estimate of the position of the walk is quite accurate.

Random Walk with Changing Drift

In the third example, one tracks a random walk that has changing drift. This system is modeled by the following equations:

$$X_1(n + 1) = X_1(n) + X_2(n) + V_1(n) \tag{9.27}$$

$$X_2(n + 1) = X_2(n) + V_2(n) \tag{9.28}$$

$$Y(n) = X_1(n) + W(n) \tag{9.29}$$

$$\text{var}(V_1(n)) = 1, \text{var}(V_2(n)) = 0.01, \tag{9.30}$$

$$\text{var}(W(n)) = 0.25. \tag{9.31}$$

In this model, $X_2(n)$ is the varying drift and $X_1(n)$ is the value of the "random walk." Figure 9.17 shows a simulation of the filter. It shows that the filter tries to track the drift and that the estimate of the position of the walk is quite accurate.

Falling Object

In the fourth example, one tracks a falling object. The elevation $Z(n)$ of that falling object follows the equation

$$Z(n) = Z(0) + S(0)n - gn^2/2 + V(n), n \geq 0,$$

where $S(0)$ is the initial vertical velocity of the object and g is the gravitational constant at the surface of the earth. In this expression, $V(n)$ is some noise that perturbs the motion. We observe $\eta(n) = Z(n) + W(n)$, where $W(n)$ is some noise.

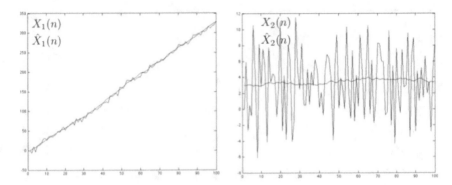

Fig. 9.17 The Kalman Filter for (9.27)–(9.31)

Fig. 9.18 The Kalman Filter
for (9.32)–(9.35)

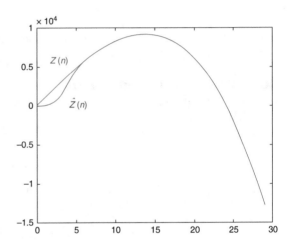

Since the term $-gn^2/2$ is known, we consider

$$X_1(n) = Z(n) + gn^2/2 \text{ and } Y(n) = \eta(n) + gn^2/2.$$

With this change of variables, the system is described by the following equations:

$$X_1(n + 1) = X_1(n) + X_2(n) + V(n) \tag{9.32}$$

$$X_2(n + 1) = X_2(n) \tag{9.33}$$

$$Y(n) = X_1(n) + W(n) \tag{9.34}$$

$$\text{var}(V_1(n)) = 100 \text{ and } \text{var}(W(n)) = 1600. \tag{9.35}$$

Figure 9.18 shows a simulation of the filter that computes $\hat{X}_1(n)$ from which we
subtract $gt^2/2$ to get an estimate of the actual altitude $Z(n)$ of the object.

9.9 Summary

- LLSE, linear regression, and MMSE;
- Projection characterization;
- MMSE of jointly Gaussian is linear;
- Kalman Filter.

9.9.1 Key Equations and Formulas

LLSE	$L[X\|Y] = E(X) + \mathrm{cov}(X, Y)\mathrm{var}(Y)^{-1}(Y - E(Y))$	Theorem 9.1
Orthogonality	$X - L[X\|Y] \perp a + bY$	(9.6)
Linear Regression	converges to $L[X\|Y]$	Theorem 9.2
Conditional Expectation	$E[X\|Y] = \ldots$	Definition 9.2
Orthogonality	$X - E[X\|Y] \perp g(Y)$	Lemma 9.4
MMSE = CE	$MMSE[X\|Y] = E[X\|Y]$	Theorem 9.3
Properties of CE	Linearity, smoothing, etc...	Theorem 9.5
CE for J.G.	If X, Y J.G., then $E[X\|Y] = L[X\|Y] = \cdots$	Theorem 9.6
LLSE vectors	$L[\mathbf{X}\|\mathbf{Y}] = E(\mathbf{X}) + \Sigma_{\mathbf{X,Y}}\Sigma_{\mathbf{Y}}^{-1}(\mathbf{Y} - E(\mathbf{Y}))$	Theorem 9.7
Kalman Filter	$\hat{X}(n) = A\hat{X}(n-1) + K_n[Y(n) - CA\hat{X}(n-1)]$	Theorem 9.8

9.10 References

LLSE, MMSE, and linear regression are covered in Chapter 4 of Bertsekas and Tsitsiklis (2008). The Kalman filter was introduced in Kalman (1960). The text (Brown and Hwang 1996) is an easy introduction to Kalman filters with many examples.

9.11 Problems

Problem 9.1 Assume that $X_n = Y_n + 2Y_n^2 + Z_n$ where the Y_n and Z_n are i.i.d. $U[0, 1]$. Let also $X = X_1$ and $Y = Y_1$.

(a) Calculate $L[X|Y]$ and $E((X - L[X|Y])^2)$;
(b) Calculate $Q[X|Y]$ and $E((X - Q[X|Y])^2)$ where $Q[X|Y]$ is the quadratic least squares estimate of X given Y.

(c) Design a stochastic gradient algorithm to compute $Q[X|Y]$ and implement it in Python.

Problem 9.2 We want to compare the off-line and on-line methods for computing $L[X|Y]$. Use the setup of the previous problem.

(a) Generate $N = 1,000$ samples and compute the linear regression of X given Y. Say that this is $X = aY + b$
(b) Using the same samples, compute the linear fit recursively using the stochastic gradient algorithm. Say that you obtain $X = cY + d$
(c) Evaluate the quality of the two estimates your obtained by computing $E((X - aY - b)^2)$ and $E((X - cY - d)^2)$.

Problem 9.3 The random variables X, Y, Z are jointly Gaussian,

$$(X, Y, Z)^T \sim N\left((0, 0, 0)^T, \begin{bmatrix} 2 & 2 & 1 \\ 2 & 4 & 2 \\ 1 & 2 & 1 \end{bmatrix}\right).$$

(a) Find $E[X|Y, Z]$;
(b) Find the variance of error.

Problem 9.4 You observe three i.i.d. samples X_1, X_2, X_3 from the distribution $f_{X|\theta}(x) = \frac{1}{2}e^{-|x-\theta|}$, where $\theta \in \mathbb{R}$ is the parameter to estimate. Find $MLE[\theta|X_1, X_2, X_3]$.

Problem 9.5

(a) Given three independent $N(0, 1)$ random variables X, Y, and Z, find the following minimum mean square estimator:

$$E[X + 3Y|2Y + 5Z].$$

(b) For the above, compute the mean squared error of the estimator.

Problem 9.6 Given two independent $N(0, 1)$ random variables X and Y, find the following linear least square estimator:

$$L[X|X^2 + Y].$$

Hint: The characteristic function of a $N(0, 1)$ random variable X is as follows:

$$E(e^{isX}) = e^{-\frac{1}{2}s^2}.$$

Problem 9.7 Consider a sensor network with n sensors that are making observations $\mathbf{Y}^n = (Y_1, \ldots, Y_n)$ of a signal X where

$$Y_i = aX + Z_i, i = 1, \ldots, n.$$

In this expression, $X =_D N(0, 1)$, $Z_i =_D N(0, \sigma^2)$, for $i = 1, \ldots, n$ and these random variables are mutually independent.

(a) Compute the MMSE estimator of X given $\mathbf{Y}^n$.
(b) Compute the mean squared error σ_n^2 of the estimator.
(c) Assume each measurement has a cost C and that we want to minimize

$$nC + \sigma_n^2.$$

Find the best value of n.
(d) Assume that we can decide at each step whether to make another measurement or to stop. Our goal is to minimize the expected value of

$$\nu C + \sigma_\nu^2,$$

where ν is the random number of measurements. Do you think there is a decision rule that will do better than the deterministic value n derived in (c)? Explain.

Problem 9.8 We want to use a Kalman filter to detect a change in the popularity of a word in twitter messages. To do this, we create a model of the number Y_n of times that particular word appears in twitter messages on day n. The model is as follows:

$$X(n + 1) = X(n)$$
$$Y(n) = X(n) + W(n),$$

where the $W(n)$ are zero-mean and uncorrelated. This model means that we are observing numbers of occurrences with an unknown mean $X(n)$ that is supposed to be constant. The idea is that if the mean actually changes, we should be able to detect it by noticing that the errors between $\hat{Y}(n)$ and $Y(n)$ are large. Propose an algorithm for detecting that change and implement it in Python.

Problem 9.9 The random variable X is exponentially distributed with mean 1. Given X, the random variable Y is exponentially distributed with rate X.

(a) Calculate $E[Y|X]$.
(b) Calculate $E[X|Y]$.

Problem 9.10 The random variables X, Y, Z are i.i.d. $\mathcal{N}(0, 1)$.

(a) Find $L[X^2 + Y^2 | X + Y]$;
(b) Find $E[X + 2Y | X + 3Y + 4Z]$;
(c) Find $E[(X + Y)^2 | X - Y]$.

Problem 9.11 Let $(V_n, \; n \geq 0)$ be i.i.d. $N(0, \sigma^2)$ and independent of $X_0 = N(0, u^2)$. Define

$$X_{n+1} = aX_n + V_n, \; n \geq 0.$$

1. What is the distribution of X_n for $n \geq 1$?
2. Find $E[X_{n+m} | X_n]$ for $0 \leq n < n + m$.
3. Find u so that the distribution of X_n is the same for all $n \geq 0$.

Problem 9.12 Let $\theta =_D U[0, 1]$, and given θ, the random variable X is uniformly distributed in $[0, \theta]$. Find $E[\theta | X]$.

Problem 9.13 Let $(X, Y)^T \sim N([0; 0], [3, 1; 1, 1])$. Find $E[X^2 | Y]$.

Problem 9.14 Let $(X, Y, Z)^T \sim N([0; 0; 0], [5, 3, 1; 3, 9, 3; 1, 3, 1])$. Find $E[X | Y, Z]$.

Problem 9.15 Consider arbitrary random variables X and Y. Prove the following property:

$$\text{var}(Y) = E(\text{var}[Y | X]) + \text{var}(E[Y | X]).$$

Problem 9.16 Let the joint p.d.f. of two random variables X and Y be

$$f_{X,Y}(x, y) = \frac{1}{4}(2x + y)1\{0 \leq x \leq 1\}1\{0 \leq y \leq 2\}.$$

First show that this is a valid joint p.d.f. Suppose you observe Y drawn from this joint density. Find $\text{MMSE}[X | Y]$.

Problem 9.17 Given four independent $N(0, 1)$ random variables X, Y, Z, and V, find the following minimum mean square estimate:

$$E[X + 2Y + 3Z | Y + 5Z + 4V].$$

Find the mean squared error of the estimate.

Problem 9.18 Assume that X, Y are two random variables that are such that $E[X | Y] = L[X | Y]$. Then, it must be that (choose the correct answers, if any)

☐ X and Y are jointly Gaussian;
☐ X can be written as $X = aY + Z$ where Z is a random variable that is independent of Y;
☐ $E((X - L[X|Y])Y^k) = 0$ for all $k \geq 0$;
☐ $E((X - L[X|Y]) \sin(3Y + 5)) = 0$.

Problem 9.19 In a linear system with independent Gaussian noise, with state X_n and observation Y_n, the Kalman filter computes (choose the correct answers, if any)

☐ $MLE[Y_n|X^n]$;
☐ $MLE[X_n|Y^n]$;
☐ $MAP[Y_n|X^n]$;
☐ $MAP[X_n|Y^n]$;
☐ $E[X_n|Y^n]$;
☐ $E[Y_n|X^n]$;
☐ $E[X_n|Y_n]$;
☐ $E[Y_n|X_n]$.

Problem 9.20 Let $(X, \mathbf{Y})$ where $\mathbf{Y}' = [Y_1, Y_2, Y_3, Y_4]$ be $N(\mu, \Sigma)$ with $\mu' = [2, 1, 3, 4, 5]$ and

$$\Sigma = \begin{bmatrix} 3 & 4 & 6 & 12 & 8 \\ 4 & 6 & 9 & 18 & 12 \\ 6 & 9 & 14 & 28 & 18 \\ 12 & 18 & 28 & 56 & 36 \\ 8 & 12 & 18 & 36 & 24 \end{bmatrix}.$$

Find $E[X|\mathbf{Y}]$.

Problem 9.21 Let $\mathbf{X} = A\mathbf{V}$ and $\mathbf{Y} = C\mathbf{V}$ where $\mathbf{V} = N(\mathbf{0}, \mathbf{I})$.
Find $E[\mathbf{X}|\mathbf{Y}]$.

Problem 9.22 Given $\theta \in \{0, 1\}$, $\mathbf{X} = N(\mathbf{0}, \Sigma_\theta)$ where

$$\Sigma_0 = \begin{bmatrix} 1 & 0 \\ 0 & 1 \end{bmatrix} \text{ and } \Sigma_1 = \begin{bmatrix} 1 & \rho \\ \rho & 1 \end{bmatrix},$$

where $\rho > 0$ is given.
Find $MLE[\theta|\mathbf{X}]$.

Problem 9.23 Given two independent $N(0, 1)$ random variables X and Y, find the following linear least square estimator:

$$L[X|X^3 + Y].$$

Hint: The characteristic function of a $N(0, 1)$ random variable X is as follows:

$$E(e^{isX}) = e^{-\frac{1}{2}s^2}.$$

Problem 9.24 Let X, Y, Z be i.i.d. $\mathcal{N}(0, 1)$. Find

$$E[X|X + Y, X + Z, Y - Z].$$

Hint: Argue that the observation $Y - Z$ is redundant.

Problem 9.25 Let X, Y_1, Y_S, Y_3 be zero-mean with covariance matrix

$$\Sigma = \begin{bmatrix} 10 & 6 & 5 & 16 \\ 6 & 9 & 6 & 21 \\ 5 & 6 & 6 & 18 \\ 16 & 21 & 18 & 57 \end{bmatrix}.$$

Find $L[X|Y_1, Y_2, Y_3]$. *Hint:* You will observe that $\Sigma_{\mathbf{Y}}$ is singular. This means that at least one of the observations Y_1, Y_2, or Y_3 is redundant, i.e., is a linear combination of the others. This implies that $L[X|Y_1, Y_2, Y_3] = L[X|Y_1, Y_2]$.

Topics: Derivation and properties of Kalman filter; Extended Kalman filter

10.1 Updating LLSE

In many situations, one keeps making observations and one wishes to update the estimate accordingly, hopefully without having to recompute everything from scratch. That is, one hopes for a method that enables to calculate $L[\mathbf{X}|\mathbf{Y}, \mathbf{Z}]$ from $L[\mathbf{X}|\mathbf{Y}]$ and $\mathbf{Z}$.

The key idea is in the following result.

Theorem 10.1 (LLSE Update—Orthogonal Additional Observation) *Assume that* $\mathbf{X}$, $\mathbf{Y}$, *and* $\mathbf{Z}$ *are zero-mean and that* $\mathbf{Y}$ *and* $\mathbf{Z}$ *are orthogonal. Then*

$$L[\mathbf{X}|\mathbf{Y}, \mathbf{Z}] = L[\mathbf{X}|\mathbf{Y}] + L[\mathbf{X}|\mathbf{Z}]. \tag{10.1}$$

∎

Proof Figure 10.1 shows why the result holds. To be convinced mathematically, we need to show that the error

$$\mathbf{X} - (L[\mathbf{X}|\mathbf{Y}] + L[\mathbf{X}|\mathbf{Z}])$$

is orthogonal to $\mathbf{Y}$ and to $\mathbf{Z}$. To see why it is orthogonal to $\mathbf{Y}$, note that the error is

© The Author(s) 2021
J. Walrand, *Probability in Electrical Engineering and Computer Science*,
https://doi.org/10.1007/978-3-030-49995-2_10

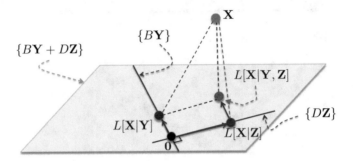

Fig. 10.1 The LLSE is easy to update after an additional orthogonal observation

$$(\mathbf{X} - L[\mathbf{X}|\mathbf{Y}]) - L[\mathbf{X}|\mathbf{Z}].$$

Now, the term between parentheses is orthogonal to $\mathbf{Y}$, by the projection property of $L[\mathbf{X}|\mathbf{Y}]$. Also, the second term is linear in $\mathbf{Z}$, and is therefore orthogonal to $\mathbf{Y}$ since $\mathbf{Z}$ is orthogonal to $\mathbf{Y}$. One shows that the error is orthogonal to $\mathbf{Z}$ in the same way.   □

A simple consequence of this result is the following fact.

Theorem 10.2 (LLSE Update—General Additional Observation) *Assume that* $\mathbf{X}$, $\mathbf{Y}$, *and* $\mathbf{Z}$ *are zero-mean. Then*

$$L[\mathbf{X}|\mathbf{Y}, \mathbf{Z}] = L[\mathbf{X}|\mathbf{Y}] + L[\mathbf{X}|\mathbf{Z} - L[\mathbf{Z}|\mathbf{Y}]]. \tag{10.2}$$

■

Proof The idea here is that one considers the *innovation* $\tilde{\mathbf{Z}} := \mathbf{Z} - L[\mathbf{Z}|\mathbf{Y}]$, which is the information in the new observation $\mathbf{Z}$ that is orthogonal to $\mathbf{Y}$.

To see why the result holds, note that any linear combination of $\mathbf{Y}$ and $\mathbf{Z}$ can be written as a linear combination of $\mathbf{Y}$ and $\tilde{\mathbf{Z}}$. For instance, if $L[\mathbf{Z}|\mathbf{Y}] = C\mathbf{Y}$, then

$$A\mathbf{Y} + B\mathbf{Z} = A\mathbf{Y} + B(\mathbf{Z} - C\mathbf{Y}) + BC\mathbf{Y} = (A + BC)\mathbf{Y} + B\tilde{\mathbf{Z}}.$$

Thus, the set of linear functions of $\mathbf{Y}$ and $\mathbf{Z}$ is the same as the set of linear functions of $\mathbf{Y}$ and $\tilde{\mathbf{Z}}$, so that

$$L[\mathbf{X}|\mathbf{Y}, \mathbf{Z}] = L[\mathbf{X}|\mathbf{Y}, \tilde{\mathbf{Z}}].$$

Thus, (10.2) follows from Theorem 10.1 since $\mathbf{Y}$ and $\tilde{\mathbf{Z}}$ are orthogonal.   □

10.2   Derivation of Kalman Filter

We derive the equations for the Kalman filter, as stated in Theorem 9.8. For convenience, we repeat those equations here:

$$\hat{X}(n) = A\hat{X}(n-1) + K_n[Y(n) - CA\hat{X}(n-1)] \tag{10.16}$$

$$K_n = S_n C'[CS_n C' + \Sigma_W]^{-1} \tag{10.17}$$

$$S_n = A\Sigma_{n-1}A' + \Sigma_V \tag{10.18}$$

$$\Sigma_n = (I - K_n C)S_n \tag{10.19}$$

and

$$S_n = \text{cov}(X(n) - A\hat{X}(n-1)) \text{ and } \Sigma_n = \text{cov}(X(n) - \hat{X}(n)). \tag{10.20}$$

In the algebra, we repeatedly use the fact that

$$\text{cov}(BV, DW) = B \text{ cov}(V, W)D'$$

and also that if V and W are orthogonal, then

$$\text{cov}(V + W) = \text{cov}(V) + \text{cov}(W).$$

The algebra is a bit tedious, but the key steps are worth noting.
Let

$$Y^n = (Y(0), \dots, Y(n)).$$

Note that

$$L\left[X(n)|Y^{n-1}\right] = L\left[AX(n-1) + V(n-1)|Y^{n-1}\right] = A\hat{X}(n-1).$$

Hence,

$$L\left[Y(n)|Y^{n-1}\right] = L\left[CX(n) + W(n)|Y^{n-1}\right] = CL\left[X(n)|Y^{n-1}\right] = CA\hat{X}(n-1),$$

so that, by Theorem 10.2,

$$Y(n) - L\left[Y(n)|Y^{n-1}\right] = Y(n) - CA\hat{X}(n-1).$$

Thus,

$$\hat{X}(n) = L[X(n)|Y^n] = L\left[X(n)|Y^{n-1}\right] + L\left[X(n)|Y(n) - L\left[Y(n)|Y^{n-1}\right]\right]$$

$$= A\hat{X}(n-1) + K_n\left[Y(n) - CA\hat{X}(n-1)\right].$$

This derivation shows that (10.16) is a fairly direct consequence of the formula in Theorem 10.2 for updating the LLSE.

The calculation of the gain K_n is a bit more complex. Let

$$\tilde{Y}(n) = Y(n) - L\left[Y(n)|Y^{n-1}\right] = Y(n) - CA\hat{X}(n-1).$$

Then

$$K_n = \text{cov}\left(X(n), \tilde{Y}(n)\right)\text{cov}\left(\tilde{Y}(n)\right)^{-1}.$$

Now,

$$\text{cov}\left(X(n), \tilde{Y}(n)\right) = \text{cov}\left(X(n) - L\left[X(n)|Y^{n-1}\right], \tilde{Y}(n)\right),$$

because $\tilde{Y}(n)$ is orthogonal to Y^{n-1}. Also,

$$\text{cov}(X(n) - L\left[X(n)|Y^{n-1}\right], \tilde{Y}(n))$$

$$= \text{cov}(X(n) - A\hat{X}(n-1), Y(n) - CA\hat{X}(n-1))$$

$$= \text{cov}(X(n) - A\hat{X}(n-1), CX(n) + W(n) - CA\hat{X}(n-1))$$

$$= S_n C',$$

by (10.20).

To calculate $\text{cov}(\tilde{Y}(n))$, we note that

$$\text{cov}(\tilde{Y}(n)) = \text{cov}\left(CX(n) + W(n) - CL\left[X(n)|Y^{n-1}\right]\right) = CS_n C' + \Sigma_W.$$

Thus,

$$K_n = S_n C'\left[CS_n C' + \Sigma_W\right]^{-1}.$$

To show (10.18), we note that

$$S_n = \text{cov}\left(X(n) - L\left[X(n)|Y^{n-1}\right]\right)$$

$$= \text{cov}\left(AX(n-1) + V(n-1) - A\hat{X}(n-1)\right)$$

$$= A\Sigma_{n-1}A' + \Sigma_V.$$

Finally, to derive (10.19), we calculate

$$\Sigma_n = \text{cov}\left(X(n) - \hat{X}(n)\right).$$

We observe that

$$X(n) - L[X(n)|Y^n] = X(n) - A\hat{X}(n-1) - K_n\left[Y(n) - CA\hat{X}(n-1)\right]$$

$$= X(n) - A\hat{X}(n-1) - K_n\left[CX(n) + W(n) - CA\hat{X}(n-1)\right]$$

$$= [I - K_nC]\left[X(n) - A\hat{X}(n-1)\right] - K_nW(n),$$

so that

$$\Sigma_n = [I - K_nC]S_n[I - K_nC]' + K_n\Sigma_W K_n'$$

$$= S_n - 2K_nCS_n + K_n\left[CS_nC' + \Sigma_W\right]K_n'$$

$$= S_n - 2K_nCS_n + K_n\left[CS_nC' + \Sigma_W\right]\left[CS_nC' + \Sigma_W\right]^{-1}CS_n \text{ by (10.17)}$$

$$= S_n - K_nCS_n,$$

as we wanted to show.

$$\square$$

10.3   Properties of Kalman Filter

The goal of this section is to explain and justify the following result. The terms *observable* and *reachable* are defined after the statement of the theorem.

Theorem 10.3 (Properties of the Kalman Filter)

(a) *If (A, C) is observable, then Σ_n is bounded. Moreover, if $\Sigma_0 = 0$, then*

$$\Sigma_n \to \Sigma \text{ and } K_n \to K, \tag{10.37}$$

where Σ is a finite matrix.

(b) *Also, if in addition, $(A, \Sigma_V^{1/2})$ is reachable, then the filter with $K_n = K$ is such that the covariance of the error also converges to Σ.*

$$\blacksquare$$

We explain these properties in the subsequent sections. Let us first make a few comments.

- For some systems, the errors grow without bound. For instance, if one does not observe anything (e.g., $C = 0$) and if the system is unstable (e.g., $X(n) = 2X(n-1) + V(n)$), then Σ_n goes to infinity. However, (a) says that "if the observations are rich enough," this does not happen: one can track $X(n)$ with an error that has a bounded covariance.
- Part (b) of the theorem says that in some cases, one can use the filter with a constant gain K without having a bigger error, asymptotically. This is very convenient as one does not have to compute a new gain at each step.

10.3.1 Observability

Are the observations good enough to track the state with a bounded error covariance? Before stating the result, we need a precise notion of good observations.

Definition 10.1 (Observability) We say that (A, C) is *observable* if the null space of

$$
\begin{bmatrix}
C \\
CA \\
\vdots \\
CA^d
\end{bmatrix}
$$

is $\{\mathbf{0}\}$. Here, d is the dimension of $X(n)$. A matrix M has null space $\{\mathbf{0}\}$ if $\{\mathbf{0}\}$ is the only vector $\mathbf{v}$ such that $M\mathbf{v} = \mathbf{0}$.

$\diamond$

The key result is the following.

Lemma 10.4 (Observability Implies Bounded Error Covariance)

(a) If the system is observable, then Σ_n is bounded.
(b) If in addition, $\Sigma_0 = 0$, then Σ_n converges to some finite Σ.

Proof

(a) Observability implies that there is only one $X(0)$ that corresponds to $(Y(0), \ldots, Y(d))$ if the system has no noise. Indeed, in that case,

$$X(n) = AX(n-1) \text{ and } Y(n) = CX(n).$$

Then,

$$X(1) = AX(0), X(2) = A^2 X(0), \ldots, X(d) = A^{d-1} X(0),$$

so that

$$Y(0) = CX(0), Y(1) = CAX(0), \ldots, Y(d-1) = CA^{d-1} X(0).$$

Consequently,

$$\begin{bmatrix} Y(0) \\ Y(1) \\ \vdots \\ Y(d) \end{bmatrix} = \begin{bmatrix} C \\ CA \\ \vdots \\ CA^{d-1} \end{bmatrix} X(0).$$

Now, imagine that there are two different initial states, say $X(0)$ and $\mathring{X}(0)$ that give the same outputs $Y(0), \ldots, Y(d)$. Then,

$$\begin{bmatrix} Y(0) \\ Y(1) \\ \vdots \\ Y(d) \end{bmatrix} = \begin{bmatrix} C \\ CA \\ \vdots \\ CA^{d-1} \end{bmatrix} X(0) = \begin{bmatrix} C \\ CA \\ \vdots \\ CA^{d-1} \end{bmatrix} \mathring{X}(0),$$

so that

$$\begin{bmatrix} C \\ CA \\ \vdots \\ CA^{d-1} \end{bmatrix} (X(0) - \mathring{X}(0)) = \mathbf{0}.$$

The observability property implies that $X(0) - \mathring{X}(0) = \mathbf{0}$.

Thus, if (A, C) is observable, one can identify the initial condition $X(0)$ uniquely after $d + 1$ observations of the output, when there is no noise. Hence, when there is no noise, one can then determine $X(1), X(2), \ldots$ exactly. Thus, when (A, C) is observable, one can determine the state $X(n)$ precisely from the outputs.

However, our system has some noise. If (A, C) is observable, we are able to identify $X(0)$ from $Y(0), \ldots, Y(d)$, up to some linear function of the noise that has affected those outputs, i.e., up to a linear function of $\{V(0), \ldots, V(d - 1), W(0), \ldots, W(d)\}$. Consequently, we can determine $X(d)$ from $Y(0), \ldots, Y(d)$, up to some linear function of $\{V(0), \ldots, V(d - 1), W(0), \ldots, W(d)\}$. Similarly, we can determine $X(n)$ from $Y(n - d), \ldots, Y(n)$, up to some linear function of $\{V(n), \ldots, V(n + d - 1), W(n - d), \ldots, W(n)\}$.

This implies that the error between $X(n)$ and $\hat{X}(n)$ is a linear combination of d noise contributions, so that Σ_n is bounded.

(b) One can show that if $\Sigma_0 = 0$, i.e., if we know $X(0)$, then Σ_n increases in the sense that $\Sigma_n - \Sigma_{n-1}$ is nonnegative definite. Being bounded and increasing implies that Σ_n converges, and so does K_n.

$\square$

10.3.2 Reachability

Assume that $\Sigma_V = QQ'$. We say that (A, Q) is reachable if the rank of

$$[Q, AQ, \ldots, A^{d-1}Q]$$

is full. To appreciate the meaning of this property, note that we can write the state equations as

$$X(n) = AX(n-1) + Q\eta_n,$$

where $\text{cov}(\eta_n) = \mathbf{I}$. That is, the components of η are orthogonal. In the Gaussian case, the components of η are $N(0, 1)$ and independent. If (A, Q) is reachable, this means that for any $\mathbf{x} \in \mathfrak{R}^d$, there is some sequence $\eta_0, \ldots, \eta_d$ such that if $X(0) = 0$, then $X(d) = \mathbf{x}$. Indeed,

$$X(d) = \sum_{k=0}^{d} A^k Q\eta_{d-k} = \left[Q, AQ, \ldots, A^{d-1}Q \right] \begin{bmatrix} \eta_d \\ \eta_{d-1} \\ \vdots \\ \eta_0 \end{bmatrix}.$$

Since the matrix is full rank, the span of its columns is $\mathfrak{R}^d$, which means precisely that there is a linear combination of these columns that is equal to any given vector in $\mathfrak{R}^d$.

The proof of part (b) of the theorem is a bit too involved for this course.

10.4   Extended Kalman Filter

The Kalman filter is often used for nonlinear systems. The idea is that if the system is almost linear over a few steps, then one may be able to use the Kalman filter locally and change the matrices A and C as the estimate of the state changes.

The model is as follows:

$$X(n+1) = f(X(n)) + V(n)$$
$$Y(n+1) = g(X(n+1)) + W(n+1).$$

The *extended Kalman filter* is then

$$\hat{X}(n+1) = f\left(\hat{X}(n)\right) + K_n \left[Y(n+1) - g\left(f(\hat{X}(n))\right)\right]$$

$$K_n = S_n C_n' \left[C_n S_n C_n' + \Sigma_W\right]^{-1}$$

$$S_n = A_n \Sigma_n A_n' + \Sigma_V$$

$$\Sigma_{n+1} = [I - K_n C_n]S_n,$$

where

$$[A_n]_{ij} = \frac{\partial}{\partial x_j} f_i\left(\hat{X}(n)\right) \text{ and } [C_n]_{ij} = \frac{\partial}{\partial x_j} g_i\left(\hat{X}(n)\right).$$

Thus, the idea is to linearize the system around the estimated state value and then apply the usual Kalman filter.

Note that we are now in the realm of heuristics and that very little can be said about the properties of this filter. Experiments show that it works well when the nonlinearities are small, whatever this means precisely, but that it may fail miserably in other conditions.

10.4.1 Examples

Tracking a Vehicle
In this example, borrowed from "Eric Feron, Notes for AE6531, Georgia Tech.", the goal is to track a vehicle that moves in the plane by using noisy measurements of distances to 9 points $p_i \in \mathfrak{R}^2$. Let $p(n) \in \mathfrak{R}^2$ be the position of the vehicle and $u(n) \in \mathfrak{R}^2$ be its velocity at time $n \geq 0$.

We assume that the velocity changes accruing to a known rule, except for some random perturbation. Specifically, we assume that

$$p(n+1) = p(n) + 0.1u(n) \tag{10.38}$$

$$u(n+1) = \begin{bmatrix} 0.85 & 0.15 \\ -0.1 & 0.85 \end{bmatrix} u(n) + w(n), \tag{10.39}$$

where the $w(n)$ are i.i.d. $N(0, \mathbf{I})$. The measurements are

$$y_i(n) = \|p(n) - p_i\| + v_i(n), i = 1, 2, \ldots, 9,$$

where the $v_i(n)$ are i.i.d. $N(0, 0.3^2)$.

Figure 10.2 shows the result of the extended Kalman filter for $X(n) = (p(n), u(n))$ initialized with $\hat{x}(0) = 0$ and $\Sigma_0 = \mathbf{I}$.

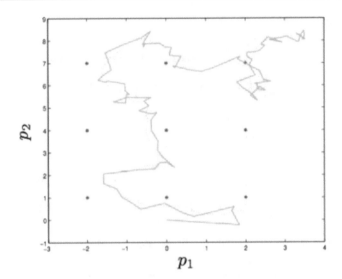

Fig. 10.2 The Extended Kalman Filter for the system (10.38)–(10.39)

Fig. 10.3 The chemical
reactions

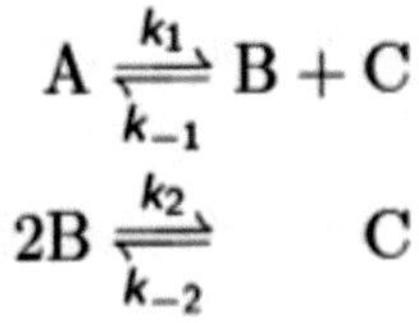

$$A \xrightleftharpoons[k_{-1}]{k_1} B + C$$

$$2B \xrightleftharpoons[k_{-2}]{k_2} \quad C$$

Tracking a Chemical Reaction

This example concerns estimating the state of a chemical reactor from measurements of the pressure. This example is borrowed from James B. Rawlings and Fernando V. Lima, U. Wisconsin, Madison. There are three components A, B, C in the reactions and they are modeled as shown in Fig. 10.3 where the k_i are the kinetic constants.

Let C_A, C_B, C_C be the concentrations of the A, B, C, respectively. The model is

$$\frac{d}{dt} \begin{bmatrix} C_A \\ C_B \\ C_C \end{bmatrix} = \begin{bmatrix} -1 & 0 \\ 1 & -2 \\ 1 & 1 \end{bmatrix} \begin{bmatrix} k_1 C_A - k_{-1} C_B C_C \\ k_2 C_B^2 - k_{-2} C_C \end{bmatrix}$$

and

$$y = RT(C_A + C_B + C_C).$$

As shown in the top part of Fig. 10.4, this filter does not track the concentrations correctly. In fact, some concentrations that the filter estimates are negative!

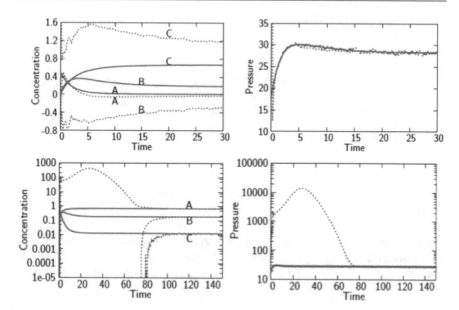

Fig. 10.4 The top two graphs show that the extended Kalman filter does not track the concentrations correctly. The bottom two graphs show convergence after modifying the equations

The bottom graphs show that the filter tracks the concentrations converge after modifying the equations and replacing negative estimates by 0.

The point of this example is that the extended Kalman filter is not guaranteed to converge and that, sometimes, a simple modification makes it converge.

10.5 Summary

- Updating LLSE;
- Derivation of Kalman Filter;
- Observability and Reachability;
- Extended Kalman Filter.

10.5.1 Key Equations and Formulas

Updating LLSE & zero-mean	$\Rightarrow L[\mathbf{X}|\mathbf{Y}, \mathbf{Z}] = L[\mathbf{X}|\mathbf{Y}] + L[\mathbf{X}|\mathbf{Z} - L[\mathbf{Z}|\mathbf{Y}]]$	T. 10.2
Observability	$\Rightarrow$ bounded error covariance	L.10.4
Observability + Reachability	$\Rightarrow$ asymptotic filter is good enough	T.9.8
Extended Kalman Filter	Linearize equations	S.10.4

10.6   References

The book Goodwin and Sin (2009) survey filtering and applications to control. The textbook Kumar and Varaiya (1986) is a comprehensive yet accessible presentation of control theory, filtering, and adaptive control. It is available online.

Speech Recognition: A
11

Application: Recognizing Speech
Topics: Hidden Markov chain, Viterbi decoding, EM Algorithms

11.1   Learning: Concepts and Examples

In artificial intelligence, "learning" refers to the process of discovering the relation-ship between related items, for instance between spoken words and sounds heard (Fig. 11.1).

As a simple example, consider the binary symmetric channel example of Problem 7.5 in Chap. 7. The inputs X_n are i.i.d. $B(p)$ and, given the inputs, the output Y_n is equal to X_n with probability $1 - \epsilon$, for $n \geq 0$. In this example, there is a probabilistic relationship between the inputs and the outputs described by ϵ. Learning here refers to estimating ϵ.

There are two basic situations. In *supervised learning*, one observes the inputs $\{X_n, n = 0, \ldots, N\}$ and the outputs $\{Y_n, n = 0, \ldots, N\}$. One can think of this form of learning as a *training* phase for the system. Thus, one observes the channel with a set of known input values. Once one has "learned" the channel, i.e., estimated ϵ, one can then design the best receiver and use it on unknown inputs. In *unsupervised learning*, one observes only the outputs. The benefit of this form of learning is that it takes place while the system is operational and one does not "waste" time with a training phase. Also, the system can adapt automatically to slow changes of ϵ without having to re-train it with a new training phase.

As you can expect, there is a trade-off when choosing supervised versus unsupervised learning. A training phase takes time but the learning is faster than

© The Author(s) 2021
J. Walrand, *Probability in Electrical Engineering and Computer Science*,
https://doi.org/10.1007/978-3-030-49995-2_11

Fig. 11.1 Can you hear me?

in unsupervised learning. The best method to use depends on characteristics of the practical situation, such as the likely rate of change of the system parameters.

11.2   Hidden Markov Chain

A hidden Markov chain is a Markov chain together with a state observation model. The Markov chain is $\{X(n), n \geq 0\}$ and it has its transition matrix P on the state space $\mathscr{X}$ and its initial distribution π_0. The state observation model specifies that when the state of the Markov chain is x, one observes a value y with probability $Q(x, y)$, for $y \in \mathscr{Y}$. More precisely, here is the definition (Fig. 11.2).

Definition 11.1 (Hidden Markov Chain)   A hidden Markov chain is a random sequence $\{(X(n), Y(n)), n \geq 0\}$ such that $X(n) \in \mathscr{X} = \{1, \ldots, N\}$ and $Y(n) \in \mathscr{Y} = \{1, \ldots, M\}$ and

$$P(X(0) = x_0, Y(0) = y_0, \ldots, X(n) = x_n, Y(n) = y_n)$$
$$= \pi_0(x_0) Q(x_0, y_0) P(x_0, x_1) Q(x_1, y_1) \times \cdots \times P(x_{n-1}, x_n) Q(x_n, y_n),$$
$$\text{for all } n \geq 0, x_m \in \mathscr{X}, y_m \in \mathscr{Y}. \tag{11.1}$$

◇

In the speech recognition application, the X_n are "parts of speech," i.e., segments of sentences, and the Y_n are sounds. The structure of the language determines relationships between the X_n that can be approximated by a Markov chain. The relationship between X_n and Y_n is speaker-dependent.

The recognition problem is the following. Assume that you have observed that $\mathbf{Y}^n := (Y_0, \ldots, Y_n) = \mathbf{y}^n := (y_0, \ldots, y_n)$. What is the most likely sequence $\mathbf{X}^n := (X_0, \ldots, X_n)$? That is, in the terminology of Chap. 7, we want to compute

$$MAP[\mathbf{X}^n \mid \mathbf{Y}^n = \mathbf{y}^n].$$

Thus, we want to find the sequence $\mathbf{x}^n \in \mathscr{X}^{n+1}$ that maximizes

Fig. 11.2 The hidden
Markov chain

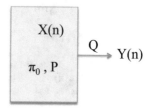

$$P[\mathbf{X}^n = \mathbf{x}^n \mid \mathbf{Y}^n = \mathbf{y}^n].$$

Note that

$$P[\mathbf{X}^n = \mathbf{x}^n \mid \mathbf{Y}^n = \mathbf{y}^n] = \frac{P(\mathbf{X}^n = \mathbf{x}^n, \mathbf{Y}^n = \mathbf{y}^n)}{P(\mathbf{Y}^n = \mathbf{y}^n)}.$$

The MAP is the value of $\mathbf{x}^n$ that maximizes the numerator. Now, by (11.1), the logarithm of the numerator is equal to

$$\log(\pi_0(x_0)Q(x_0, y_0)) + \sum_{m=1}^{n} \log(P(x_{m-1}, x_m)Q(x_m, y_m)).$$

Define

$$d(x_0) = -\log(\pi_0(x_0)Q(x_0, y_0))$$

and

$$d_m(x_{m-1}, x_m) = -\log(P(x_{m-1}, x_m)Q(x_m, y_m)).$$

Then, the MAP is the sequence $\mathbf{x}^n$ that minimizes

$$d(x_0) + \sum_{m=1}^{n} d_m(x_{m-1}, x_m). \tag{11.2}$$

The expression (11.2) can be viewed as the length for a path in the graph shown in Fig. 11.3. Finding the MAP is then equivalent to solving a shortest path problem. There are a few standard algorithms for solving such problems. We describe the *Bellman–Ford Algorithm* due to Bellman (Fig. 11.4) and Ford.

For $m = 0, \ldots, n$ and $x \in \mathcal{X}$, let $V_m(x)$ be the length of the shortest path from $X(m) = x$ to the column $X(n)$ in the graph. Also, let $V_n(x) = 0$ for all $x \in \mathcal{X}$. Then, one has

$$V_m(x) = \min_{x' \in \mathcal{X}} \{d_{m+1}(x, x') + V_{m+1}(x')\}, x \in \mathcal{X}, m = 0, \ldots, n-1. \tag{11.3}$$

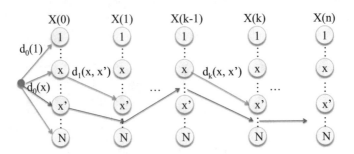

Fig. 11.3 The MAP as a shortest path

Fig. 11.4 Richard Bellman,
1920–1984

Finally, let

$$V = \min_{x \in \mathcal{X}} \{d_0(x) + V_0(x)\}. \tag{11.4}$$

Then, V is the minimum value of expression (11.2).
The algorithm is then as follows:

Step (1): Calculate $\{V_m(x), x \in \mathcal{X}\}$ recursively for $m = n - 1, n - 2, \ldots, 0$, using
 (11.3). At each step, note the arc out of each x that achieves the minimum. Say
 that the arc out of $x_m = x$ goes to $x_{m+1} = s(m, x)$ for $x \in \mathcal{X}$.
Step (2): Find the value x_0 that achieves the minimum in (11.4).
Step (3): The MAP is then the sequence

$$x_0, x_1 = s(0, x_0), x_2 = s(1, x_1), \ldots, x_n = s(n - 1, x_{n-1}).$$

Equations (11.3) are the *Bellman–Ford Equations*. They are a particular version
of *Dynamic Programming Equations (DPE)* for the shortest path problem.

Note that the essential idea was to define the length of the shortest remaining path
starting from every node in the graph and to write recursive expressions for those
quantities. Thus, one solves the DPE backwards and then one finds the shortest path
forward. This application of the shortest path algorithm for finding a MAP is called
the *Viterbi Algorithm* due to Andrew Viterbi (Fig. 11.5).

Fig. 11.5 Andrew Viterbi, b.
1934

11.3 Expectation Maximization and Clustering

Expectation maximization is a class of algorithms to estimate parameters of distributions. We first explain these algorithms on a simple clustering problem. We apply expectation maximization to the HMC model in the next section.

The *clustering problem* consists in grouping sample points into clusters of "similar" values. We explain a simple instance of this problem and we discuss the expectation maximization algorithm.

11.3.1 A Simple Clustering Problem

You look at set of N exam results $\{X(1), \ldots, X(N)\}$ in your probability course and you must decide who are the A and the B students. To study this problem, we assume that the results of A students are i.i.d. $\mathcal{N}(a, \sigma^2)$ and those of B students are $\mathcal{N}(b, \sigma^2)$ where $a > b$.

For simplicity, assume that we know σ^2 and that each student has probability 0.5 of being an A student. However, we do not know the parameters (a, b).

(The same method applies when one does not know the variances of the scores of A and B students, nor the prior probability that a student is of type A.)

One heuristic is as follows (see Fig. 11.6). Start with a guess (a_1, b_1) for (a, b). Student n with score $X(n)$ is more likely to be of type A if $X(n) > (a_1 + b_1)/2$. Let us declare that such students are of type A and the others are of type B. Let then a_2 be the average score of the students declared to be of type A and b_2 that of the other students. We repeat the procedure after replacing (a_1, b_1) by (a_2, b_2) and we keep doing this until the values seem to converge. This heuristic is called the *hard expectation maximization algorithm*.

A slightly different heuristic is as follows (see Fig. 11.7). Again, we start with a guess (a_1, b_1).

Using Bayes' rule, we calculate the probability $p(n)$ that student n with score $X(n)$ is of type A. We then calculate

$$a_2 = \frac{\sum_n X(n) p(n)}{\sum_n p(n)} \text{ and } b_2 = \frac{\sum_n X(n)(1 - p(n))}{\sum_n (1 - p(n))}.$$

Fig. 11.6 Clustering with
hard EM. The initial guess is
(a_1, b_1), which leads to the
MAP of the types and the
next guess (a_2, b_2), and so on

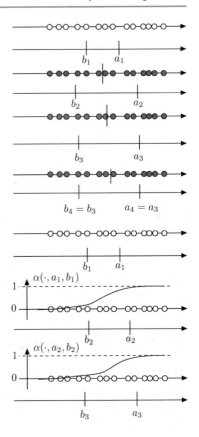

Fig. 11.7 Clustering with
soft EM. The initial guess is
(a_1, b_1), which leads to the
probabilities of the types and
the next guess (a_2, b_2), and
so on

We then repeat after replacing (a_1, b_1) by (a_2, b_2). Thus, the calculation of a_2
weighs the scores of the students by the likelihood that they are of type A, and
similarly for the calculation of b_2.

This heuristic is called the *soft expectation maximization algorithm*.

11.3.2 A Second Look

In the previous example, one attempts to estimate some parameter $\theta = (a, b)$ based
on some observations $\mathbf{X} = (X_1, \ldots, X_N)$. Let $\mathbf{Z} = (Z_1, \ldots, Z_N)$ where $Z_n = A$ if
student n is of type A and $Z_n = B$ otherwise.

We would like to maximize $f[\mathbf{x}|\theta]$ over θ, to find $MLE[\theta|\mathbf{X} = \mathbf{x}]$. One has

$$f[\mathbf{x}|\theta] = \sum_{\mathbf{z}} f[\mathbf{x}|\mathbf{z}, \theta] P[\mathbf{z}|\theta],$$

where the sum is over the 2^N possible values of $\mathbf{Z}$. This is computationally too
difficult.

Fig. 11.8 Hard and soft
EM?

Hard EM (Fig. 11.8) replaces the sum over **z** by

$$f[\mathbf{x}|\mathbf{z}^*, \theta] P[\mathbf{z}^*|\theta],$$

where $\mathbf{z}^*$ is the most likely value of $\mathbf{Z}$ given the observations and a current guess for
θ. That is, if the current guess is θ_k, then

$$\mathbf{z}^* = MAP[\mathbf{Z}|\mathbf{X} = \mathbf{x}, \theta_k] = \arg\max_{\mathbf{z}} P[\mathbf{Z} = \mathbf{z}|\mathbf{X} = \mathbf{x}, \theta_k].$$

The next guess is then

$$\theta_{k+1} = \arg\max_{\theta} f[\mathbf{x}|\mathbf{z}^*, \theta] P[\mathbf{z}^*|\theta].$$

Soft EM makes a different approximation. First, it replaces

$$\log(f[\mathbf{x}|\theta]) = \log\left(\sum_{\mathbf{z}} f[\mathbf{x}|\mathbf{z}, \theta] P[\mathbf{z}|\theta]\right)$$

by

$$\sum_{\mathbf{z}} \log(f[\mathbf{x}|\mathbf{z}, \theta]) P[\mathbf{z}|\theta].$$

That is, it replaces the logarithm of an expectation by the expectation of the
logarithm.

Second, it replaces the expression above by

$$\sum_{\mathbf{z}} \log(f[\mathbf{x}|\mathbf{z}, \theta]) P[\mathbf{z}|\mathbf{x}, \theta_k]$$

and the new guess θ_{k+1} is the maximizer of that expression over θ. Thus, it replaces
the distribution of $\mathbf{Z}$ by the conditional distribution given the current guess and the
observations.

If this heuristic did not work in practice, nobody would mention it. Surprisingly, it seems to work for some classes of problems. There is some theoretical justification for the heuristic. One can show that it converges to a local maximum of $f[\mathbf{x}|\theta]$. Generally, this is little comfort because most problems have many local maxima. See Roche (2012).

11.4   Learning: Hidden Markov Chain

Consider once again a hidden Markov chain model but assume that (π, P, Q) are functions of some parameter θ that we wish to estimate. We write this explicitly as $(\pi_\theta, P_\theta, Q_\theta)$. We are interested in the value of θ that makes the observed sequence $\mathbf{y}^n$ most likely.

Recall that MLE of θ given that $\mathbf{Y}^n = \mathbf{y}^n$ is defined as

$$MLE[\theta|\mathbf{Y}^n = \mathbf{y}^n] = \arg\max_\theta P[\mathbf{Y}^n = \mathbf{y}^n \mid \theta].$$

As in the discussion of clustering, we have

$$P[\mathbf{Y}^n = \mathbf{y}^n \mid \theta] = \sum_{\mathbf{x}^n} P[\mathbf{Y}^n = \mathbf{y}^n \mid \mathbf{X}^n = \mathbf{x}^n, \theta] P[\mathbf{X}^n = \mathbf{x}^n|\theta]. \qquad (11.5)$$

11.4.1 HEM

The HEM algorithm replaces the sum over $\mathbf{x}^n$ by

$$P[\mathbf{Y}^n = \mathbf{y}^n \mid \mathbf{X}^n = \mathbf{x}^n_*, \theta] P[\mathbf{X}^n = \mathbf{x}^n_*|\theta]$$

and then $P[\mathbf{X}^n = \mathbf{x}^n_*|\theta]$ by

$$P[\mathbf{X}^n = \mathbf{x}^n_*|Y^n, \theta_0],$$

where

$$\mathbf{x}^n_* = MAP[\mathbf{x}^n|\mathbf{Y}^n, \theta_0].$$

Recall that one can find $\mathbf{x}^n_*$ by using Viterbi's algorithm. Also,

$$P[\mathbf{Y}^n = \mathbf{y}^n \mid \mathbf{X}^n = \mathbf{x}^n, \theta]$$
$$= \pi_\theta(x_0) Q_\theta(x_0, y_0) Q_\theta(x_1, y_1) \times \cdots \times P_\theta(x_{n-1}, x_n) Q_\theta(x_n, y_n).$$

11.4.2 Training the Viterbi Algorithm

The Viterbi algorithm requires knowing P and Q. In practice, Q depends on the speaker and P may depend on the local dialect. (Valley speech uses more "likes" than Berkeley speakers.) We explained that if a parametric model is available, then one can use HEM.

Without a parametric model, a simple *supervised training* approach where one knows both $\mathbf{x}^n$ and $\mathbf{y}^n$ is to estimate P and Q by using empirical frequencies. For instance, the number of pairs (x_m, x_{m+1}) that are equal to (a, b) in $\mathbf{x}^n$ divided by the number of times that $x_m = a$ provides an estimate of $P(a, b)$. The estimation of Q is similar.

11.5   Summary

- Hidden Markov Chain;
- Viterbi Algorithm for $MAP[\mathbf{X}|\mathbf{Y}]$;
- Clustering and Expectation Maximization;
- EM for HMC.

11.5.1 Key Equations and Formulas

Definition of HMC	$X(n) = \text{MC} \ \& \ P[Y_n|X_n]$	D.11.1
Bellman–Ford Equations	$V_n(x) = \min_y\{d(x, y) + V_{n+1}(y)\}$	(11.3)
EM, Soft and Hard	$\theta \to \mathbf{z} \to \mathbf{x}$; Heuristics to compute $MAP[\theta|\mathbf{x}]$	S.11.3

11.6   References

The text Wainwright and Jordan (2008) is great presentation of graphical models. It covers expectation maximization and many other useful techniques.

11.7   Problems

Problem 11.1 Let (X_n, Y_n) be a hidden Markov chain. Let $Y^n = (Y_0, \ldots, Y_n)$ and $X^n = (X_0, \ldots, X_n)$. The Viterbi algorithm computes

☐ $MLE[Y^n | X^n]$;
☐ $MLE[X^n | Y^n]$;
☐ $MAP[Y^n | X^n]$;
☐ $MAP[X^n | Y^n]$.

Problem 11.2 Assume that the Markov chain X_n is such that $\mathscr{X} = \{a, b\}$, $\pi_0(a) = \pi_0(b) = 0.5$ and $P(x, x') = \alpha$ for $x \neq x'$ and $P(x, x) = 1 - \alpha$. Assume also that X_n is observed through a BSC with error probability ϵ, as shown in Fig. 11.9. Implement the Viterbi algorithm and evaluate its performance.

Problem 11.3 Suppose that the grades of students in a class are distributed as a mixture of two Gaussian distribution, $N(\mu_1, \sigma_1^2)$ with probability p and $N(\mu_2, \sigma_2^2)$ with probability $1 - p$. All the parameters $\theta = (\mu_1, \sigma_1, \mu_2, \sigma_2, p)$ are unknown.

(a) You observe n i.i.d. samples, $y_1, \ldots, y_n$ drawn from the mixed distribution. Find $f(y_1, \ldots, y_n | \theta)$.
(b) Let the type random variable X_i be 0 if $Y_i \sim N(\mu_1, \sigma_1^2)$ and 1 if $Y_i \sim N(\mu_2, \sigma_2^2)$. Find $MAP[X_i | Y_i, \theta]$.
(c) Implement Hard EM algorithm to approximately find $MLE[\theta | Y_1, \ldots, Y_n]$. To this end, use MATLAB to generate 1000 data points $(y_1, \ldots, y_{1000})$, according to $\theta = (10, 4, 30, 6, 0.4)$. Use your data to estimate θ. How well is your algorithm working?

Fig. 11.9 A simple hidden Markov chain

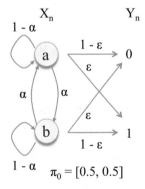

Speech Recognition: B

12

> **Topics:** Stochastic Gradient, Matching Pursuit, Compressed Sensing, Recommendation Systems

12.1 Online Linear Regression

This section explains the stochastic gradient descent algorithm, which is a technique used in many learning schemes.

Recall that a linear regression finds the parameters a and b that minimize the error

$$\sum_{k=1}^{K}(X_k - a - bY_k)^2,$$

where the (X_k, Y_k) are observed samples that are i.i.d. with some unknown distribution $f_{X,Y}(x, y)$.

Assume that, instead of calculating the linear regression based on K samples, we keep updating the parameters (a, b) every time we observe a new sample.

Our goal is to find a and b that minimize

$$E\left((X - a - bY)^2\right)$$
$$= E\left(X^2\right) + a^2 + b^2 E\left(Y^2\right) - 2aE(X) - 2bE(XY) + 2abE(Y)$$
$$=: h(a, b).$$

© The Author(s) 2021
J. Walrand, *Probability in Electrical Engineering and Computer Science*,
https://doi.org/10.1007/978-3-030-49995-2_12

One idea is to use a gradient descent algorithm to minimize $h(a, b)$. Say that at step k of the algorithm, one has calculated $(a(k), b(k))$. The gradient algorithm would update $(a(k), b(k))$ in the direction opposite of the gradient, to make $h(a(k), b(k))$ decrease. That is, the algorithm would compute

$$a(k + 1) = a(k) - \alpha \frac{\partial}{\partial a} h(a(k), b(k))$$

$$b(k + 1) = b(k) - \alpha \frac{\partial}{\partial b} h(a(k), b(k)),$$

where α is a small positive number that controls the step size. Thus,

$$a(k + 1) = a(k) - \alpha[2a(k) - 2E(X) + 2b(k)E(Y)]$$

$$b(k + 1) = b(k) - \alpha[2b(k)E(Y^2) - 2E(XY) + 2a(k)E(Y)].$$

However, we do not know the distributions and cannot compute the expected values. Instead, we replace the mean values by the values of the new samples. That is, we compute

$$a(k + 1) = a(k) - \alpha[2a(k) - 2X(k + 1) + 2b(k)Y(k + 1)]$$

$$b(k + 1) = b(k) - \alpha[2b(k)Y^2(k + 1)$$

$$- 2X(k + 1)Y(k + 1) + 2a(k)Y(k + 1)].$$

That is, instead of using the gradient algorithm we use a *stochastic gradient algorithm* where the gradient is replaced by a noisy version. The intuition is that, if the step size is small, the errors between the true gradient and its noisy version average out.

The top part of Fig. 12.1 shows the updates of this algorithm for the example (9.4) with $\alpha = 0.002$, $E(X^2) = 1$, and $E(Z^2) = 0.3$. In this example, we know that the LLSE is

$$L[X|Y] = a + bY = \frac{1}{1.3} Y = 0.77Y.$$

The figure shows that (a_k, b_k) approaches $(0, 0.77)$.

The bottom part of Fig. 12.1 shows the coefficients for (9.5) with $\gamma = 0.05$, $\alpha = 1$, and $\beta = 6$. We see that (a_k, b_k) approaches $(-1, 7)$, which are the values for the LLSE.

Fig. 12.1 The coefficients
"learned" with a stochastic
gradient algorithm for (9.4)
(top) and (9.5) (bottom)

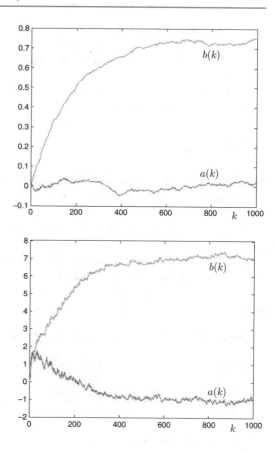

12.2   Theory of Stochastic Gradient Projection[1]

In this section, we explain the theory of the stochastic gradient algorithm that
we illustrated in the case of online regression. We start with a discussion of the
deterministic gradient projection algorithm.

Consider a smooth convex function on a convex set, such as a soup bowl. A
standard algorithm to minimize that function, i.e., to find the bottom of the bowl,
is the *gradient projection algorithm*. This algorithm is similar to going downhill by
making smaller and smaller jumps along the steepest slope. The projection makes
sure that one remains in the acceptable set. The step size of the algorithm decreases
over time so that one does not keep on overshooting the minimum.

---

[1]This algorithm is also called 'stochastic gradient descent'.

The stochastic gradient projection algorithm is similar except that one has access only to a noisy version of the gradient. As the step size gets small, the errors in the gradient tend to average out and the algorithm converges to the minimum of the function.

We first review the gradient projection algorithm and then discuss the stochastic gradient projection algorithm.

12.2.1 Gradient Projection

Consider the problem of minimizing a convex differentiable function $f(\mathbf{x})$ on a closed convex subset $\mathscr{C}$ of $\mathfrak{R}^d$. By definition, $\mathscr{C}$ is a *convex set* if

$$\theta\mathbf{x} + (1 - \theta)\mathbf{y} \in \mathscr{C}, \forall \mathbf{x}, \mathbf{y} \in \mathscr{C} \text{ and } \theta \in (0, 1). \tag{12.1}$$

That is, $\mathscr{C}$ contains the line segment between any two of its points. That is, there are no holes or kinks in the set boundary (Fig. 12.2).

Also (see Fig. 12.3), recall that a function $f : \mathscr{C} \to \mathfrak{R}$ is a *convex function* if (Fig. 12.3)

$$f(\theta\mathbf{x} + (1 - \theta)\mathbf{y}) \leq \theta f(\mathbf{x}) + (1 - \theta)f(\mathbf{y}), \forall \mathbf{x}, \mathbf{y} \in \mathscr{C} \text{ and } \theta \in (0, 1). \tag{12.2}$$

A standard algorithm is *gradient projection* (GP):

$$\mathbf{x}_{n+1} = [\mathbf{x}_n - \alpha_n \nabla f(\mathbf{x}_n)]_{\mathscr{C}}, \text{ for } n \geq 0.$$

Fig. 12.2 A non-convex set (left) and a convex set (right)

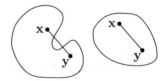

Fig. 12.3 A non-convex function (top) and a convex function (bottom)

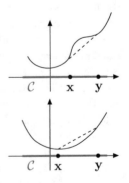

Fig. 12.4 The gradient projection algorithm (12.4) and (12.5)

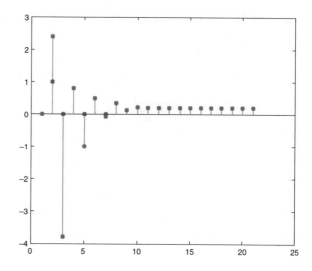

Here,

$$\nabla f(\mathbf{x}) := \left[\frac{\partial}{\partial x_1} f(\mathbf{x}), \ldots, \frac{\partial}{\partial x_d} f(\mathbf{x}) \right]'$$

is the *gradient* of $f(\cdot)$ at $\mathbf{x}$ and $[\mathbf{y}]_{\mathscr{C}}$ indicates the closest point to $\mathbf{y}$ in $\mathscr{C}$, also called the *projection* of y onto $\mathscr{C}$. The constants $\alpha_n > 0$ are called the *step sizes* of the algorithm.

As a simple example, let $f(x) = 6(x - 0.2)^2$ for $x \in \mathscr{C} := [0, 1]$. The factor 6 is there only to have big steps initially and show the necessity of projecting back into the convex set. With $\alpha_n = 1/n$ and $x_0 = 0$, the algorithm is

$$x_{n+1} = \left[x_n - \frac{12}{n}(x_n - 0.2) \right]_{\mathscr{C}}. \tag{12.3}$$

Equivalently,

$$y_{n+1} = x_n - \frac{12}{n}(x_n - 0.2) \tag{12.4}$$

$$x_{n+1} = \max\{0, \min\{1, y_{n+1}\}\} \tag{12.5}$$

with $y_0 = x_0$.

As the Fig. 12.4 shows, when the step size is large, the update y_{n+1} falls outside the set $\mathscr{C}$ and it is projected back into that set. Eventually, the updates fall into the set $\mathscr{C}$.

There are many known sufficient conditions that guarantee that the algorithm converges to the unique minimizer of $f(\cdot)$ on $\mathscr{C}$. Here is an example.

Theorem 12.1 *Assume that $f(\mathbf{x})$ is convex and differentiable on the convex set $\mathscr{C}$*
and such

$$f(x) \text{ has a unique minimizer } x^* \text{ in } \mathscr{C} \tag{12.6}$$

$$||\nabla f(x)||^2 \leq K, \forall x \in \mathscr{C} \tag{12.7}$$

$$\sum_n \alpha_n = \infty \text{ and } \sum_n \alpha_n^2 < \infty. \tag{12.8}$$

Then

$$x_n \to x^* \text{ as } n \to \infty.$$

■

Proof The idea of the proof is as follows. Let $d_n = \frac{1}{2}||x_n - x^*||^2$. Fix $\epsilon > 0$. One
shows that there is some $n_0(\epsilon)$ so that, when $n \geq n_0(\epsilon)$,

$$d_{n+1} \leq d_n - \gamma_n, \text{ if } d_n \geq \epsilon \tag{12.9}$$

$$d_{n+1} \leq 2\epsilon, \text{ if } d_n < \epsilon. \tag{12.10}$$

Moreover, in (12.9), $\gamma_n > 0$ and $\sum_n \gamma_n = \infty$.

It follows from (12.9) that, eventually, for some $n = n_1(\epsilon) \geq n_0(\epsilon)$, one has
$d_n < \epsilon$. But then, because of (12.9) and (12.10), $d_n < 2\epsilon$ for all $n \geq n_1(\epsilon)$. Since
$\epsilon > 0$ is arbitrary, this proves that $x_n \to x^*$.

To show (12.9) and (12.10), we first claim that

$$d_{n+1} \leq d_n + \alpha_n(x^* - x_n)^T \nabla f(x_n) + \frac{1}{2}\alpha_n^2 K. \tag{12.11}$$

To see this, note that

$$d_{n+1} = \frac{1}{2}||[x_n - \alpha_n \nabla f(x_n)]_{\mathscr{C}} - x^*||^2$$

$$\leq \frac{1}{2}||x_n - \alpha_n \nabla f(x_n) - x^*||^2 \tag{12.12}$$

$$\leq d_n + \alpha_n(x^* - x_n)^T \nabla f(x_n) + \frac{1}{2}\alpha_n^2 K. \tag{12.13}$$

The inequality in (12.12) comes from the fact that projection on a convex set is
non-expansive. That is,

$$||x_{\mathscr{C}} - y_{\mathscr{C}}|| \leq ||x - y||.$$

Fig. 12.5 Projection on a convex set is non-expansive

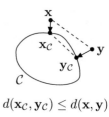

$$d(\mathbf{x}_C, \mathbf{y}_C) \le d(\mathbf{x}, \mathbf{y})$$

Fig. 12.6 The inequality (12.14)

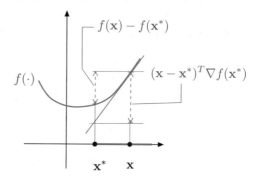

This property is clear from a picture (see Fig. 12.5) and is not difficult to prove.

Observe that $\alpha_n \to 0$, because $\sum_n \alpha_n^2 < \infty$. Hence, (12.13) and (12.7) imply (12.10).

It remains to show (12.9). As Fig. 12.6 shows, the convexity of $f(\cdot)$ implies that

$$(x^* - x)^T \nabla f(x) \le f(x^*) - f(x). \tag{12.14}$$

Also, if $d_n \ge \epsilon$, one has $f(x^*) - f(x_n) \le -\delta(\epsilon)$, for some $\delta(\epsilon) > 0$. Thus, whenever $d_n \ge \epsilon$, one has

$$(x^* - x_n)^T \nabla f(x_n) \le -\delta(\epsilon).$$

Together with (12.11), this implies

$$d_{n+1} \le d_n - \alpha_n \delta(\epsilon) + \frac{1}{2}\alpha_n^2 K.$$

Now, let

$$\gamma_n = \alpha_n \delta(\epsilon) - \frac{1}{2}\alpha_n^2 K. \tag{12.15}$$

Since $\alpha_n \to 0$, there is some $n_2(\epsilon)$ such that $\gamma_n > 0$ for $n \geq n_2(\epsilon)$. Moreover, (12.8) is seen to imply that $\sum_n \gamma_n = \infty$. This proves (12.9) after replacing $n_0(\epsilon)$ by $\max\{n_0(\epsilon), n_2(\epsilon)\}$. $\square$

12.2.2 Stochastic Gradient Projection

There are many situations where one cannot measure directly the gradient $\nabla f(\mathbf{x}_n)$ of the function. Instead, one has access to a random estimate of that gradient, $\nabla f(\mathbf{x}_n) + \eta_n$, where η_n is a random variable. One hopes that, if the error η_n is small enough, GP still converges to x^* when one uses $\nabla f(\mathbf{x}_n) + \eta_n$ instead of $\nabla f(\mathbf{x}_n)$. The point of this section is to justify this hope.

The algorithm is as follows (see Fig. 12.7):

$$\mathbf{x}_{n+1} = [\mathbf{x}_n - \alpha_n \mathbf{g}_n]_{\mathscr{C}}, \tag{12.16}$$

where

$$\mathbf{g}_n = \nabla f(\mathbf{x}_n) + \mathbf{z}_n + \mathbf{b}_n \tag{12.17}$$

is a noisy estimate of the gradient. In (12.17), z_n is a zero-mean random variable that models the estimation noise and b_n is a constant that models the estimation bias.

As a simple example, let $f(x) = 6(x - 0.2)^2$ for $x \in \mathscr{C} := [0, 1]$. With $\alpha_n = 1/n$, $b_n = 0$, and $x_0 = 0$, the algorithm is

$$x_{n+1} = \left[x_n - \frac{12}{n}(x_n - 0.2 + z_n)\right]_{\mathscr{C}}. \tag{12.18}$$

In this expression, the z_n are i.i.d. $U[-0.5, 0.5]$. Figure 12.8 shows the values that the algorithm produces.

Fig. 12.7 The figure shows level curves of $f(\cdot)$ and the convex set $\mathscr{C}$. It also shows the first few iterations of GPA in red and of SGPA in blue

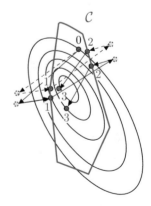

Fig. 12.8 The stochastic gradient projection algorithm (12.18)

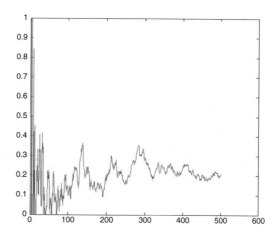

This algorithm converges to the minimum $x^* = 0.2$ of the function, albeit slowly. For the algorithm (12.16) and (12.17) to converge, one needs the estimation noise z_n and bias b_n to be small. Specifically, one has the following result.

Theorem 12.2   *Assume that $\mathscr{C}$ is bounded and*

$$f(.) \text{ has a unique minimizer } \mathbf{x}^* \text{ in } \mathscr{C}; \tag{12.19}$$

$$||\nabla f(\mathbf{x})||^2 \leq K, \forall x \in \mathscr{C}; \tag{12.20}$$

$$\alpha_n > 0, \sum_n \alpha_n = \infty, \sum_n \alpha_n^2 < \infty. \tag{12.21}$$

In addition, assume that

$$\sum_{n=0}^{\infty} \alpha_n ||\mathbf{b}_n|| < \infty; \tag{12.22}$$

$$E[z_{n+1} \mid z_0, z_1, \ldots, z_n] = 0; \tag{12.23}$$

$$E(||z_n||^2) \leq A, n \geq 0. \tag{12.24}$$

Then $\mathbf{x}_n \to \mathbf{x}^*$ with probability one.

■

Proof   The proof is essentially the same as for the deterministic case. The inequality (12.11) becomes

$$d_{n+1} \leq d_n + \alpha_n (\mathbf{x}^* - \mathbf{x}_n)^T [\nabla f(\mathbf{x}_n) + \mathbf{z}_n + \mathbf{b}_n] + \frac{1}{2} \alpha_n^2 K. \tag{12.25}$$

Accordingly, γ_n in (12.15) is replaced by

$$\gamma_n = \alpha_n \left[\delta(\epsilon) + (\mathbf{x}^* - \mathbf{x}_n)^T (\mathbf{z}_n + \mathbf{b}_n) \right] - \frac{1}{2} \alpha_n^2 K. \qquad (12.26)$$

Now, (12.23) implies that $\mathbf{v}_n := \sum_{m=0}^{n} \alpha_m \mathbf{z}_m$ is a martingale.[2] Because of (12.24) and (12.21), one has $E(||\mathbf{v}_n||^2) \leq A \sum_{m=0}^{\infty} \alpha_m < \infty$ for all n. This implies, by the *Martingale Convergence Theorem* 12.3, that v_n converges to a finite random variable. Combining this fact with (12.22) shows that[3] $\sum_{m=n}^{\infty} \alpha_m [\mathbf{z}_m + \mathbf{b}_m] \rightarrow 0$. Since $||\mathbf{x}_n - \mathbf{x}^*||$ is bounded, this implies that the effect of the estimation error is asymptotically negligible and that argument used in the proof of GP applies here.   □

12.2.3 Martingale Convergence

We discuss the theory of martingales in Sect. 15.9. Here are the ideas we needed in the proof of Theorem 12.2.

Let $\{x_n, y_n, n \geq 0\}$ be random variables such that $E(x_n)$ is well-defined for all n. The sequence x_n is said to be a *martingale* with respect to $\{(x_m, y_m), m \geq 0\}$ if

$$E[x_{n+1} | x_m, y_m, m \leq n] = x_n, \forall n.$$

Theorem 12.3 (Martingale Convergence Theorem) *If a martingale x_n is such that $E(x_n^2) \leq B < \infty$ for all n, then it converges with probability one to a finite random variable.*

■

For a proof, see Theorem 15.13.

12.3   Big Data

The web makes it easy to collect a vast amount of data from many sources. Examples include books, movie, and restaurants that people like, website that they visit, their mobility patterns, their medical history, and measurements from sensors. This data

---

[2] See the next section.

[3] Recall that if a series $\sum_n w_n$ converges, then the tail $\sum_{m \geq n} w_m$ of the series converges to zero as $n \rightarrow \infty$.

Fig. 12.9 The web provides access to vast amounts of data. How does one extract useful knowledge from that data?

can be useful to recommend items that people will probably like, treatments that are likely to be effective, people you might want to commute with, to discover who talks to who, efficient management techniques, and so on. Moreover, new technologies for storage, databases, and cloud computing make it possible to process huge amounts of data. This section explains a few of the formulations of such problems and algorithms to solve them (Fig. 12.9).

12.3.1 Relevant Data

Many factors potentially affect an outcome, but what are the most relevant ones? For instance, the success in college of a student is correlated with her high-school GPA, her scores in advanced placement courses and standardized tests. How does one discover the factors that best predict her success? A similar situation occurs for predicting the odds of getting a particular disease, the likelihood of success of a medical treatment, and many other applications.

Identifying these important factors can be most useful to improve outcomes. For instance, if one discovers that the odds of success in college are most affected by the number of books that a student has to read in high-school and by the number of hours she spends playing computer games, then one may be able to suggest strategies for improving the odds of success.

One formulation of the problem is that the outcome Y is correlated with a collection of factors that we represent by a vector $\mathbf{X}$ with $N \gg 1$ components. For instance, if Y is the GPA after 4 years in college, the first component X_1 of $\mathbf{X}$ might indicate the high-school GPA, the second component X_2 the score on a specific standardized test, X_3 the number of books the student had to write reports on, and so on. Intuition suggests that, although $N \gg 1$, only relatively few of the components of $\mathbf{X}$ really affect the outcome Y in a significant way. However, we do not want to presume that we know what these components are.

Say that you want to predict Y on the basis of six components of $\mathbf{X}$. Which ones should you consider? This problem turns out to be hard because there are many (about $N^6/6!$) subsets with 6 elements in $\mathcal{N} = \{1, 2, \ldots, N\}$, and this combinatorial aspect of the problem makes it intractable when N is large. To

Fig. 12.10

make progress, we change the formulation slightly and resort to some heuristic (Fig. 12.10).[4]

The change in formulation is to consider the problem of minimizing

$$J(\mathbf{b}) = E\left((Y - \sum_n b_n X_n)^2\right)$$

over $\mathbf{b} = (b_1, \ldots, b_N)$, subject to a bound on

$$C(\mathbf{b}) = \sum_n |b_n|.$$

This is called the *LASSO* problem, for "least absolute shrinkage and selection operator." Thus, the hard constraint on the number of components is replaced by a cost for using large coefficients. Intuitively, the problem is still qualitatively similar. Also, the constraint is such that the solution of the problem has many b_n equal to zero. Intuitively, if a component is less useful than others, its coefficient is probably equal to zero in the solution.

One interpretation of this problem as follows. In order to simplify the algebra, we assume that Y and $\mathbf{X}$ are zero-mean. Assume that

$$Y = \sum_n B_n X_n + Z,$$

where Z is $\mathcal{N}(0, \sigma^2)$ and the coefficients B_n are random and independent with a prior distribution of B_n given by

$$f_n(b) = \frac{\lambda}{2} \exp\{-\lambda |b|\}.$$

Then

---

[4]If you cannot crack a nut, look for another one. (A difference between Engineering and Mathematics?)

$$MAP[\mathbf{B}|\mathbf{X} = \mathbf{x}, Y = y] = \arg\max_{\mathbf{b}} f_{\mathbf{B}|\mathbf{X},Y}[\mathbf{b}|\mathbf{x}, y]$$

$$= \arg\max_{\mathbf{b}} f_{\mathbf{B}}(\mathbf{b}) f_{Y|\mathbf{X}}[y|\mathbf{x}]$$

$$= \arg\max_{\mathbf{b}} \exp\left\{ -\frac{1}{2\sigma^2}\left(y - \sum_n b_n x_n \right)^2 \right\}$$

$$\times \exp\{-\lambda \sum_n |b_n|\}$$

$$= \arg\min_{\mathbf{b}} \left\{ \left(y - \sum_n b_n x_n \right)^2 + \mu \sum_n |b_n| \right\}$$

with $\mu = 2\lambda\sigma^2$. This formulation is the Lagrange multiplier formulation of the LASSO problem where the constraint on the cost $C(\mathbf{b})$ is replaced by a penalty $\mu C(\mathbf{b})$. Thus, the LASSO problem is equivalent to finding $MAP[\mathbf{B}|\mathbf{X}, Y]$ under the assumptions stated above.

We explain a greedy algorithm that selects the components one by one, trying to maximize the progress that it makes with each selection. First assume that we can choose only one component X_n among the N elements in $\mathbf{X}$. We know that

$$L[Y|X_n] = \frac{\text{cov}(Y, X_n)}{\text{var}(X_n)} X_n =: b_n X_n$$

and

$$E((Y - L[Y|X_n])^2) = \text{var}(Y) - \frac{\text{cov}(Y, X_n)^2}{\text{var}(X_n)}$$

$$= \text{var}(Y) - |\text{cov}(Y, X_n)| \times |b_n|.$$

Thus, one unit of "cost" $C(b_n) = |b_n|$ invested in b_n brings a reduction $|\text{cov}(Y, X_n)|$ in the objective $J(b_n)$. It then makes sense to choose the first component with the largest value of "reward per unit cost" $|\text{cov}(Y, X_n)|$. Say that this component is X_1 and let $\hat{Y}_1 = L[Y|X_1]$.

Second, assume that we stick to our choice of X_1 with coefficient b_1 and that we look for a second component X_n with $n \neq 1$ to add to our estimate. Note that

$$E((Y - b_1 X_1 - b_n X_n)^2)$$

$$= E((Y - b_1 X_1)^2) - 2b_n \text{cov}(Y - b_1 X_1, X_n) + b_n^2 \text{var}((X_n)).$$

This expression is minimized over b_n by choosing

$$b_n = \frac{\text{cov}(Y - b_1 X_1, X_n)}{\text{var}(X_n)}$$

and it is then equal to

$$E((Y - b_1 X_1)^2) - \frac{\text{cov}(Y - b_1 X_1, X_n)^2}{\text{var}(X_n)}.$$

Thus, as before, one unit of additional cost in $C(b_1, b_n)$ invested in b_n brings a reduction

$$|\text{cov}(Y - b_1 X_1, X_n)|$$

in the cost $J(b_1, b_n)$. This suggests that the second component X_n to pick should be the one with the largest covariance with $Y - b_1 X_1$.

These observations suggest the following algorithm, called the *stepwise regression algorithm*. At each step k, the algorithm finds the component X_n that is most correlated with the residual error $Y - \hat{Y}_k$, where $\hat{Y}_k$ is the current estimate. Specifically, the algorithm is as follows:

Step 0 : $\hat{Y}_0 = E(Y)$ and $S_0 = \emptyset$;

Step $k + 1$: Find $n \notin S_k$ that maximizes $E((Y - \hat{Y}_k) X_n)$

        Let $S_{k+1} = S_k \cup \{n\}$, $Y_{k+1} = L[Y | X_n, n \in S_{k+1}]$, $k = k + 1$;

Repeat until $E((Y - \hat{Y}_k)^2) \leq \epsilon$.

In practice, one is given a collection of outcomes $\{Y^m, m = 1, \ldots, M\}$ of with factors $\mathbf{X}^m = (X_1^m, X_2^m, \ldots, X_N^m)$. Here, each m corresponds to one sample, say one student in the college success example. From those samples, one can estimate the mean values by the sample means. Thus, in step k, one has calculated coefficients $(b_1, \ldots, b_k)$ to calculate

$$\hat{Y}_k^m = b_1 X_1^m + \cdots + b_k X_k^m.$$

One then estimates $E((Y - \hat{Y}_k) X_n)$ by

$$\frac{1}{M} \sum_{m=1}^{M} (Y^m - \hat{Y}_k^m) X_n^m.$$

Also, one approximates $L[Y | X_n, n \in S_{k+1}]$ by the linear regression.

It is useful to note that, by the Law of Large Numbers, the number M of samples needed to estimate the means and covariances is not necessarily very large. Thus, although one may have data about millions of students, a reasonable estimate may be obtained from a few thousand. Recall that one can use the sample moments to compute confidence intervals for these estimates.

Signal processing uses a similar algorithm called *matching pursuit* introduced in Mallat and Zhang (1993). In that context, the problem is to find a compact representation of a signal, such as a picture or a sound. One considers a representation of the signal as a linear combination of basis functions. The matching pursuit algorithm finds the most important basis functions to use in the representation.

An Example
Our example is very small, so that we can understand the steps. We assume that all the random variables are zero-mean and that $N = 3$ with

$$
\Sigma_{\mathbf{Z}} =
\begin{bmatrix}
4 & 3 & 2 & 2 \\
3 & 4 & 2 & 2 \\
2 & 2 & 4 & 1 \\
2 & 2 & 1 & 4
\end{bmatrix},
$$

where $\mathbf{Z}' = (Y, X_1, X_2, X_3) = (Y, \mathbf{X}')$.

We first try the stepwise regression. The component X_n most correlated with Y is X_1. Thus,

$$
\hat{Y}_1 = L[Y|X_1] = \frac{\mathrm{cov}(Y, X_1)}{\mathrm{var}(X_1)} X_1 = \frac{3}{4} X_1 =: b_1 X_1.
$$

The next step is to compute the correlations $E(X_n(Y - \hat{Y}_1))$ for $n = 2, 3$. We find

$$
E(X_2(Y - \hat{Y}_1)) = E(X_2(Y - b_1 X_1)) = 2 - 2b_1 = 0.5
$$
$$
E(X_3(Y - \hat{Y}_1)) = E(X_3(Y - b_1 X_1)) = 2 - 2b_1 = 0.5.
$$

Hence, the algorithm selects X_2 as the next components and one finds

$$
\hat{Y}_2 = L[Y|X_1, X_2] = \begin{bmatrix} 3 & 2 \end{bmatrix} \begin{bmatrix} 4 & 2 \\ 2 & 4 \end{bmatrix}^{-1} \begin{bmatrix} X_1 \\ X_2 \end{bmatrix} = \frac{2}{3} X_1 + \frac{1}{6} X_2.
$$

The resulting error variance is

$$
E\left((Y - \hat{Y}_2)^2\right) = \frac{5}{3}.
$$

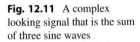

Fig. 12.11 A complex looking signal that is the sum of three sine waves

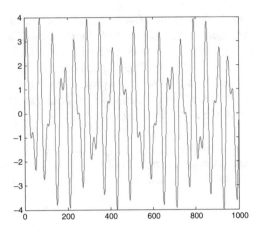

12.3.2 Compressed Sensing

Complex looking objects may have a simple hidden structure. For example, the signal $s(t)$ shown in Fig. 12.11 is the sum of three sine waves. That is,

$$s(t) = \sum_{i=1}^{3} b_i \sin(2\pi \phi_i t), t \geq 0. \tag{12.27}$$

A classical result, called the *Nyquist sampling theorem*, states that one can reconstruct a signal exactly from its values measured every T seconds, provided that $1/T$ is at least twice the largest frequency in the signal. According to that result, we could reconstruct $s(t)$ by specifying its value every T seconds if $T < 1/(2\phi_i)$ for $i = 1, 2, 3$. However, in the case of (12.27), one can describe $s(t)$ completely by specifying the values of the six parameters $\{b_i, \phi_i, i = 1, 2, 3\}$. Also, it seems clear in this particular case that one does not need to know many sample values $s(t_k)$ for different times t_k to be able to reconstruct the six parameters and therefore the signal $s(t)$ for all $t \geq 0$. Moreover, one expects the reconstruction to be unique if we choose a few sampling times t_k randomly. The same is true if the representation is in terms of different functions, such as polynomials or wavelets.

This example suggests that if a signal has a simple representation in terms of some basis functions (e.g., sine waves), then it is possible to reconstruct it exactly from a small number of samples.

Computing the parameters of (12.27) from a number of samples $s(t_k)$ is highly nontrivial, so that the fact that it is possible does not seem very useful. However, a slightly different perspective shows that the problem can be solved. Assume that we have a collection of functions (Fig. 12.12)

$$g_n(t) = \sin(2\pi f_n t), t \geq 0, n = 1, \ldots, N.$$

Fig. 12.12 A tough nut to crack!

Assume also that the frequencies $\{\phi_1, \phi_2, \phi_3\}$ in $s(t)$ are in the collection $\{f_n, n = 1, \ldots, N\}$. We can then try to find the vector $\mathbf{a} = \{a_n, n = 1, \ldots, N\}$ such that

$$s(t_k) = \sum_{n=1}^{N} a_n g_n(t_k), \text{ for } k = 1, \ldots, K.$$

We should be able to do this with three functions, by choosing the appropriate coefficients. How do we do this systematically? A first idea is to formulate the following problem:

$$\text{Minimize } \sum_{n} 1\{a_n \neq 0\}$$

$$\text{such that } s(t_k) = \sum_{n} a_n g_n(t_k), \text{ for } k = 1, \ldots, K.$$

That is, one tries to find the most economical representation of $s(t)$ as a linear combination of functions in the collection.

Unfortunately, this problem is intractable because of the number of choices of sets of nonzero coefficients a_n, a difficulty we already faced in the previous section. The key trick is, as before, to convert the problem into a much easier one that retains the main goal.

The new problem is as follows:

$$\text{Minimize } \sum_{n} |a_n|$$

$$\text{such that } s(t_k) = \sum_{n} a_n g_n(t_k), \text{ for } k = 1, \ldots, K.$$

$$(12.28)$$

Trying to minimize the sum of the absolute values of the coefficients a_n is a relaxation of limiting the number of nonzero coefficients. (Simple examples show that choosing $\sum_n |a_n|^2$ instead of $\sum_n |a_n|$ often leads to bad reconstructions.) The result is that if K is large enough, then the solution is exact with a high probability.

Theorem 12.4 (Exact Recovery from Random Samples)   *The signal $s(t)$ can be recovered exactly with a very high probability from K samples by solving (12.28) if*

$$K \geq C \times B \times \log(N).$$

In this expression, C is a small constant, B is the number of sine waves that make up $s(t)$, and N is the number of sine waves in the collection.

■

Note that this is a probabilistic statement. Indeed, one could be unlucky and choose sampling times t_k, where $s(t_k) = 0$ (see Fig. 12.11) and these samples would not enable the reconstruction of $s(t)$. More generally, the samples could be chosen so that they do not enable an exact reconstruction. The theorem says that the probability of poor samples is very small.

Thus, in our example, where $B = 3$, one can expect to recover the signal $s(t)$ exactly from about $3 \log(100) \approx 14$ samples if $N \leq 100$.

Problem (12.28) is equivalent to the following linear programming problem, which implies that it is easy to solve:

$$\text{Minimize } \sum_n b_n$$

$$\text{such that } s(t_k) = \sum_n a_n g_n(t_k), \text{ for } k = 1, \ldots, K$$

$$\text{and } -b_n \leq a_n \leq b_n, \text{ for } n = 1, \ldots, N. \tag{12.29}$$

Assume that

$$s(t) = \sin(2\pi t) + 2\sin(2.4\pi t) + 3\sin(3.2\pi t), t \in [0, 1]. \tag{12.30}$$

The frequencies in $s(t)$ are $\phi_1 = 1, \phi_2 = 1.2$, and $\phi_3 = 1.6$. The collection of functions is

$$\{g_n(t) = \sin(2\pi f_n t), n = 1, \ldots, 100\},$$

where $f_n = n/10$.

The frequencies of the sine waves in the collection are $0.1, 0.2, \ldots, 10$. Thus, the frequencies in $s(t)$ are contained in the collection, so that perfect reconstruction is possible as

$$s(t) = \sum_n a_n g_n(t)$$

with $a_{10} = 1$, $a_{12} = 2$, and $a_{16} = 3$, and all the other coefficients a_n equal to zero. The theory tells us that reconstruction should be possible with about 14 samples. We choose 15 sampling times t_k randomly and uniformly in [0, 1]. We then ask Python to solve (12.29). The solution is shown in Fig. 12.13.

Another Example

Figure 12.14, from Candes and Romberg (2007), shows another example. The image on top has about one million pixels. However, it can be represented as a linear combination of 25,000 functions called wavelets. Thus, the compressed sensing results tell us that one should be able to reconstruct the picture exactly from a small

Fig. 12.13 Exact reconstruction of the signal (12.30) with 15 samples chosen uniformly in [0, 1]. The signal is in green and the reconstruction in blue

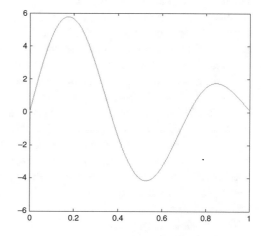

Fig. 12.14 Original image with 10^6 pixels (top) and reconstruction from 96,000 randomly chosen pixels (bottom)

multiple of 25,000 randomly chosen pixels. It turns out that this is indeed the case with about 96,000 pixels.

12.3.3 Recommendation Systems

Which movie would you like to watch? One formulation of the problem is as follows. There is a $K \times N$ matrix Y. The entry $Y(k, n)$ of the matrix indicates how much user k likes movie n. However, one does not get to observe the complete matrix. Instead, one observes a number of entries, when users actually watch movies and one gets to record their rankings. The problem is to complete the matrix to be able to recommend movies to users.

This *matrix completion* is based on the idea that the entries of the matrix are not independent. For instance, assume that Bob and Alice have seen the same five movies and gave them the same ranking. Assume that Bob has seen another movie he loved. Chances are that Alice would also like it.

To formulate this dependency of the entries of the matrix Y, one observes that even though there are thousands of movies, a few factors govern how much users like them. Thus, it is reasonable to expect that many columns of the matrix are combinations of a few common vectors that correspond to the hidden factors that influence the rankings by users. Thus, a few independent vectors get combined into linear combinations that form the columns. Consequently the matrix Y has a small number of linearly independent columns, i.e., it is a *low rank* matrix.[5] This observation leads to the question of whether one can recover a low rank matrix Y from observed entries?

One possible formulation is

$$\text{Minimize rank}(X) \text{ s.t. } X(k, n) = M(k, n), \forall (k, n) \in \Omega.$$

Here, $\{M(k, n), (k, n) \in \Omega\}$ is the set of observed entries of the matrix. Thus, one wishes to find the lowest-rank matrix X that is consistent with the observed entries.

As before, such a problem is hard. To simplify the problem, one replaces the rank by the *nuclear norm* $||X||_*$ where

$$||X||_* = \sum_i \sigma_i,$$

where the σ_i are the singular values of the matrix X. The rank of the matrix counts the number of nonzero singular values. The nuclear norm is a convex function of the entries of the matrix, which makes the problem a convex programming problem that is easy to solve. Remarkably, as in the case of compressed sensing, the solution of the modified problem is very good.

---

[5]The rank of a matrix is the number of linearly independent columns.

Theorem 12.5 (Exact Matrix Completion from Random Entries) *The solution of the problem*

$$Minimize \; ||X||_* \; s.t. \; X(k, n) = M(k, n), \forall (k, n) \in \Omega$$

is the matrix Y with a very high probability if the observed entries are chosen uniformly at random and if there are at least

$$Cn^{1.25}r \log(n)$$

observations. In this expression, C is a small constant, $n = \max\{K, N\}$, and r is the rank of Y.

■

This result is useful in many situations where this number of required observations is much smaller than $K \times N$, which is the number of entries of Y. The reference contains many extensions of these results and details on numerical solutions.

12.4   Deep Neural Networks

Deep neural networks (DNN) are electronic processing circuits inspired by the structure of the brain. For instance, our vision system consists of layers. The first layer is in the retina that captures the intensity and color of zones in our field of vision. The next layer extracts edges and motion. The brain receives these signals and extracts higher level features. A simplistic model of this processing is that the neurons are arranged in successive layers, where each neuron in one layer gets inputs from neurons in the previous layer through connections called synapses. Presumably, the weights of these connections get tuned as we grow up and learn to perform tasks, possibly by trial and errors. The figure sketches a DNN. The inputs at the left of the DNN are the features X from which the system produces the probability that X corresponds to a dog, or the estimate of some quantity (Fig. 12.15).

Fig. 12.15 A neural network

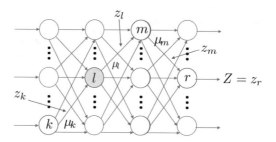

Fig. 12.16 The logistic function

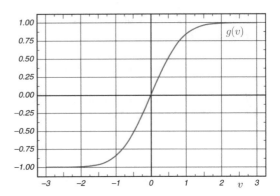

Each circle is a circuit that we call a neuron. In the figure, z_k is the output of neuron k. It is multiplied by θ_k to contribute the quantity $\theta_k z_k$ to the total input V_l of neuron l. The parameter θ_k represents the strength of the connection between neuron k and neuron l. Thus, $V_l = \sum_n \theta_n z_n$, where the sum is over all the neurons n of the layer to the immediate left of neuron l, including neuron k. The output z_l of neuron l is equal to $f(a_l, V_l)$, where a_l is a parameter specific to that neuron and f is some function that we discuss later.

With this structure, it is easy to compute the derivative of some output Z with respect to some weight, say θ_k. We do it in the last section of this chapter.

What should be the functions $f(a, V)$? Inspired by the idea that a neuron fires if it is excited enough, one may use a function $f(a, V)$ that is close to 1 if $V > a$ and close to -1 if $V < a$. To make the function differentiable, one may use $f(a, V) = g(V - a)$ with

$$g(v) = \frac{2}{1 + e^{-\beta v}} - 1,$$

where β is a positive constant. If β is large, then $e^{-\beta v}$ goes from a very large to a very small value when v goes from negative to positive. Consequently, $g(v)$ goes from -1 to $+1$ (Fig. 12.16).

The DNN is able to model many functions by adjusting its parameters. To see why, consider neuron l. The output of this neuron indicates whether the linear combination $V_l = \sum_n \theta_n z_n$ is larger or smaller than the thresholds a_l of the neurons. Consequently, the first layer divides the set of inputs into regions separated by hyperplanes. The next layer then further divides these regions. The number of regions that can be obtained by this process is exponential in the number of layers. The final layer then assigns values to the regions, thus approximating a complex function of the input vector by an almost piecewise constant function.

The missing piece of the puzzle is that, unfortunately, the cost function is not a nice convex function of the parameters of the DNN. Instead, it typically has many local minima. Consequently, by using the SGD algorithm, the tuning of the DNN may get stuck in a local minimum. Also, to reduce the number of parameters to tune, one usually selects a few layers with fixed parameters, such as edge detectors in vision systems. Thus, the selection of the DNN becomes somewhat of an art, like cooking.

Thus, it remains impossible to predict whether the DNN will be a good technique for machine learning in a specific application. The answer of the practitioners is to try and see. If it works, they publish a paper. We are far from the proven convergence results of adaptive systems. Ah, nostalgia. . . .

There is a worrisome aspect to these black-box approaches. When the DNN has been tuned and seems to perform well on many trials, not only one does not understand what it really does, but one has no guarantee that it will not seriously misbehave for some inputs. Imagine then a killer drone with a DNN target recognition system. . . . It is not surprising that a number of serious scientists have raised concerns about "artificial stupidity" and the need to build safeguards into such systems. "Open the pod bay doors, Hal."

12.4.1 Calculating Derivatives

Let's compute the derivative of Z with respect to θ_k.

See you increase θ_k by ϵ. This increases V_l by ϵz_k. In turn, this increases z_l by $\delta z_l := \epsilon z_k f'(a_l, V_l)$, where $f'(a_l, V_l)$ is the derivative of $f(a_l, V_l)$ with respect to V_l. Consequently, this increases V_m by $\theta_l \delta z_l$. The result is an increase of z_m by $\delta z_m = \theta_l \delta z_l f'(a_m, V_m)$. Finally, this increase V_r by $\theta_m \delta z_m$ and Z by $\theta_m \delta z_m f'(a_r, V_r)$. We conclude that

$$\frac{dZ}{d\theta_k} = z_k f'(a_l, V_l)\theta_l f'(a_m, V_m)\theta_m f'(a_r, V_r).$$

The details do not matter too much. The point is that the structure of the network makes the calculation of the derivatives straightforward.

12.5   Summary

- Online Linear Regression;
- Convex Sets and Functions;
- Gradient Projection Algorithm;
- Stochastic Gradient Projection Algorithm;
- Deep Neural Networks;
- Martingale Convergence Theorem;
- Big Data: Relevant Data, Compressed Sensing, Recommendation Systems.

12.5.1 Key Equations and Formulas

Convex Set	if it contains its chords	(12.1)
Convex Function	if it is above its tangents	(12.2)
Convergence of GP	if unique minimizer and bounded gradient	T.12.1
Convergence of SGP	if bounded drift and noise variance	T.12.2
Martingale CT	L^1 or L^2-bounded MG converges w.p. 1	T.12.3

12.6   References

Online linear regression algorithms are discussed in Strehl and Littman (2007). The book Bertsekas and Tsitsiklis (1989) is an excellent presentation of distributed optimization algorithms. It explains the gradient projection algorithm and distributed implementations. The LASSO algorithm and many other methods are clearly explained in Hastie et al. (2009), together with applications. The theory of martingales is nicely presented by its father in Doob (1953). Theorem 12.4 is from Candes and Romberg (2007).

12.7   Problems

Problem 12.1 Let $\{Y_n, n \geq 1\}$ be i.i.d. $U[0, 1]$ random variables and $\{Z_n, n \geq 1\}$ be i.i.d. $\mathcal{N}(0, 1)$ random variables. Define $X_n = 1\{Y_n \geq a\} + Z_n$ for some constant a. The goal of the problem is to design an algorithm that "learns" the value of a from the observation of pairs (X_n, Y_n). We construct a model

Fig. 12.17 The logistic function (12.31) with $\lambda = 10$

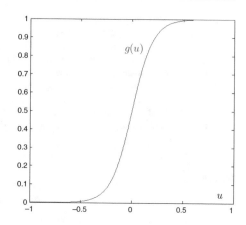

$$X_n = g(Y_n - \theta),$$

where

$$g(u) = \frac{1}{1 + \exp\{-\lambda u\}} \tag{12.31}$$

with $\lambda = 10$. Note that when $u > 0$, the denominator of $g(u)$ is close to 1, so that $g(u) \approx 1$. Also, when $u < 0$, the denominator is large and $g(u) \approx 0$. Thus, $g(u) \approx 1\{u \geq 0\}$. The function $g(\cdot)$ is called the *logistic function*. Use SGD in Python to estimate θ (Fig. 12.17).

Problem 12.2 Implement the stepwise regression algorithm with

$$\Sigma_{\mathbf{Z}} = \begin{bmatrix} 10 & 5 & 6 & 7 \\ 5 & 6 & 5 & 2 \\ 6 & 5 & 11 & 5 \\ 7 & 2 & 5 & 6 \end{bmatrix},$$

where $\mathbf{Z}' = (Y, X_1, X_2, X_3) = (Y, \mathbf{X}')$.

Problem 12.3 Implement the compressed sensing algorithm with

$$s(t) = 3\sin(2\pi t) + 2\sin(3\pi t) + 4\sin(4\pi t), t \in [0, 1],$$

where you choose sampling times t_k independently and uniformly in $[0, 1]$. Assume that the collection of sine waves has the frequencies $\{0.1, 0.2, \ldots, 3\}$.

What is the minimum number of samples that you need for exact reconstruction?

Route Planning: A

13

Application: Choosing a fast route given uncertain delays, Controlling a Markov chain

Topics: Stochastic Dynamic Programming, Markov Decision Problems

13.1   Model

One is given a finite connected directed graph. Each edge (i, j) is associated with a travel time $T(i, j)$. The travel times are *independent* and have known *distributions*. There are a start node s and a destination node d. The goal is to choose a fast route from s to d. We consider a few different formulations (Fig. 13.1).

To make the situation concrete, we consider the very simple example illustrated in Fig. 13.2.

The goal is to choose the fastest path from s to d. In this example, the possible paths are sd, sad, and $sabd$. We assume that the delays $T(i, j)$ on the edges (i, j) are as follows:

$$T(s, a) =_D U[5, 13], T(a, d) = 10, T(a, b) =_D U[2, 10],$$

$$T(b, d) = 4, T(s, d) = 20.$$

Thus, the delay from s to a is uniformly distributed in $[5, 13]$, the delay from a to d is equal to 10, and so on. The delays are assumed to be independent, which is an unrealistic simplification.

© The Author(s) 2021
J. Walrand, *Probability in Electrical Engineering and Computer Science*,
https://doi.org/10.1007/978-3-030-49995-2_13

Fig. 13.1 Road network. How to select a path?

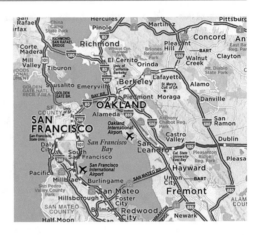

Fig. 13.2 A simple graph

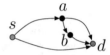

13.2   Formulation 1: Pre-planning

In this formulation, one does not observe anything and one plans the journey ahead of time. In this case, the solution is to look at the average travel times $E(T(i, j)) = c(i, j)$ and to run a shortest path algorithm.

For our example, the average delays are $c(s, a) = 9$, $c(a, d) = 10$, and so on, as shown in the top part of Fig. 13.3.

Let $V(i)$ be the minimum average travel time from node i to the destination d. The *Bellman–Ford Algorithm* calculates these values as follows. Let $V_n(i)$ be an estimate of the shortest average travel time from i to d, as calculated after the n-th iteration of the algorithm. The algorithm starts with $V_0(d) = 0$ and $V_0(i) = \infty$ for $i \neq d$. Then, the algorithm calculates

$$V_{n+1}(i) = \min_j\{c(i, j) + V_n(j)\}, n \geq 0. \tag{13.1}$$

The interpretation is that $V_n(i)$ is the minimum expected travel time from i to d over all paths that go through at most n edges. The distance is infinite if no path with at most n edges reaches the destination d. This is exactly the same algorithm we discussed in Sect. 11.2 to develop the Viterbi algorithm.

These relations are justified by the fact that the mean value of a sum is the sum of the mean values. For instance, say that the minimum average travel time from a to d using a path that has at most 2 edges is $V_2(a, d)$ and it corresponds to a path with

Fig. 13.3 The average
delays (top) and the
successive steps of the
Bellman–Ford algorithm to
calculate the minimum
expected times (shown in red)
from the nodes to the
destination

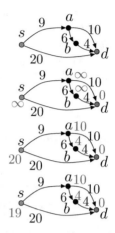

random travel time $W_2(a, d)$. Then, the minimum average travel time from s to d using a path that has at most 3 edges follows either the direct path sd, that has travel time $T(s, d)$, or the edge sa followed by the fastest path from a to d that uses at most 2 edges with travel time $W_2(a, d)$. Accordingly, the minimum expected travel time $V_3(s)$ from s to d using at most three edges is the minimum of $E(T(s, d)) = c(s, d)$ and the mean value of $T(s, a) + W_2(a, d)$. Thus,

$$V_3(s) = \min\{c(s, d), E(T(s, a) + W_2(a, d))\}$$
$$= \min\{c(s, d), c(s, a) + V_2(a, d)\}.$$

Since the graph is finite, V_n converges to V in at most N steps, where N is the length of the longest cycle-free path to node d. The limit is such that $V(i)$ is the shortest average travel time from i to d. Note that V satisfies the following fixed-point equations:

$$V(i) = \min_j\{c(i, j) + V(j)\}, \forall j \text{ and } V(d) = 0. \tag{13.2}$$

These are called the *dynamic programming equations* (DPE). Thus, (13.1) is an algorithm for solving (13.2).

13.3   Formulation 2: Adapting

We now assume that when we get to a node i, we see the actual travel times along the edges out of i. However, we do not see beyond those edges. How should we modify our path planning? If the travel times are in fact deterministic, then nothing changes. However, if they are random, we may notice that the actual travel times on some edges out of i are smaller than their mean value, whereas others may be larger. Clearly, we should use that information.

Here is a systematic procedure for calculating the best path. Let $V(i)$ be the minimum average time to get to d starting from node i, for $i \in \{s, a, b, d\}$. We see that $V(b) = T(b, d) = 4$.

To calculate $V(a)$, define $W(a)$ to be the minimum expected time from a to d given the observed delays along the edges out of a. That is,

$$W(a) = \min\{T(a, b) + V(b), T(a, d)\}.$$

Hence, $V(a) = E(W(a))$. Thus,

$$V(a) = E(\min\{T(a, b) + V(b), T(a, d)\}). \tag{13.3}$$

For this example, we see that $T(a, b) + V(b) =_D U[6, 14]$. Since $T(a, d) = 10$, if $T(a, b) + V(b) < 10$, which occurs with probability $1/2$, we choose the path abd that has a travel time uniformly distributed in $[6, 10]$ with a mean value 8. Also, if $T(a, b) + V(b) > 10$, then we choose the travel time $T(a, d) = 10$, also with probability $1/2$. Thus, the minimum expected travel time $V(a)$ from a to d is equal to 8 with probability $1/2$ and to 10 with probability $1/2$, so that its average value is $8(1/2) + 10(1/2) = 9$. Hence, $V(a) = 9$.

Similarly,

$$V(s) = E(\min\{T(s, a) + V(a), T(s, d)\}),$$

where $T(s, a) + V(a) =_D U[14, 22]$ and $T(s, d) = 20$. Thus, if $T(s, a) + V(a) < 20$, which occurs with probability $(20 - 14)/(22 - 14) = 3/4$, then we choose a path that goes from s to a and has a delay that is uniformly distributed in $[14, 20]$, with mean value 17. If $T(s, a) + V(a) > 20$, which occurs with probability $1/4$, we choose the direct path sd that has delay 20. Hence $V(s) = 17(3/4) + 20(1/4) = 71/4 = 17.75$.

Note that by observing the delays on the next edges and making the appropriate decisions, we reduce the expected travel time from s to d from 19 to 17.5. Not surprisingly, more information helps. Observe also that the decisions we make depend on the observed delays. For instance, starting in node s, we go along edge sd if $T(s, a) + V(a) > T(s, d)$, i.e., if $T(s, a) + 9 > 20$, or $T(s, a) > 11$. Otherwise, we follow the edge sa.

Let us now go back to the general model. The key relationships are as follows:

$$V(i) = E(\min_j\{T(i, j) + V(j)\}), \forall i. \tag{13.4}$$

The interpretation is simple: starting from i, one can choose to go next to j. In that case, one faces a travel time $T(i, j)$ from i to j and a subsequent minimum average time from j to d equal to $V(j)$. Since the path from i to d must necessarily go to a next node j, the minimum expected travel time from i to d is given by the expression above. As before, these equations are justified by the fact that the expected value of a sum is the sum of the expected values.

An algorithm for solving these fixed-point equations is

$$V_{n+1}(i) = E(\min_{j}\{T(i, j) + V_n(j)\}), n \geq 0, \qquad (13.5)$$

where $V_0(i) = 0$ for all i. The interpretation of $V_n(i)$ is the same as before: it is the minimum expected time from i to d using a path with at most n edges, given that at each step along the path one observes the delays along the edges out of the current node.

Equations (13.4) are the *stochastic dynamic programming equations* for the problem. Equations (13.5) are called the *value iteration equations*.

13.4   Markov Decision Problem

A more general version of the path planning problem is the control of a Markov chain. At each step, one looks at the state and one chooses an action that determines the transition probabilities and also the cost for the next step.

More precisely, to define a controlled Markov chain $X(n)$ on some state space $\mathscr{X}$, one specifies, for each $x \in \mathscr{X}$, a set $A(x)$ of possible actions. For each state $x \in \mathscr{X}$ and each action $a \in A(x)$, one has transition probabilities $P(x, x'; a) \geq 0$ with $\sum_{x' \in \mathscr{X}} P(x, x'; a) = 1$. One also specifies a cost $c(x, a)$ of taking the action a when in state x.

The sequence $X(n)$ is then defined by

$$P[X(1) = x_1, X(2) = x_2, \ldots, X(n) = x_n | X(0) = x_0, a_0, \ldots, a_{n-1}]$$
$$= P(x_0, x_1; a_0) P(x_1, x_2; a_1) \times \cdots \times P(x_{n-1}, x_n; a_{n-1}).$$

The goal is to choose the actions to minimize the average total cost

$$E\left[\sum_{m=0}^{n} c(X(m), a(m)) | X(0) = x\right]. \qquad (13.6)$$

For each $m = 0, \ldots, n$, the action $a(m) \in A(X(m))$ is determined from the knowledge of $X(m)$ and also of the previous states $X(0), \ldots, X(m-1)$ and previous actions $a(0), \ldots, a(m-1)$.

This problem is called a *Markov decision problem (MDP)*.

To solve this problem, we follow a procedure identical to the path planning problem where we think of the state as the node that has been reached during the travel. Let $V_m(x)$ be the minimum value of the cost (13.6) when n is replaced by m. That is, $V_m(x)$ is the minimum average cost of the next $m + 1$ steps, starting from $X(0) = x$. The function $V_m(\cdot)$ is called the *value function*.

The DPE are

$$V_m(x) = \min_{a \in A(x)} \{c(x, a) + E[V_{m-1}(x')|X(0) = x, a(0) = a]\}$$

$$= \min_{a \in A(x)} \left\{ c(x, a) + \sum_{x'} P(x, x'; a) V_{m-1}(x') \right\}. \tag{13.7}$$

Let $a = g_m(x)$ be the value of $a \in A(x)$ that achieves the minimum in (13.7). Then the choices $a(m) = g_{n-m}(X(m))$ achieve the minimum of (13.6).

The existence of the minimizing a in (13.7) is clear if $\mathscr{X}$ and each $A(x)$ are finite and also under weaker assumptions.

13.4.1 Examples

Guess a Card
Here is a simple example. One is given a perfectly shuffled deck of 52 cards. The cards are turned over one at a time. Before one turns over a new card, you have the option of saying "Stop." If the next card is an ace, you win $1.00. If not, the game stops and you lose. The problem is for you to decide when to stop (Fig. 13.4).

Assume that there are still x aces in a deck with m remaining cards. Then, if you say stop, you win with probability x/m. If you do not say stop, then after the next card is turned over, $x - 1$ aces remain with probability x/m and x remain otherwise.

Let $V(m, x)$ be the maximum expected probability that you win if there are still x aces in the deck with m remaining cards.

The DPE are

$$V(m, x) = \max \left\{ \frac{x}{m}, \frac{x}{m} V(m-1, x-1) + \frac{m-x}{m} V(m-1, x) \right\}.$$

Interestingly, the solution of these equations is $V(m, x) = x/m$, as you can verify. Also, the two terms in the maximum are equal if $x > 0$. The conclusion is that you can stop at any time, as long as there is still at least one ace in the deck.

Scheduling Jobs
You have two sets of jobs to perform. Jobs of type i (for $i = 1, 2$) have a waiting cost equal to c_i per unit of waiting time until they are completed. Also, when you work on a job of type i, it completes with probability μ_i in the next time unit,

Fig. 13.4 Guessing if the next card is an Ace

independently of how long you have worked on it. That is, the job processing times are geometrically distributed with parameter μ_i. The problem is to decide which job to work on to minimize the total waiting cost of the jobs.

Let $V(x_1, x_2)$ be the minimum expected total remaining waiting cost given that there are x_1 jobs to type 1 and x_2 jobs of type 2. The DPE are

$$V(x_1, x_2) = x_1 c_1 + x_2 c_2 + \min\{V_1(x_1, x_2), V_2(x_1, x_2)\},$$

where

$$V_1(x_1, x_2) = \mu_1 V((x_1 - 1)^+, x_2) + (1 - \mu_1)V(x_1, x_2)$$

and

$$V_1(x_1, x_2) = \mu_2 V(x_1, (x_2 - 1)^+) + (1 - \mu_1)V(x_1, x_2).$$

As can be verified directly, the solution of the DPE is as follows. Assume that $c_1\mu_1 > c_2\mu_2$. Then

$$V(x_1, x_2) = c_1 \frac{x_1(x_1 + 1)}{2\mu_1} + c_2 \frac{x_2(x_2 + 1)}{2\mu_2} + c_2 \frac{x_1 x_2}{\mu_1}.$$

Moreover, this minimum expected cost is achieved by performing all the jobs of type 1 first and then the jobs of type 2. This strategy is called the $c\mu$ rule. Thus, although one might be tempted to work on the longest queue first, this is not optimal.

There is a simple interchange argument to confirm the optimality of the $c\mu$ rule. Say that you decide to work on the jobs in the following order: 1221211. Thus, you work on a job of type 1 until it completes, then a job of type 2, then another job of type 2, and so on. Modify the strategy as follows. Instead of working on the second job of type 2, work on the second job of type 1, until it completes. Then work on the second job of type 2 and continue as you would have. Thus, the processings of two jobs have been interchanged: the second job of type 2 and the second job of type 1. Only the waiting times of these two jobs change. The waiting time of the job of type 1 is reduced by $1/\mu_2$, on average, since this is the average completion time of the job of type 2 that was previously processed before the job of type 1. Thus, the waiting cost of the job of type 1 is reduced by c_1/μ_2. Similarly, the waiting cost of the job of type 2 is increased by c_2/μ_1, on average. Thus, the average cost decreases by $c_1/\mu_2 - c_2/\mu_1$ which is a positive amount since $c_1\mu_1 > c_2\mu_2$. By induction, it is optimal to process all the jobs of type 1 first.

Of course, there are very few examples of control problems where the optimal policy can be proved by a simple argument. Nevertheless, keep this possibility in mind because it can yield elegant results simply. For instance, assume that jobs arrive at the queues shown in Fig. 13.5 according to independent Bernoulli processes. That is, with probability λ_i, a job of type i arrives during each time step, independently of the past, for $i = 1, 2$. The same interchange argument shows that

Fig. 13.5 What job to work on next?

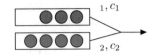

the $c\mu$ rule minimizes the long-term average expected waiting cost of the jobs (a cost that we have not defined, but you may be able to imagine what it means). This is useful because the DPE can no longer be solved explicitly and proving the optimality of this rule analytically is quite complicated.

Hiring a Helper

Jobs arrive at random times and you must decide whether to work on them yourself or hire some helper. Intuition suggests that you should get some help if the backlog of jobs to be performed exceeds some threshold. We examine a model of this situation.

At time $n = 0, 1, \ldots$, a job arrives with probability $\lambda \in (0, 1)$. If you work alone, you complete a job with probability $\mu \in (0, 1)$ in one time unit, independently of the past. If you hire a helper, then together you complete a job with probability $\alpha\mu \in (0, 1)$ in one unit of time, where $\alpha > 1$. Let the cost at time n be $c(n) = \beta > 0$ if you hire a helper at time step n and $c(n) = 0$ otherwise. The goal is to minimize

$$E\left[\sum_{n=0}^{N}(X(n) + c(n))\right],$$

where $X(n)$ is the number of jobs yet to be processed at time n. This cost measures the waiting cost of the jobs plus the cost of hiring the helper. The waiting cost is minimized if you hire the helper all the time and the helper cost is minimized if you never hire him. The goal of the problem is to figure out when to hire a helper to achieve the best trade-off between these two costs.

The state of the system is $X(n)$ at time n. Let

$$V_m(x) = \min E\left[\sum_{n=0}^{m}(X(n) + c(n))|X(0) = x\right],$$

where the minimum is over the possible choices of actions (hiring or not) that depend on the state up to that time. The stochastic dynamic programming equations are

$$V_m(x) = x + \min_{a \in \{0,1\}} \{\beta 1\{a = 1\} + (1 - \lambda)(1 - \mu(a))V_{m-1}(x)$$

$$+ \lambda(1 - \mu(a))V_{m-1}(\min\{x + 1, K\})$$

$$+ (1 - \lambda)\mu(a)V_{m-1}(\max\{x - 1, 0\})$$

$$+ \lambda\mu(a)V_{m-1}(x)\}, n \geq 0,$$

Fig. 13.6 One should hire a helper at time n if the backlog exceeds $\gamma(N - n)$

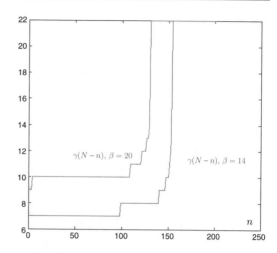

where we defined $\mu(0) = \mu$ and $\mu(1) = \alpha\mu$ and $V_{-1}(x) = 0$. Also, we limit the backlog of jobs to K, so that if one job arrives where there are already K, we discard the new arrival.

We solve these equations using Python. As expected, the solution shows that one should hire a helper at time n if $X(n) > \gamma(N - n)$, where $\gamma(m)$ is a constant that decreases with m. As the time to go m increases, the cost of holding extra jobs increases and so does the incentive to hire a helper. Figure 13.6 shows the values of $\gamma(n)$ for $\beta = 14$ and $\beta = 20$. The figure corresponds to $\lambda = 0.5, \mu = 0.6, \alpha = 1.5, K = 20$, and $N = 200$. Not surprisingly, when the helper is more expensive, one waits until the backlog is larger before hiring him.

Which Queue to Join?

After shopping in the supermarket, you get to the cashiers and have to choose a queue to join. Naturally, you try to identify the queue with the shortest expected waiting time, and you join that queue. Everyone does the same, and it seems quite natural that this strategy should minimize the expected waiting time of all the customers. Your friend, who has taken this class before, tells you that this is not necessarily the case. Let us try to understand this apparent paradox.

Assume that there are two queues and customers arrive with probability λ at each time step. The service times in queue i are geometrically distributed with parameter μ_i in queue i, for $i = 1, 2$.

Say that when you arrive, there are x_i customers in queue i, for $i = 1, 2$. You should join queue 1 if

$$\frac{x_1 + 1}{\mu_1} < \frac{x_2 + 1}{\mu_2},$$

Fig. 13.7 The socially
optimal policy is shown in
blue and the selfish policy is
shown in green

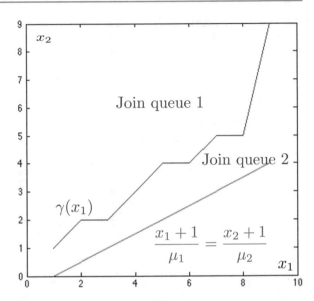

as this will minimize the expected time until you are served. However, if we consider
the problem of minimizing the total average waiting time of customers in the two
queues, we find that the optimal policy does not agree with the selfish choice of
individual customers. Figure 13.7 shows an example with $\mu_2 < \mu_1$. It indicates that
under the socially optimal policy some customers should join queue 2, even though
they will then incur a longer delay than under the selfish policy.

This example corresponds to minimizing the total cost

$$\sum_{n=0}^{N} \beta^n E(X_1(n) + X_2(n)).$$

In this expression, $X_i(n)$ is the number of customers in queue i at time n. The
capacity of each queue is K. To prevent the system from discarding too many
customers, one imposes the constraint that if only one queue is full when a customer
arrives, he should join the non-full queue. In the expression for the total cost, one
uses a discount factor $\beta \in (0, 1)$ to keep the cost bounded. The figure corresponds
to $K = 8, \lambda = 0.3, \mu_1 = 0.4, \mu_2 = 0.2, N = 100$, and $\beta = 0.95$. (The graphs are
in fact for $x_1 + 1$ and $x_2 + 2$ as Python does not like the index value 0.)

13.5 Infinite Horizon

The problem of minimizing (13.6) involves a finite horizon. The problem stops at time n. We have seen that the minimum cost to go when there are m more steps is $V_m(x)$ when in state x. Thus, not surprisingly, the cost to go depends on the time to go and, consequently, the best action to choose in a given state x generally depends on the time to go.

The problem is simpler when one considers an infinite horizon because the time to go remains the same at each step. To make the total cost finite, one discounts the future costs. That is, one considers the problem of minimizing the expected total *discounted* cost:

$$E\left[\sum_{m=0}^{\infty} \beta^m c(X(m), a(m))|X(0) = x\right]. \tag{13.8}$$

In this expression, $0 < \beta < 1$ is the discount rate. Intuitively, if β is small, then future costs do not matter much and one tends to be short-sighted. However, if β is close to 1, then one pays a lot of attention to the long term.

Define $V(x)$ to be the minimum value of the cost (13.8), where the minimum is over all the possible choices of the actions at each step. Arguing as before, one can show that

$$V(x) = \min_{a \in A(x)} \{c(x, a) + \beta E[V(X(1))|X(0) = x, a(0) = a]\}$$

$$= \min_{a \in A(x)} \left\{c(x, a) + \beta \sum_{y} P(x, y; a)V(y)\right\}. \tag{13.9}$$

These equations are similar to (13.7), with two differences: the discount factor and the fact that the value function does not depend on time. Note that these equations are fixed-point equations. A standard method to solve them is to consider the equations

$$V_{n+1}(x) = \min_{a \in A(x)} \left\{c(x, a) + \beta \sum_{y} P(x, y; a)V_n(y)\right\}, n \geq 0, \tag{13.10}$$

where one chooses $V_0(x) = 0, \forall x$. Note that these equations correspond to

$$V_n(x) = \min E\left[\sum_{m=0}^{n} \beta^m c(X(m), a(m))|X(0) = x\right]. \tag{13.11}$$

One can show that the solution $V_n(x)$ of (13.10) is such that $V_n(x) \to V(x)$ as $n \to \infty$, where $V(x)$ is the solution of (13.9).

13.6   Summary

- Dynamic Programming Equations;
- Controlled Markov Chain;
- Markov Decision Problem.

13.6.1 Key Equations and Formulas

MDP	$P(x, y; a)$	S.13.4
SDPE	$V_{m+1}(x) = \min_a \{c(x, a) + \sum_y P(x, y; a) V_m(y)\}$	(13.7)

13.7   References

The book Ross (1995) is a splendid introduction to stochastic dynamic programming. We borrowed the "guess a card" example from it. It explains the key ideas simply and the many variations of the theory illustrated by carefully chosen examples. The textbook Bertsekas (2005) is a comprehensive presentation of the algorithms for dynamic programming. It contains many examples and detailed discussions of the theory and practice.

13.8   Problems

Problem 13.1 Consider a single queue with one server in discrete time. At each time, a new customer arrives to the queue with probability $\lambda < 1$, and if the server works on the queue at rate $\mu \in [0, 1]$, it serves one customer in one unit of time with probability μ. Due to energy constraints, you want your server to work with the smallest rate as possible without making the queue unstable. Thus, you want your server to work at rate $\mu^* = \lambda$. Unfortunately, you do not know the value of λ. All you can observe is the queue length. We try to design an algorithm based on stochastic gradient to learn μ^* in the following steps:

(a) Minimize the function $V(\mu) = \frac{1}{2}(\lambda - \mu)^2$ over μ using gradient descent.
(b) Find $E[Q(n+1) - Q(n) | Q(n) = q]$, for some $q > 0$, given that server allocates capacity μ_n during time slot n. $Q(n)$ is the queue length at time n. What happens if $q = 0$?
(c) Use the stochastic gradient projection algorithm and write a Python code based on parts (a) and (b) to learn μ^*. Note that $0 \le \mu \le 1$.

Hint To avoid the case when the queue length is 0, start with a large initial queue length.

Problem 13.2 Consider a routing network with three nodes: the start node s, the destination node d, and an intermediate node r. There is a direct path from s to d with travel time 20. The travel time from s to r is 7. There are two paths from r to d. They have independent travel times that are uniformly distributed between 8 and 20.

(a) If you want to do pre-planning, which path should be chosen to go from s to d?
(b) If the travel times from r to d are revealed at r which path should be chosen?

Problem 13.3 Consider a single queue in discrete time with Bernoulli arrival process of rate λ. The queue can hold K jobs, and there is a fee γ when its backlog reaches K. There is one server dedicated to the queue with service rate $\mu(0)$. You can decide to allocate another server to the queue that increases the rate to $\mu(1) \in (\mu(0), 1)$. However, using the additional server has some cost. You want to minimize the cost

$$\sum_{n=0}^{\infty} \beta^n E(X(n) + \alpha H(n) + \gamma 1\{X(n) = K\}),$$

where $H(n)$ is equal to one if you use an extra helper at time n and is zero otherwise.

(a) Write the dynamic programming equations.
(b) Solve the DPE with MATLAB for $\lambda = 0.4$, $\mu(0) = 0.35$, $\mu(1) = 0.5$, $\alpha = 2.5$, $\beta = 0.95$, and $\gamma = 30$.

Problem 13.4 We want to plan routing from node 1 to 5 in the graph of Fig. 13.8. The travel times on the edges of the graph are as follows: $T(1, 2) = 2$, $T(1, 3) \sim U[2, 4]$, $T(2, 4) = 1$, $T(2, 5) \sim U[4, 6]$, $T(4, 5) \sim U[3, 5]$, and $T(3, 5) = 4$. Note that $X \sim U[a, b]$ means X is a random variable uniformly distributed between a and b.

(a) If you want to do pre-planning, which path would you choose? What is the expected travel time?

Fig. 13.8 Route planning

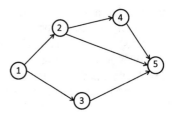

(b) Now suppose that at each node, the travel times of <u>two</u> steps ahead are revealed. Thus, at node 1 all the travel times are revealed except $T(4, 5)$. Write the dynamic programming equations that solve the route planning problem and solve them. That is, let $V(i)$ be the minimum expected travel time from i to 5, and $1 \le i \le 5$. Find $V(i)$ for $1 \le i \le 5$.

Problem 13.5 Consider a factory, DilBox, that stores boxes. At the beginning of year k, they have x_k boxes in storage. Now at the end of every year k they are mandated by contracts to provide d_k boxes. However, the number of boxes d_k is unknown until the year actually ends.

At the beginning of the year, they can request u_k boxes. Using very shoddy Elbonian labor each box has costs A to produce. At the end of the year DilBox is able to borrow y_k boxes from BoxR'Us at the cost $s(y_k)$ to meet the contract.

The boxes remaining after meeting the demand are carried over to the next year $x_{k+1} = x_k + u_k + y_k - d_k$. Sadly, they need to pay to store the boxes at a cost given by a function $r(x_{k+1})$.

Now your job is to provide a box creation and storage plan for the upcoming 20 years. Your goal is to minimize the total cost for the 20 years. You can treat costs as being paid at the end of the year and there is no inflation. Also, you get your pension after 20 years so you do not care about costs beyond those paid in the 20th year. (Assume you start with zero boxes, of course, it does not really matter).

(a) Formulate the problem as a Markov decision problem;
(b) Write the dynamic programming equations;
(c) Use Python to solve the equations with the following parameters:

  - $r(x_k) = 5x_k$;
  - $s(y_k) = 20y_k$;
  - $A = 1$;
  - $d_k =_D U\{1, \ldots, 10\}$.

Problem 13.6 Consider a video game duel where Bob starts at time 0 at distance $T = 10$ from Alice and gets closer to her at speed 1. For instance, Alice is at location $(0, 0)$ in the plane and Bob starts at location $(0, T)$ and moves toward Alice, so that after t seconds, Bob is at location $(0, T - t)$. Alice has picked a random time, uniformly distributed in $[0, T]$, when she will shoot Bob. If Alice shoots first, Bob is dead. Alice never misses. [This is only a video game.]

(a) Bob has to find at what time t he should shoot Alice to maximize the probability of killing her. If Bob shoots from a distance x, the probability that he hits (and kills) Alice is $1/(1 + x)^2$. Bob has only one bullet.
(b) What is the maximum probability that Bob wins the duel?
(c) Assume now that Bob has two bullets. You must find the times t_1 and t_2 when Bob should shoot Alice to maximize the probability that he wins the duel. Again,

for each bullet that Bob shoots from distance x, the probability of success is $1/(1 + x)^2$, independently for each bullet.

Problem 13.7 You play a game where you win the amount you bet with probability $p \in (0, 0.5)$ and you lose it with probability $1 - p$. Your initial fortune is 16 and you gamble a fixed amount γ at each step, where $\gamma \in \{1, 2, 4, 8, 16\}$. Find the probability that you reach a fortune equal to 256 before you go broke. What is the gambling amount that maximizes that probability?

Route Planning: B

14

Topics: LQG Control, incomplete observations

14.1 LQG Control

The ideas of dynamic programming that we explained for a controlled Markov chain apply to other controlled systems. We discuss the case of a *linear system with quadratic cost and Gaussian noise*, which is called the *LQG* problem. For simplicity, we consider only the scalar case.

The system is

$$X(n+1) = aX(n) + U(n) + V(n), n \geq 0. \tag{14.1}$$

Here, $X(n)$ is the state, $U(n)$ is a control value, and $V(n)$ is the noise. We assume that the random variables $V(n)$ are i.i.d. and $N(0, \sigma^2)$.

The problem is to choose, at each time n, the control value $U(n)$ in $\Re$ based on the observed state values up to time n to minimize the expected cost

$$E\left[\sum_{n=0}^{N} \left(X(n)^2 + \beta U(n)^2 \right) | X(0) = x \right]. \tag{14.2}$$

Thus, the goal of the control is to keep the state value close to zero, and one pays a cost for the control.

The problem is then to trade-off the cost of a large value of the state and that of the control that can bring the state back close to zero. To get some intuition for the solution, consider a simple form of this trade-off: minimizing

© The Author(s) 2021

J. Walrand, *Probability in Electrical Engineering and Computer Science*,
https://doi.org/10.1007/978-3-030-49995-2_14

Fig. 14.1 The optimal
control is linear in the state

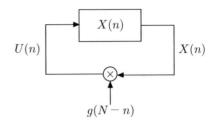

$$(ax + u)^2 + \beta u^2.$$

In this simple version of the problem, there is no noise and we apply the control only once. To minimize this expression over u, we set the derivative with respect to u equal to zero and we find

$$2(ax + u) + 2\beta u = 0,$$

so that

$$u = -\frac{a}{1 + \beta}x.$$

Thus, the value of the control that minimizes the cost is linear in the state. We should use a large control value when the state is far from the desired value 0. The following result shows that the same conclusion holds for our problem (Fig. 14.1).

Theorem 14.1 *Optimal LQG Control   The control values $U(n)$ that minimize (14.2) for the system (14.1) are*

$$U(n) = g(N - n)X(n),$$

where

$$g(m) = -\frac{ad(m - 1)}{\beta + d(m - 1)}, m \geq 0; \qquad (14.3)$$

$$d(m) = 1 + \frac{a^2\beta d(m - 1)}{\beta + d(m - 1)}, m \geq 0 \qquad (14.4)$$

with $d(-1) = 0$.

That is, the optimal control is linear in the state and the coefficient depends on the time-to-go. These coefficients can be pre-computed at time 0 and they do not depend on the noise variance. Thus, the control values would be calculated in the same way if $V(n) = 0$ for all n.

■

Proof Let $V_m(x)$ be the minimum value of (14.2) when N is replaced by m. The stochastic dynamic programming equations are

$$V_m(x) = \min_u \left\{ x^2 + \beta u^2 + E(V_{m-1}(ax + u + V)) \right\}, m \geq 0, \qquad (14.5)$$

where $V = N(0, \sigma^2)$. Also, $V_{-1}(x) := 0$.

We claim that the solution of these equations is

$$V_m(x) = c(m) + d(m)x^2$$

for some constants $c(m)$ and $d(m)$ where $d(m)$ satisfies (14.4).

That is, we claim that

$$\min_u \{x^2 + \beta u^2 + E[c(m-1) + d(m-1)(ax + u + V)^2]\} = c(m) + d(m)x^2, \qquad (14.6)$$

where $d(m)$ is given by (14.4) and the minimizer is $u = g(m)x$ where $g(m)$ is given by (14.3).

The verification is a simple algebraic exercise that we leave to the reader.   □

14.1.1 Letting $N \to \infty$

What happens if N becomes very large in (14.2)? Proceeding formally, we examine (14.4) and observe that if $|a| < 1$, then $d(m) \to d$ as $m \to \infty$ where d is the solution of the fixed-point equation

$$d = f(d) := 1 + \frac{a^2 \beta d}{\beta + d}.$$

To see why this is the case, note that

$$f'(d) = \frac{a^2 \beta^2}{(\beta + d)^2},$$

so that $0 < f'(d) < a^2$ for $d \geq 0$. Also, $f(d) > 0$ for $d \geq 0$. Hence, $f(d)$ is a *contraction*. That is,

$$|f(d_1) - f(d_2)| \leq \alpha |d_1 - d_2|, \forall d_1, d_2 \geq 0$$

for some $\alpha \in (0, 1)$. (Here, $\alpha = a^2$.) In particular, choosing $d_1 = d$ and $d_2 = d(m)$, we find that

$$|d - d(m + 1)| \leq \alpha |d - d(m)|, \forall m \geq 0.$$

Fig. 14.2 The optimal
control for the average cost

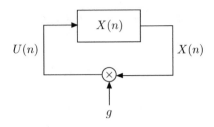

Thus,

$$|d - d(m)| \leq \alpha^m |d - d(0)|,$$

which shows that $d(m) \to d$, as claimed. Consequently, (14.3) shows that $g(m) \to g$ as $m \to \infty$, where

$$g = -\frac{ad}{\beta + d}.$$

Thus, when the time-to-go m is very large, the optimal control approaches $U(N - m) = gX(N - m)$. This suggests that this control may minimize the cost (14.2) when N tends to infinity (Fig. 14.2).

The formal way to study this problem is to consider the *long-term average cost* defined by

$$\lim_{N \to \infty} \frac{1}{N} E \left[\sum_{n=0}^{N} \left(X(n)^2 + \beta U(n)^2 \right) | X(0) = x \right].$$

This expression is the average cost per unit time. One can show that if $|a| < 1$, then the control $U(n) = gX(n)$ with g defined as before indeed minimizes that average cost.

14.2   LQG with Noisy Observations

In the previous section, we controlled a linear system with Gaussian noise assuming that we observed the state. We now consider the case of noisy observations.

The system is

$$X(n + 1) = aX(n) + U(n) + V(n), n \geq 0; \tag{14.7}$$

$$Y(n) = X(n) + W(n), \tag{14.8}$$

where the random variables $W(n)$ are i.i.d. $\mathcal{N}(0, w^2)$ and are independent of the $V(n)$.

Fig. 14.3 The optimal
control is linear in the
estimate of the state

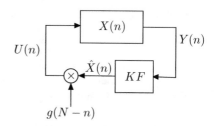

The problem is to find, for each n, the value of $U(n)$ based on the values of $Y^n := \{Y(0), \ldots, Y(n)\}$ that minimize the expected total cost (14.2).

The following result gives the solution of the problem (Fig. 14.3).

Theorem 14.2 *Optimal LQG Control with Noisy Observations The solution of the problem is*

$$U(n) = g(N - n)\hat{X}(n),$$

where

$$\hat{X}(n) = E[X(n)|Y(0), \ldots, Y(n), U(0), \ldots, U(n - 1)]$$

can be computed by using the Kalman filter and the constants $g(m)$ are given by (14.3)–(14.4).

Thus, the control values are the same as when $X(n)$ is observed exactly, except that $X(n)$ is replaced by $\hat{X}(n)$. This feature is called certainty equivalence.

∎

Proof The fact that the values of $g(n)$ do not depend on the noise $V(n)$ gives us some inkling as to why the result in the theorem can be expected: given Y^n, the state $X(n)$ is $\mathcal{N}(\hat{X}(n), v^2)$ for some variance v^2. Thus, we can view the noisy observation as increasing the variance of the state, as if the variance of $V(n)$ were increased.

Instead of providing the complete algebra, let us sketch why the result holds. Assume that the minimum expected cost-to-go at time $N - m + 1$ given Y^{N-m+1} is

$$c(m - 1) + d(m - 1)\hat{X}(N - m + 1)^2.$$

Then, at time $N - m$, the expected cost-to-go given Y^{N-m} and $U(N - m) = u$ is the expected value of

$$X(N - m)^2 + \beta u^2 + c(m - 1) + d(m - 1)\hat{X}(N - m + 1)^2$$

given Y^{N-m} and $U(N-m) = u$. Now,

$$X(N-m) = \hat{X}(N-m) + \eta,$$

where η is a Gaussian random variable independent of Y^{N-m}. Also, as we saw when we discussed the Kalman filter,

$$\hat{X}(N-m+1) = a\hat{X}(N-m) + u$$
$$+ K(N-m+1)\{Y(N-m+1) - E[Y(N-m+1)|Y^{N-m}]\}.$$

Moreover, we know from our study of conditional expectation of jointly Gaussian random variables, that $Y(N-m+1) - E[Y(N-m+1)|Y^{N-m}]$ is a Gaussian random variable that has mean zero and is independent of Y^{N-m}. Hence,

$$\hat{X}(N-m+1) = a\hat{X}(N-m) + u + Z$$

for some independent zero-mean Gaussian random variable Z.

Thus, the expected cost-to-go at time $N-m-1$ is the expected value of

$$(\hat{X}(N-m) + \eta)^2 + \beta u^2 + c(m-1)$$
$$+ d(m-1)(a\hat{X}(N-m) + Z)^2,$$

i.e., of

$$\hat{X}(N-m)^2 + \beta u^2 + c(m-1) + d(m-1)(a\hat{X}(N-m) + u + Z)^2.$$

This expression is identical to (14.6), except that x is replaced by $\hat{X}(N-m)$ and V is replaced by Z. Since the variance of V does not affect the calculations of $c(m)$ and $d(m)$, this concludes the proof.   □

14.2.1 Letting $N \to \infty$

As when $X(n)$ is observed exactly, one can show that, if $|a| < 1$, the control

$$U(n) = g\hat{X}(n)$$

minimizes the average cost per unit time. Also, in this case, we know that the Kalman filter becomes stationary and has the form (Fig. 14.4)

$$\hat{X}(n+1) = a\hat{X}(n) + u + K[Y(n+1) - a\hat{X}(n) - U(n)].$$

Fig. 14.4 The optimal
control for the average cost
with noisy observations.
Here, the Kalman filter is
stationary

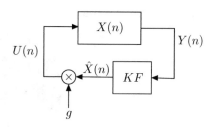

14.3   Partially Observed MDP

In the previous chapter, we considered a controlled Markov chain and the action
is based on the knowledge of the state. In this section, we look at problems where
the state of the Markov chain is not observed exactly. In other words, we look at
a controlled hidden Markov chain. These problems are called *partially observed
Markov decision problems (POMDPs)*.

Instead of discussing the general version of this problem, we look at one concrete
example to convey the basic ideas.

14.3.1 Example: Searching for Your Keys

The example is illustrated in Fig. 14.5. You have misplaced your keys but you
know that they are either in bag A, with probability p, or in bag B, otherwise.
Unfortunately, your bags are cluttered and if you spend one unit of time (say 10 s)
looking in bag A, you find your keys with probability α if they are there. Similarly,
the probability for bag B is β. Every time unit, you choose which bag to explore.
Your objective is to minimize the expected time until you find your keys.

The state of the system is the location A or B of your keys. However, you do
not observe that state. The key idea (excuse the pun) is to consider the conditional
probability p_n that the keys are in bag A given all your observations up to time n. It
turns out that p_n is a controlled Markov chain, as we explain shortly. Unfortunately,
the set of possible value of p_n is $[0, 1]$, which is not finite, nor even countable. Let
us not get discouraged by this technical issue.

Assume that at time n, when the keys are in bag A with probability p_n, you look
in bag A for one unit of time and you do not see the keys. What is then p_{n+1}? We
claim that

$$p_{n+1} = \frac{p_n(1 - \alpha)}{p_n(1 - \alpha) + (1 - p_n)} =: f(A, p_n).$$

Indeed, this is the probability that the keys are in bag A and we do not see them,
divided by the probability that we do not see the keys (either when they are there or
when they are not). Of course, if we see the keys, the problem stops.

Fig. 14.5 Where to look for
your keys?

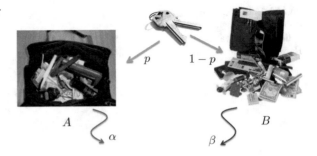

Similarly, say that we look in bag B and we do not see the keys. Then

$$p_{n+1} = \frac{p_n}{p_n + (1 - p_n)(1 - \beta)} =: f(B, p_n).$$

Thus, we control p_n with our actions. Let $V(p)$ be the minimum expected time until we find the keys, given that they are in bag A with probability p. Then, the DPE are

$$V(p) = 1 + \min\{(1 - p\alpha)V(f(A, p)), (1 - (1 - p)\beta)V(f(B, p))\}. \quad (14.9)$$

The constant 1 is the duration of the first step. The first term in the minimum is what happens when you look in bag A. With probability $1 - p\alpha$, you do not find your keys and you will then have to wait a minimum expected time equal to $V(f(A, p))$ to find your keys, because the probability that they are in bag A is now $f(A, p)$. The other term corresponds to first looking in bag B.

These equations look hopeless. However, they are easy to solve in Python. One discretizes $[0, 1]$ into K intervals and one rounds off the updates $f(A, p)$ and $f(B, p)$.

Thus, the updates are for a finite vector $\mathbf{V} = (V(1/K), V(2/K), \ldots, V(1))$. With this discretization, the equations (14.9) look like

$$\mathbf{V} = \phi(\mathbf{V}),$$

where $\phi(\cdot)$ is the right-hand side of (14.9). These are fixed-point equations. To solve them, we initialize $\mathbf{V}_0 = \mathbf{0}$ and we iterate

$$\mathbf{V}_{t+1} = \phi(\mathbf{V}_t), t \geq 0.$$

With a bit of luck, that can be justified mathematically, this algorithm converges to $\mathbf{V}$, the solution of the DPE. The solution is shown in Fig. 14.6, for different values of α and β. The figure also shows the optimum action as a function of p. The discretization uses $K = 1000$ values in $[0, 1]$ and the iteration is performed 100 times.

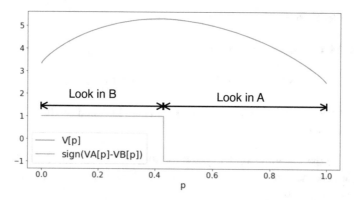

Fig. 14.6 Numerical solution of (14.9)

14.4   Summary

- LQG Control Problem with State Observations;
- LQG Control Problem with Noisy Observations;
- Partially Observed MDP.

14.4.1 Key Equations and Formulas

LQG problem	Formulation	(14.1)–(14.2)
Solution of LQG	$U_n = g_{N-n} X_n$	T.14.1
Noisy observations	$Y_n = X_n + W_n$	(14.8)
Solution with noisy observations	$U_n = g_{N-n} \hat{X}_n$	T.14.2
Partially observed MDP	Replace X_n by $P[X_n = x|Y^n]$	S.14.3

14.5   References

The texts Bertsekas (2005), Kumar and Varaiya (1986) and Goodwin and Sin (2009) cover LQG control. The first two texts discuss POMDP.

14.6   Problems

Problem 14.1 Consider the system

$$X(n+1) = 0.8X(n) + U(n) + V(n), n \geq 0,$$

where $X(0) = 0$ and the random variables $V(n)$ are i.i.d. and $\mathcal{N}(0, 0.2)$. The $U(n)$ are control values.

(a) Simulate the system when $U(n) = 0$ for all $n \geq 0$.
(b) Implement the control given in Theorem 14.1 with $N = 100$ and simulate the controlled system.
(c) Implement the control with the constant gain $g = \lim_{n \to \infty} g(n)$ and simulate the system.

Problem 14.2 Consider the system

$$X(n+1) = 0.8X(n) + U(n) + V(n), n \geq 0$$
$$Y(n) = X(n) + W(n), n \geq 0,$$

where $X(0) = 0$ and the random variables $V(n), W(n)$ are independent with $V(n) =_D \mathcal{N}(0, 0.2)$ and $W(n) =_D \mathcal{N}(0, \sigma^2)$.

(a) Implement the control described in Theorem 14.2 for $\sigma^2 = 0.1$ and $\sigma^2 = 0.4$ and simulate the controlled system.
(b) Implement the limiting control with the limiting gain and the stationary Kalman filter for $\sigma^2 = 0.1$ and $\sigma^2 = 0.4$. Simulate the system.
(c) Compare the systems with the time-varying and the limiting controls.

Problem 14.3 There are two coins. One is fair and the other one has a probability of "head" equal to 0.6. You cannot tell which is which by looking at the coins. At each step $n \geq 1$, you must choose which coin to flip. The goal is to maximize the expected number of "heads."

(a) Formulate the problem as a POMDP.
(b) Discretize the state of the system as we did in the "searching for your keys" example and write the SDPEs.
(c) Implement the SDPEs in Python and simulate the resulting system.

Perspective and Complements

15

Topics: Inference, Sufficient Statistic, Infinite Markov Chains, Poisson, Boosting, Multi-Armed Bandits, Capacity, Bounds, Martingales, SLLN

15.1 Inference

One key concept that we explored is that of *inference*. The general problem of inference can be formulated as follows. There is a pair of random quantities (X, Y). One observes Y and one wants to guess X (Fig. 15.1).

Thus, the goal is to find a function $g(\cdot)$ such that $\hat{X} := g(Y)$ is close to X, in a sense to be made precise. Here are a few sample problems:

- X is the weight of a person and Y is her height;
- $X = 1$ is a house is on fire, $X = 0$ otherwise, and Y is a measurement of the CO density at a sensor;
- $X \in \{0, 1\}^N$ is a bit string that a transmitter sends and $Y \in \Re^{[0,T]}$ is a signal that the receiver receives;
- Y is one woman's genome and $X = 1$ if she develops a specific form of breast cancer and $X = 0$ otherwise;
- Y is a vector of characteristics of a movie and of one person and X is the number of stars that the person gives to the movie;
- Y is the photograph of a person's face and $X = 1$ if it is that of a man and $X = 0$ otherwise;
- X is a sentence and Y is the signal that a microphone picks up.

© The Author(s) 2021
J. Walrand, *Probability in Electrical Engineering and Computer Science*,
https://doi.org/10.1007/978-3-030-49995-2_15

$$Y \overset{?}{\to} X$$

Fig. 15.1 The inference problem is to guess the value of X from that of Y

We explained a few different formulations of this problem in Chaps. 7 and 9:

- **Known Distribution**: We know the joint distribution of (X, Y);
- **Off-Line**: We observe a set of sample values of (X, Y);
- **On-Line**: We observe successive values of samples of (X, Y);
- **Maximum Likelihood Estimate**: We do not want to assume a distribution for X, only the conditional distribution of Y given X; the goal is to find the value of X that makes the observed Y most likely;
- **Maximum A Posteriori Estimate**: We know a prior distribution for X and the conditional distribution of Y given X; the goal is to find the value of X that is most likely given Y;
- **Hypothesis Test**: We do not want to assume a distribution for $X \in \{0, 1\}$, only a conditional distribution of Y given X; the goal is to maximize the probability of correctly deciding that $X = 1$ while keeping the probability that we decide that $X = 1$ when in fact $X = 0$ below some given β.
- **MMSE**: Given the joint distribution of X and Y, we want to find the function $g(Y)$ that minimizes $E((X - g(Y))^2)$.
- **LLSE**: Given the joint distribution of X and Y, we want to find the linear function $a + bY$ that minimizes $E((X - a - bY)^2)$.

15.2   Sufficient Statistic

A useful notion for inference problems is that of a *sufficient statistic*. We have not discussed this notion so far. It is time to do it.

Definition 15.1 (Sufficient Statistic)   We say that $h(Y)$ is a *sufficient statistic* for X if

$$f_{Y|X}[y|x] = f(h(y), x)g(y),$$

or, equivalently, it

$$f_{Y|h(Y),X}[y|s, x] = f_{Y|h(Y)}[y|s].$$

◇

We leave the verification of this equivalence to the reader.

Before we discuss the meaning of this definition, let us explore some implications. First note that if we have a prior $f_X(x)$ and we want to calculate $MAP[X|Y = y]$, we have

$$MAP[X|Y = y] = \arg\max_x f_X(x) f_{Y|X}[y|x]$$

$$= \arg\max_x f_X(x) f(h(y), x) g(y)$$

$$= \arg\max_x f_X(x) f(h(y), x).$$

Consequently, the maximizer is some function of $h(y)$. Hence,

$$MAP[X|Y] = g(h(Y)),$$

for some function $g(\cdot)$. In words, the information in Y that is useful to calculate $MAP[X|Y]$ is contained in $h(Y)$.

In the same way, we see that $MLE[X|Y]$ is also a function of $h(Y)$.

Observe also that

$$f_{X|Y}[x|y] = \frac{f_X(x) f_{Y|X}[y|x]}{f_Y(y)} = \frac{f_X(x) f(h(y), x) g(y)}{f_Y(y)}.$$

Now,

$$f_Y(y) = \int_{-\infty}^{\infty} f_X(x) f(h(y), x) g(y) dx = g(y) \int_{-\infty}^{\infty} f_X(x) f(h(y), x) dx$$

$$= g(y) \phi(h(y)),$$

where

$$\phi(h(y)) = \int_{-\infty}^{\infty} f_X(x) f(h(y), x) dx.$$

Hence,

$$f_{X|Y}[x|y] = \frac{f_X(x) f(h(y), x)}{\phi(h(Y))}.$$

Thus, the conditional density of X given Y depends only on $h(Y)$. Consequently,

$$E[X|Y] = \psi(h(Y)).$$

Now, consider the hypothesis testing problem when $X \in \{0, 1\}$. Note that

$$L(y) = \frac{f_{Y|X}[y|1]}{f_{Y|X}[y|0]} = \frac{f(h(y), 1)g(y)}{f(h(y), 0)g(y)} = \psi(h(y)).$$

Thus, the likelihood ratio depends only on $h(y)$ and it follows that the solution of the hypothesis testing problem is also a function of $h(Y)$.

15.2.1 Interpretation

The definition of sufficient statistic is quite abstract. The intuitive meaning is that if $h(Y)$ is sufficient for X, then Y is some function of $h(Y)$ and a random variable Z that is independent of X and Y. That is,

$$Y = g(h(Y), Z). \tag{15.1}$$

For instance, say that $Y = (Y_1, \ldots, Y_n)$ where the Y_m are i.i.d. and Bernoulli with parameter $X \in [0, 1]$. Let $h(Y) = Y_1 + \cdots + Y_n$. Then we can think of Y as being constructed from $h(Y)$ by selecting randomly which $h(Y)$ random variables among $(Y_1, \ldots, Y_n)$ are equal to one. This random choice is some independent random variable Z. In such a case, we see that Y does not contain any information about X that is not already in $h(Y)$.

To see the equivalence between this interpretation and the definition, first assume that (15.1) holds. Then

$$P[Y \approx y | X = x] = P[h(Y) \approx h(y) | X = x] P(g(h(y), Z) \approx y)$$
$$= f(h(y), x)g(y),$$

so that $h(Y)$ is sufficient for X. Conversely, if $h(Y)$ is sufficient for X, then we can find some Z such that $g(h(y), Z)$ has the density $f_{Y|h(Y)}[y|h(y)]$.

15.3   Infinite Markov Chains

We studied Markov chains on a finite state space $\mathcal{X} = \{1, 2, \ldots, N\}$. Let us explore the countably infinite case where $\mathcal{X} = \{0, 1, \ldots\}$.

One is given an initial distribution $\pi = \{\pi(x), x \in \mathcal{X}\}$, where $\pi(x) \geq 0$ and $\sum_{x \in \mathcal{X}} \pi(x) = 1$. Also, one is given a set of nonnegative numbers $\{P(x, y), x, y \in \mathcal{X}\}$ such that

$$\sum_{y \in \mathcal{X}} P(x, y) = 1, \forall x \in \mathcal{X}.$$

The sequence $\{X(n), n \geq 0\}$ is then a Markov chain with initial distribution π and probability transition matrix P if

Fig. 15.2 An infinite
Markov chain

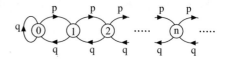

$$P(X(0) = x_0, X(1) = x_1, \ldots, X(n) = x_n)$$

$$= \pi(x_0) P(x_0, x_1) \times \cdots \times P(x_{n-1}, x_n),$$

for all $n \geq 0$ and all $x_0, \ldots, x_n$ in $\mathscr{X}$.

One defines irreducible and aperiodic as in the case of a finite Markov chain. Recall that if a finite Markov chain is irreducible, then it visits all its states infinitely often and it spends a positive fraction of time in each state.

That may not happen when the Markov chain is infinite. To see this, consider the following example (see Fig. 15.2). One has $\pi(0) = 1$ and $P(i, i + 1) = p$ for $i \geq 1$ and

$$P(i + 1, i) = 1 - p =: q = P(0, 0), \forall i.$$

Assume that $p \in (0, 1)$. Then the Markov chain is irreducible. However, it is intuitively clear that $X(n) \to \infty$ as $n \to \infty$ if $p > 0.5$. To see that this is indeed the case, let $Z(n)$ be i.i.d. random variables with $P(Z(n) = 1) = p$ and $P(Z(n) = -1) = q$. Then note that

$$X(n) = \max\{X(n - 1) + Z(n), 0\},$$

so that

$$X(n) \geq X(0) + Z(1) + \cdots + Z(n - 1), n \geq 0.$$

Also,

$$\frac{X(n)}{n} \geq \frac{X(0) + Z(1) + \cdots + Z(n - 1)}{n} \to E(Z(n)) > 0,$$

where the convergence follows by the SLLN. This implies that $X(n) \to \infty$, as claimed.

Thus, $X(n)$ eventually is larger than any given N and remains larger. This shows that $X(n)$ visits every state only finitely many times. We say that the states are *transient* because they are visited only finitely often.

We say that a state is *recurrent* if it is not transient. In that case, the state is called *positive recurrent* if the average time between successive visits is finite; otherwise it is called *null recurrent*.

Here is the result that corresponds to Theorem 1.1

Theorem 15.1 (Big Theorem for Infinite Markov Chains) *Consider an infinite Markov chain.*

(a) *If the Markov chain is irreducible, the states are either all transient, all positive recurrent, or all null recurrent. We then say that the Markov chain is transient, positive recurrent, or null recurrent, respectively.*

(b) *If the Markov chain is positive recurrent, it has a unique invariant distribution π and $\pi(i)$ is the long-term fraction of time that $X(n)$ is equal to i.*

(c) *If the Markov chain is positive recurrent and also aperiodic, then the distribution π_n of $X(n)$ converges to π.*

(d) *If the Markov chain is not positive recurrent, it does not have an invariant distribution and the fraction of time that it spends in any state goes to zero.*

∎

It turns out that the Markov chain in Fig. 15.2 is null recurrent for $p = 0.5$ and positive recurrent for $p < 0.5$. In the latter case, its invariant distribution is

$$\pi(i) = (1 - \rho)\rho^i, i \geq 0, \text{ where } \rho := \frac{p}{q}.$$

15.3.1 Lyapunov–Foster Criterion

Here is a useful sufficient condition for positive recurrence.

Theorem 15.2 (Lyapunov–Foster) *Let $X(n)$ be an irreducible Markov chain on an infinite state space $\mathscr{X}$. Assume there exists some function $V : \mathscr{X} \to [0, \infty)$ such that*

$$E[V(X(n + 1)) - V(X(n))|X(n) = x] \leq -\alpha + \beta 1\{x \in A\},$$

where A is a finite set, $\alpha > 0$ and $\beta > 0$.
  Then the Markov chain is positive recurrent.
  Such a function V is said to be a Lyapunov function for the Markov chain.

∎

The condition means that the Lyapunov function decreases by at least α on average when $X(n)$ is outside some finite set A. The intuitive reason why this makes the Markov chain positive recurrent is that, since the Lyapunov function is nonnegative, it cannot decrease forever. Thus, it must spend a positive fraction of time inside the finite set A. By the big theorem, this implies that it is positive recurrent.

15.4 Poisson Process

The Poisson process is an important model in applied probability. It is a good approximation of the arrivals of packets at a router, of telephone calls, of new TCP connections, of customers at a cashier.

15.4.1 Definition

We start with a definition of the Poisson process. (See Fig. 15.3.)

Definition 15.2 (Poisson Process) Let $\lambda > 0$ and $\{S_1, S_2, \ldots\}$ be i.i.d. $Exp(\lambda)$ random variables. Let also $T_n = S_1 + \cdots + S_n$ for $n \geq 1$. Define

$$N_t = \max\{n \geq 1 | T_n \leq t\}, t \geq 0,$$

with $N_t = 0$ if $t < T_1$. Then, $\mathbf{N} := \{N_t, t \geq 0\}$ is a Poisson process with rate λ. Note that T_n is the n-th jump time of $\mathbf{N}$.

◇

15.4.2 Independent Increments

Before exploring the properties of the Poisson process, we recall two properties of the exponential distribution.

Theorem 15.3 (Properties of Exponential Distribution) *Let τ be exponentially distributed with rate $\lambda > 0$. That is,*

$$F_\tau(t) = P(\tau \leq t) = 1 - \exp\{-\lambda t\}, t \geq 0.$$

Fig. 15.3 Poisson process: the times S_n between jumps are i.i.d. and exponentially distributed with rate λ

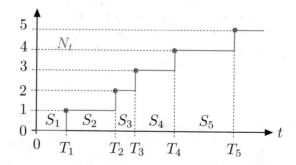

In particular, the pdf of τ is $f_\tau(t) = \lambda \exp\{-\lambda t\}$ for $t \geq 0$. Also, $E(\tau) = \lambda^{-1}$ and $var(\tau) = \lambda^{-2}$.

Then,

$$P[\tau > t + s | \tau > s] = P(\tau > t).$$

This is the memoryless property *of the exponential distribution.*

   Also,

$$P[\tau \leq t + \epsilon | \tau > t] = \lambda \epsilon + o(\epsilon).$$

∎

Proof

$$P[\tau > t + s | \tau > s] = \frac{P(\tau > t + s)}{P(\tau > s)}$$

$$= \frac{\exp\{-\lambda(t+s)\}}{\exp\{-\lambda s\}} = \exp\{-\lambda t\}$$

$$= P(\tau > t).$$

□

The interpretation of this property is that if a lightbulb has an exponentially distributed lifetime, then an old bulb is exactly as good as a new one (as long as it is still burning).

We use this property to show that the Poisson process is also memoryless, in a precise sense.

Theorem 15.4 (Poisson Process Is Memoryless) *Let* $\mathbf{N} := \{N_t, t \geq 0\}$ *is a Poisson process with rate λ. Fix $t > 0$. Given $\{N_s, s \leq t\}$, the process $\{N_{s+t} - N_t, s \geq 0\}$ is a Poisson process with rate λ.*

   As a consequence, the process has stationary and independent increments. That is, for any $0 \leq t_1 < t_2 < \cdots$, the increments $\{N_{t_{n+1}} - N_{t_n}, n \geq 1\}$ of the Poisson process are independent and the distribution of $N_{t_{n+1}} - N_{t_n}$ depends only on $t_{n+1} - t_n$.

∎

Proof Figure 15.4 illustrates that result. Given $\{N_s, s \leq t\}$, the first jump time of $\{N_{s+t} - N_t, s \geq 0\}$ is $Exp(\lambda)$, by the memoryless property of the exponential distribution. The subsequent inter-jump times are i.i.d. and $Exp(\lambda)$. This proves the theorem.   □

Fig. 15.4 Given the past of the process up to time t, the future jump times are those of a Poisson process

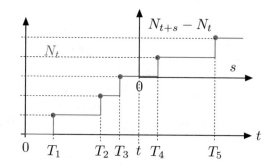

15.4.3 Number of Jumps

One has the following result.

Theorem 15.5 (The Number of Jumps Is Poisson) $\mathbf{N} := \{N_t, t \geq 0\}$ is a Poisson process with rate λ. Then N_t has a Poisson distribution with mean λt.

■

Proof There are a number of ways of showing this result. The standard way is as follows. Note that

$$P(N_{t+\epsilon} = n) = P(N_t = n)(1 - \lambda\epsilon) + P(N_t = n - 1)\lambda\epsilon + o(\epsilon).$$

Hence,

$$\frac{d}{dt}P(N_t = n) = \lambda P(N_t = n - 1) - \lambda P(N_t = n).$$

Thus,

$$\frac{d}{dt}P(N_t = 0) = -\lambda P(N_t = 0).$$

Since $P(N_0 = 0) = 1$, this shows that $P(N_t = 0) = \exp\{-\lambda t\}$ for $t \geq 0$. Now, assume that

$$P(N_t = n) = g(n, t)\exp\{-\lambda t\}, n \geq 0.$$

Then, the differential equation above shows that

$$\frac{d}{dt}[g(n, t)\exp\{-\lambda t\}] = \lambda[g(n - 1, t) - g(n, t)]\exp\{-\lambda t\},$$

i.e.,

$$\frac{d}{dt} g(n, t) = \lambda g(n - 1, t).$$

This expression shows by induction that $g(n, t) = \frac{(\lambda t)^n}{n!}$.

A different proof makes use of the density of the jumps. Let T_n be the n-th jump of the process and $S_n = T_n - T_{n-1}$, as before. Then

$$P(T_1 \in (t_1, t_1 + dt_1), \ldots, T_n \in (t_n, t_n + dt_n), T_{n+1} > t)$$
$$= P(S_1 \in (t_1, t_1 + dt_1), \ldots, S_n \in (t_n - t_{n-1}, t_n$$
$$- t_{n-1} + dt_n), S_{n+1} > t - t_n)$$
$$= \lambda \exp\{-\lambda t_1\} dt_1 \lambda \exp\{-\lambda(t_2 - t_1)\} dt_2 \cdots \exp\{-\lambda(t - t_n)\}$$
$$= \lambda^n dt_1 \cdots dt_n \exp\{-\lambda t\}.$$

To derive this expression, we used the fact that the S_n are i.i.d. $Exp(\lambda)$. The expression above shows that, given that there are n jumps in $[0, t]$, they are equally likely to be anywhere in the interval. Also,

$$P(N_t = n) = \int_S \lambda^n dt_1 \cdots dt_n \exp\{-\lambda t\},$$

where $S = \{t_1, \ldots, t_n | 0 < t_1 < \cdots < t_n < t\}$. Now, observe that S is a fraction of $[0, t]^n$ that corresponds to the times t_i being in a particular order. There are $n!$ such orders and, by symmetry, each order corresponds to a subset of $[0, t]^n$ of the same size. Thus, the volume of S is $t^n/n!$. We conclude that

$$P(N_t = n) = \frac{t^n}{n!} \lambda^n \exp\{-\lambda t\},$$

which proves the result.   □

15.5 Boosting

You follow the advice of some investment experts when you buy stocks. Their recommendations are often contradictory. How do you make your decisions so that, in retrospect, you are not doing too bad compared to the best of the experts? The intuition is that you should try to follow the leader, but randomly. To make the situation concrete, Fig. 15.5 shows three experts (B, I, T) and the profits one would make by following their advice on the successive days.

On a given day, you choose which expert to follow the next day. Figure 15.6 shows your profit if you make the sequence of selections indicated by the red circles.

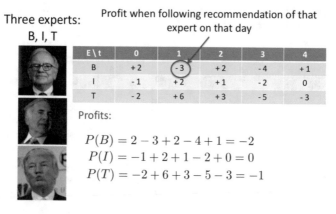

Fig. 15.5 The three experts and the profits of their recommended stocks

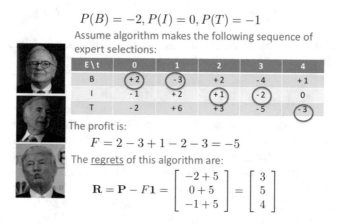

Fig. 15.6 A specific sequence of choices and the resulting profit and regrets

In these selections, you choose to follow B the first 2 days, then I the next to days, then T the last day. Of course, you have to choose the day before, and the actual profit is only known the next day. The figure also shows the *regrets* that you accumulate when comparing your profit to that of the three experts. Your total profit is -5 and the profit you would have made if you had followed B all the time would have been -2, so your *regret* compared to B is $-2 - (-5) = 3$, and similarly for the other two experts.

The problem is to make the expert selection every day so as to minimize the worst regret, i.e., the regret with respect to the most successful expert. More precisely, the goal is to minimize the rate of growth of the worst regret. Here is the result.

Theorem 15.6 (Minimum Regret Algorithm) *Generally, the worst regret grows like $O(\sqrt{n})$ with the number n of steps. One algorithm that achieves this rate of regret is to choose expert E at step $n + 1$ with probability $\pi_{n+1}(E)$ given by*

Fig. 15.7 A simulation of
the experts and the selection
algorithm

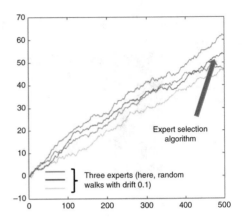

Expert selection
algorithm

Three experts (here, random
walks with drift 0.1)

$$\pi_{n+1}(E) = A_n \exp\{\alpha P_n(E)/\sqrt{n}\}, \; for \; E \in \{B, I, T\},$$

where $\eta > 0$ is a constant, A_n is such that these probabilities add up to one, and $P_n(E)$ is the profit that expert E makes in the first n days.

■

Thus, the algorithm favors successful experts. However, the algorithm makes random selections. It is easy to construct examples where a deterministic algorithm accumulates a regret that grows like n.

Figure 15.7 shows a simulation of three experts and of the selection algorithm in the theorem. The experts are random walks with drift 0.1. The simulation shows that the selection algorithm tends to fall behind the best expert by $O(\sqrt{n})$.

The proof of the theorem can be found in Cesa-Bianchi and Lugosi (2006).

15.6   Multi-Armed Bandits

Here is a classical problem. You are given two coins, both with an unknown bias (the probability of heads). At each step $k = 1, 2, \dots$ you choose a coin to flip. Your goal is to accumulate heads as fast as possible. Let X_k be the number of heads you accumulate after k steps. Let also X_k^* be the number of heads that you would accumulate if you always flipped the coin with the largest bias. The *regret* of your strategy after n steps is defined as

$$R_k = E(X_k^* - X_k).$$

Let θ_1 and θ_2 be the bias of coins 1 and 2, respectively. Then $E(X_k^*) = k \max\{\theta_1, \theta_2\}$ and the best strategy is to flip the coin with the largest bias at each step. However, since the two biases are unknown, you cannot use that strategy. We explain below that there is a strategy such that the regret grows like $\log(k)$ with the number of steps.

Any good strategy keeps on estimating the biases. Indeed, any strategy that stops estimating and then forever flips the coin that is believed to be best has a positive probability of getting stuck with the worst coin, thus accumulating a regret that grows linearly over time. Thus, a good strategy must constantly *explore*, i.e., flip both coins to learn their bias.

However, a good strategy should exploit the estimates by flipping the coin that is believed to be better more frequently than the other. Indeed, if you were to flip the two coins the same fraction of time, the regret would also grow linearly. Hence, a good strategy must *exploit* the accumulated knowledge about the biases.

The key question is how to balance exploration and exploitation. The strategy called *Thompson Sampling* does this optimally. Assume that the biases θ_1 and θ_2 of the two coins are independent and uniformly distributed in $[0, 1]$. Say that you have flipped the coins a number of times. Given the outcomes of these coin flips, one can in principle compute the conditional distributions of θ_1 and θ_2. Given these conditional distributions, one can calculate the probability that $\theta_1 > \theta_2$. The Thompson Sampling strategy is to choose coin 1 with that probability and coin 2 otherwise for the next flip. Here is the key result.

Theorem 15.7 (Minimum Regret of Thompson Sampling) *If the coins have different biases, then any strategy is such that*

$$R_k \geq O(\log k).$$

Moreover, Thompson Sampling achieves this lower bound.

■

The notation $O(\log k)$ indicates a function $g(k)$ that grows like $\log k$, i.e., such that $g(k)/\log k$ converges to a positive constant as $k \to \infty$.

Thus this strategy does not necessarily choose the coin with the largest expected bias. It is the case that the strategy favors the coin that has been more successful so far, thus exploiting the information. But the selection is random, which contributes to the exploration.

One can show that if flips of coin 1 have produced h heads and t tails, then the conditional density of θ_1 is $g(\theta; h, t)$, where

$$g(\theta; h, t) = \frac{(h+t)!}{h!t!}\theta^h(1-\theta)^t, \theta \in [0, 1].$$

The same result holds for coin 2. Thus, Thompson Sampling generates $\hat{\theta}_1$ and $\hat{\theta}_2$ according to these densities.

For a proof of this result, see Agrawal and Goyal (2012). See also Russo et al. (2018) for applications of multi-armed bandits.

A rough justification of the result goes as follows. Say that $\theta_1 > \theta_2$. One can show that after flipping coin 2 a number n of times, it takes about n steps until you flip it again when using Thompson Sampling. Your regret then grows by one at times $1, 1 + 1, 2 + 2, 4 + 4, \ldots, 2^n, 2^{n+1}, \ldots$. Thus, the regret is of order n after $O(2^n)$ steps. Equivalently, after $N = 2^n$ steps, the regret is of order $n = \log N$.

15.7 Capacity of BSC

Consider a binary symmetric channel with error probability $p \in (0, 0.5)$. Every bit that the transmitter sends has a chance of being corrupted. Thus, it is impossible to transmit any bit string fully reliably across this channel. No matter what the transmitter sends, the receiver can never be sure that it got the message right.

However, one might be able to achieve a very small probability of error. For instance, say that $p = 0.1$ and that one transmits a bit by repeating it N times, where $N \gg 1$. As the receiver gets the N bits, it uses a majority decoding. That is, if it gets more zeros than ones, it decides that transmitter sent a zero, and conversely for a one. The probability of error can be made arbitrarily small by choosing N very large. However, this scheme gets to transmit only one bit every N steps. We say that the *rate* of the channel is $1/N$ and it seems that to achieve a very small error probability, the rate has to become negligible.

It turns out that our pessimistic conclusion is wrong. Claude Shannon (Fig. 15.8), in the late 1940s, explained that the channel can transmit at any rate less than $C(p)$, where (see Fig. 15.9)

$$C(p) = 1 - H(p) \text{ with } H(p) = -p \log_2 p - (1 - p) \log_2(1 - p), \qquad (15.2)$$

with a probability less than ϵ, for any $\epsilon > 0$.

For instance, $C(0.1) \approx 0.53$. Fix a rate less than $C(0.1)$, say $R = 0.5$. Pick any $\epsilon > 0$, say $\epsilon = 10^{-8}$. Then, it is possible to transmit bits across this channel at rate $R = 0.5$, with a probability of error per bit less than 10^{-8}. The same is true if we choose $\epsilon = 10^{-12}$: it is possible to transmit at the same rate R with a probability of

Fig. 15.8 Claude Shannon. 1916–2001

Fig. 15.9 The capacity $C(p)$
of the BSC with error
probability p

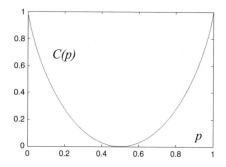

error less than 10^{-12}. The actual scheme that we use depends on ϵ, and it becomes more complex when ϵ is smaller; however, the rate R does not depend on ϵ. Quite a remarkable result! Needless to say, it baffled all the engineers who had been busily designing various ad hoc transmission schemes.

Shannon's key insight is that *long sequences are typical*. There is a statistical regularity in random sequences such as Markov chains or i.i.d. random variables and this regularity manifests itself in a characteristic of long sequences. For instance, flip many times a biased coin with $P(head) = 0.1$. The sequence that you will observe is likely to have about 10% of heads. Many other sequences are so unlikely that you will not see them. Thus, there are relatively few long sequences that are possible. In this example, although there are $M = 2^N$ possible sequences of N coin flips, only about $\sqrt{M}$ are typical when $P(head) = 0.1$. Moreover, by symmetry, these typical sequences are all equally likely. For that reason, the errors of the BSC must correspond to relatively few patterns. Say that there are only A possible patterns of errors for N transmissions. Then, any bit string of length N that the sender transmits will correspond to A possible received "output" strings: one for every typical error sequence. Thus, it might be possible to choose B different "input" strings of length N for the transmitter so that the A received "output" strings for each one of these B input strings are all distinct. However, one might worry that choosing the B input strings would be rather complex if we want their sets of output strings to be distinct.

Shannon noticed that if we pick the input strings completely randomly, this will work. Thus, Shannon scheme is as follows. Pick a large N. Choose B strings of N bits randomly, each time by flipping a fair coin N times. Call these inputs strings $\mathbf{X}_1, \ldots \mathbf{X}_B$. These are the *codewords*. Let S_1 be the set of A typical outputs that correspond to $\mathbf{X}_1$. Let $\mathbf{Y}_j$ be the output that corresponds to input $\mathbf{X}_j$. Note that the $\mathbf{Y}_j$ are sequences of fair coin flips, by symmetry of the channel. Thus, each $\mathbf{Y}_j$ is equally likely to be any one of the 2^N possible output strings. In particular, the probability that $\mathbf{Y}_j$ falls in S_1 is $A/2^N$ (Fig. 15.10).

In fact,

$$P(\mathbf{Y}_2 \in S_1 \text{ or } \mathbf{Y}_3 \in S_1 \ldots \text{ or } \mathbf{Y}_B \in S_1) \le B \times A2^{-N}.$$

Fig. 15.10 Because of the random choice of the codewords, the likelihood that one codeword produces an output that is typical for another codeword is $A2^{-N}$

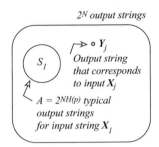

Indeed, the probability of a union of events is not larger than the sum of their probabilities. We explain below that $A = 2^{NH(p)}$. Thus, if we choose $B = 2^{NR}$, we see that the expression above is less than or equal to

$$2^{NR} \times 2^{NH(p)} \times 2^{-N}$$

and this expression goes to zero as N increases, provided that

$$R + H(p) < 1, \text{ i.e., } R < C(p) := 1 - H(p).$$

Thus, the receiver makes an error with a negligible probability if one does not choose too many codewords. Note that $B = 2^{NR}$ corresponds to transmitting NR different bits in N steps, thus transmitting at rate R.

How does the receiver recognize the bit string that the transmitter sent? The idea is to give the list of the B input strings, i.e., codewords, to the receiver. When it receives a string, the receiver looks in the list to find the codeword that is the closest to the string it received. With a very high probability, it is the string that the transmitter sent.

It remains to show that $A = 2^{NH(p)}$. Fortunately, this calculation is a simple consequence of the SLLN. Let $\mathbf{X} := \{X(n), n = 1, \dots, N\}$ be i.i.d. random variables with $P(X(n) = 1) = p$ and $P(X(n) = 0) = 1 - p$. For a given sequence $\mathbf{x} = (x(1), \dots, x(N)) \in \{0, 1\}^N$, let

$$\psi(\mathbf{x}) := \frac{1}{N} \log_2(P(\mathbf{X} = \mathbf{x})). \tag{15.3}$$

Note that, with $|\mathbf{x}| := \sum_{n=1}^{N} x(n)$,

$$\psi(\mathbf{x}) = \frac{1}{N} \log_2(p^{|\mathbf{x}|}(1 - p)^{N - |\mathbf{x}|})$$

$$= \frac{|\mathbf{x}|}{N} \log_2(p) + \frac{N - |\mathbf{x}|}{N} \log_2(1 - p).$$

Thus, the random string $\mathbf{X}$ of N bits is such that

$$\psi(\mathbf{X}) = \frac{|\mathbf{X}|}{N} \log_2(p) + \frac{N - |\mathbf{X}|}{N} \log_2(1 - p).$$

But we know from the SLLN that $|\mathbf{X}|/N \to p$ as $N \to \infty$. Thus, for $N \gg 1$,

$$\psi(\mathbf{X}) \approx p \log_2(p) + (1 - p) \log_2(1 - p) =: -H(p).$$

This calculation shows that any sequence $\mathbf{x}$ of values that $\mathbf{X}$ takes has approximately the same value of $\psi(\mathbf{x})$. But, by (15.3), this implies all the sequences $\mathbf{x}$ that occur have approximately the same probability

$$2^{-NH(p)}.$$

We conclude that there are $2^{NH(p)}$ typical sequences and that they are all essentially equally likely. Thus, $A = 2^{NH(p)}$.

Recall that for the Gaussian channel with the MLE detection rule, the channel becomes a BSC with

$$p = p(\sigma^2) := P(\mathcal{N}(0, \sigma^2) > 0.5).$$

Accordingly, we can calculate the capacity $C(p(\sigma^2))$ as a function of the noise standard deviation σ. Figure 15.11 shows the result.

These results of Shannon on the capacity, or achievable rates, of channels have had a profound impact on the design of communication systems. Suddenly, engineers had a target and they knew how far or how close their systems were to the feasible rate. Moreover, the coding scheme of Shannon, although not really practical, provided a valuable insight into the design of codes for specific channels. Shannon's theory, called *Information Theory*, is an inspiring example of how a profound conceptual insight can revolutionize an engineering field.

Fig. 15.11 The capacity of the BSC that corresponds to a $\mathcal{N}(0, \sigma^2)$ additive noise. The detector uses the MLE

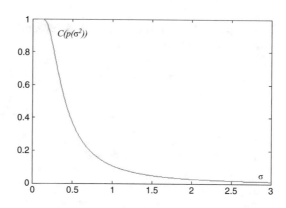

Another important part of Shannon's work concerns the coding of random objects. For instance, how many bits does it take to encode a 500-page book? Once again, the relevant notion is that of typicality. As an example, we know that to encode a string of N flips of a biased coin with $P(head) = p$, we need only $NH(p)$ bits, because this is the number of typical sequences. Here, $H(p)$ is called the *entropy* of the coin flip. Similarly, if $\{X(n), n \geq 1\}$ is an irreducible, finite, and aperiodic Markov chain with invariant distribution π and transition probabilities $P(i, j)$, then one can show that to encode $\{X(1), \ldots, X(N)\}$ one needs approximately $NH(P)$ bits, where

$$H(P) = -\sum_i \pi(i) \sum_j P(i, j) \log_2 P(i, j)$$

is called the *entropy rate* of the Markov chain. A practical scheme, called Liv–Zempel compression, essentially achieves this limit. It is the basis for most file compression algorithms (e.g., ZIP).

Shannon put these two ideas together: channel capacity and source coding. Here is an example of his source–channel coding result. How fast can one send the symbols $X(n)$ produced by the Markov chain through a BSC channel? The answer is $C(p)/H(P)$. Intuitively, it takes $H(P)$ bits per symbol $X(n)$ and the BSC can send $C(p)$ bits per unit time. Moreover, to accomplish this rate, one first encodes the source and one separately chooses the codewords for the BSC, and one then uses them together. Thus, the channel coding is independent of the source coding and vice versa. This is called the *separation theorem* of Claude Shannon.

15.8   Bounds on Probabilities

We explain how to derive estimates of probabilities using Chebyshev and Chernoff's inequalities and also using the Gaussian approximation. These methods also provide a useful insight into the likelihood of events. The power of these methods is that they can be used in very complex situations.

Theorem 15.8 (Markov, Chernoff, and Jensen Inequalities) *Let X be a random variable. Then one has*

(a) Markov's Inequality:[1]

$$P(X \geq a) \leq \frac{E(f(X))}{f(a)},\tag{15.4}$$

for all $f(\cdot)$ that is nondecreasing and positive.

---

[1]Markov's inequality is due to Chebyshev who was Markov's teacher.

Fig. 15.12 Herman
Chernoff. b. 1923

Fig. 15.13 Johan Jensen.
1859–1925

(b) Chernoff's Inequality (Fig. 15.12):[2]

$$P(X \geq a) \leq E(\exp\{\theta(X - a)\}), \qquad (15.5)$$

for all $\theta > 0$.
(c) Jensen's Inequality (Fig. 15.13):[3]

$$f(E(X)) \leq E(f(X)), \qquad (15.6)$$

for all $f(\cdot)$ that is convex.

■

These results are easy to show, so here is a proof.

---

[2]Chernoff's inequality is due to Herman Rubin (see Chernoff (2004)).
[3]Jensen's inequality seems to be due to Jensen.

Proof

(a) Since $f(\cdot)$ is nondecreasing and positive, we have

$$1\{X \geq a\} \leq \frac{f(X)}{f(a)},$$

so that (15.4) follows by taking expectations.
(b) The inequality (15.5) is a particular case of Markov's inequality (15.4) for $f(X) = \exp\{\theta X\}$ with $\theta > 0$.
(c) Let $f(\cdot)$ be a convex function. This means that it lies above any tangent. In particular,

$$f(X) \geq f(E(X)) + f'(E(X))(X - E(X)),$$

as shown in Fig. 15.14. The inequality (15.6) then follows by taking expectations.

□

15.8.1 Applying the Bounds to Multiplexing

Recall the multiplexing problem. There are N users who are independently active with probability p. Thus, the number of active users Z is $B(N, p)$. We want to find m so that $P(Z \geq m) = 5\%$.

As a first estimate of m, we use Chebyshev's inequality (2.2) which says that

$$P(|v - E(v)| > \epsilon) \leq \frac{\text{var}(v)}{\epsilon^2}.$$

Fig. 15.14 A convex function $f(\cdot)$ lies above its tangents. In particular, it lies above a tangent at $E(X)$, which implies Jensen's inequality.

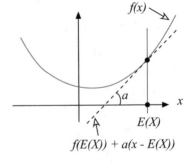

$f(E(X)) + a(x - E(X))$

Now, if $Z = B(N, p)$, one has $E(Z) = Np$ and $\text{var}(Z) = Np(1 - p)$.[4] Hence, since $\nu = B(100, 0.2)$, one has $E(\nu) = 20$ and $\text{var}(\nu) = 16$. Chebyshev's inequality gives

$$P(|\nu - 20| > \epsilon) \leq \frac{16}{\epsilon^2}.$$

Thus, we expect that

$$P(\nu - 20 > \epsilon) \leq \frac{8}{\epsilon^2},$$

because it is reasonable to think that the distribution of ν is almost symmetric around its mean, as we see in Fig. 3.4. We want to choose $m = 20 + \epsilon$ so that $P(\nu > m) \leq 5\%$. This means that we should choose ϵ so that $8/\epsilon^2 = 5\%$. This gives $\epsilon = 13$, so that $m = 33$. Thus, according to Chebyshev's inequality, it is safe to assume that no more than 33 users are active and we can choose C so that $C/33$ is a satisfactory rate for users.

As a second approach, we use Chernoff's inequality (15.5) which states that

$$P(\nu \geq Na) \leq E(\exp\{\theta(\nu - Na)\}), \forall \theta > 0.$$

To calculate the right-hand size, we note that if $Z = Bernoulli(N, p)$, then we can write as $Z = X(1) + \cdots + X(N)$, where the $X(n)$ are i.i.d. random variables with $P(X(n) = 1) = p$ and $P(X(n) = 0) = 1 - p$. Then,

$$E(\exp\{\theta Z\}) = E(\exp\{\theta X(1) + \cdots + \theta X(N)\})$$
$$= E(\exp\{\theta X(1)\} \times \cdots \times \exp\{\theta X(N)\}).$$

To continue the calculation, we note that, since the $X(n)$ are independent, so are the random variables $\exp\{\theta X(n)\}$.[5] Also, the expected value of a product of independent random variables is the product of their expected values (see Appendix A). Hence,

$$E(\exp\{\theta Z\}) = E(\exp\{\theta X(1)\}) \times \cdots \times E(\exp\{\theta X(N)\})$$
$$= E(\exp\{\theta X(1)\})^N = \exp\{N\Lambda(\theta)\}$$

where we define

$$\Lambda(\theta) = \log(E(\exp\{\theta X(1)\})).$$

---

[4]See Appendix A.

[5]Indeed, functions of independent random variables are independent. See Appendix A.

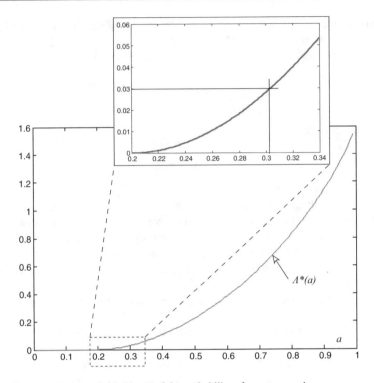

Fig. 15.15 The logarithm divided by N of the probability of too many active users

Thus, Chernoff's inequality says that

$$P(Z \geq Na) \leq \exp\{N\Lambda(\theta)\} \exp\{-\theta Na\}$$
$$= \exp\{N(\Lambda(\theta) - \theta a)\}$$

Since this inequality holds for every $\theta > 0$, let us minimize the right-hand side with respect to θ. That is, let us define

$$\Lambda^*(a) = \max_{\theta > 0}\{\theta a - \Lambda(\theta)\}.$$

Then, we see that

$$P(Z \geq Na) \leq \exp\{-N\Lambda^*(a)\}. \tag{15.7}$$

Figure 15.15 shows this function when $p = 0.2$.

We now evaluate $\Lambda(\theta)$ and $\Lambda^*(a)$. We find

$$E(\exp\{\theta X(1)\}) = 1 - p + pe^{\theta},$$

so that

$$\Lambda(\theta) = \log(1 - p + pe^{\theta})$$

and

$$\Lambda^*(a) = \max_{\theta > 0}\{\theta a - \log(1 - p + pe^{\theta})\}.$$

Setting to zero the derivative with respect to θ of the term between brackets, we find

$$a = \frac{1}{1 - p + pe^{\theta}}(pe^{\theta}),$$

which gives, for $a > p$,

$$e^{\theta} = \frac{a(1 - p)}{(1 - a)p}.$$

Substituting back in $\Lambda^*(a)$, we get

$$\Lambda^*(a) = -a \log(\frac{a}{p}) - (1 - a) \log(\frac{1 - a}{1 - p}), \forall a > p.$$

Going back to our example, we want to find $m = Na$ so that

$$P(v \geq Na) \approx 0.05.$$

Using (15.7), we need to find Na so that

$$\exp\{-N\Lambda^*(a)\} \approx 0.05 = \exp\{\log(0.05)\},$$

i.e.,

$$\Lambda^*(a) = -\frac{\log(0.05)}{N} \approx 0.03.$$

Looking at Fig. 15.15, we find $a = 0.30$. This corresponds to $m = 30$. Thus, Chernoff's estimate says that $P(v > 30) \approx 5\%$ and that we can size the network assuming that only 30 users are active at any one time.

By the way, the calculations we have performed above show that Chernoff's bound can be written as

$$P(Z \geq Na) \leq \frac{P(B(N, p) = Na)}{P(B(N, a) = Na)}.$$

15.9   Martingales

A martingale represents the sequence of fortunes of someone playing a fair game of chance. In such a game, the expected gain is always zero. A simple example is a random walk with zero-mean step size. Martingales are good models of noise and of processes discounted based on their expected value (e.g., the stock market). This theory is due to Doob (1953).

Martingales have an important property that generalizes the strong law of large numbers. It says that a martingale bounded in expectation converges almost surely. This result is used to show that fluctuations vanish and that a process converges to its mean value. The convergence of stochastic gradient algorithms and approximations of random processes by differential equations follow from that property.

15.9.1 Definitions

Let X_n be the fortune at time $n \geq 0$ when one plays a game of chance. The game is fair if

$$E[X_{n+1}|X^n] = X_n, \forall n \geq 0. \tag{15.8}$$

In this expression, $X^n := \{X_m, m \leq n\}$. Thus, in a fair game, one cannot expect to improve one's fortune. A sequence $\{X_n, n \geq 0\}$ of random variables with that property is a martingale.

This basic definition generalizes to the case where one has access to additional information and is still unable to improve one's fortune. For instance, say that the additional information is the value of other random variables Y_n. One then has the following definitions.

Definition 15.3 (Martingale, Supermartingale, Submartingale)   The sequence of random variables $\{X_n, n \geq 0\}$ is a martingale with respect to $\{X_n, Y_n, n \geq 0\}$ if

$$E[X_{n+1}|X^n, Y^n] = X_n, \forall n \geq 0 \tag{15.9}$$

with $X^n = \{X_m, m \leq n\}$ and $Y^n = \{Y_m, m \leq n\}$.

If (15.9) holds with = replaced by $\leq$, then X_n is a *supermartingale*; if it holds with $\geq$, then X_n is a *submartingale*.

◇

In many cases, we do not specify the random variables Y_n and we simply say that X_n is a martingale, or a submartingale, or a supermartingale.

Note that if X_n is a martingale, then

$$E(X_n) = E(X_0), \forall n \geq 0.$$

Indeed, $E(X_n) = E(E[X_n|X_0, Y_0])$ by the smoothing property of conditional expectation (see Theorem 9.5).

15.9.2 Examples

A few examples illustrate the definition.

Random Walk

Let $\{Z_n, n \geq 0\}$ be independent and zero-mean random variables. Then $X_n := Z_0 + \cdots + Z_n$ for $n \geq 0$ is a martingale. Indeed,

$$E[X_{n+1}|X^n] = E[Z_0 + \cdots + Z_n + Z_{n+1}|Z_0, \ldots, Z_n] = Z_0 + \cdots + Z_n = X_n.$$

Note that if $E(Z_n) \leq 0$, then X_n is a supermartingale; if $E(Z_n) \geq 0$, then X_n is a submartingale.

Product

Let $\{Z_n, n \geq 0\}$ be independent random variables with mean 1. Then $X_n := Z_0 \times \cdots \times Z_n$ for $n \geq 0$ is a martingale. Indeed,

$$E[X_{n+1}|X^n] = E[Z_0 \times \cdots \times Z_n \times Z_{n+1}|Z_0, \ldots, Z_n] = Z_0 \times \cdots \times Z_n = X_n.$$

Note that if $Z_n \geq 0$ and $E(Z_n) \leq 1$ for all n, then X_n is a supermartingale. Similarly, if $Z_n \geq 0$ and $E(Z_n) \geq 1$ for all n,, then X_n is a submartingale.

Branching Process

For $m \geq 1$ and $n \geq 0$, let X_m^n be i.i.d. random variables distributed like X that take values in $\mathbb{Z}_+ := \{0, 1, 2, \ldots\}$ and have mean μ. The branching process is defined by $Y_0 = 1$ and

$$Y_{n+1} = \sum_{m=1}^{Y_n} X_m^n, n \geq 0.$$

The interpretation is that there are Y_n individuals in a population at the n-th generation. Individual m in that population has X_m^n children.

One can see that

$$Z_n = \mu^{-n} Y_n, n \geq 0$$

is a martingale. Indeed,

$$E[Y_{n+1}|Y_0, \ldots, Y_n] = Y_n \mu,$$

so that

$$E[Z_{n+1}|Z_0, \ldots, Z_n] = E[\mu^{-(n+1)}Y_{n+1}|Y_0, \ldots, Y_n] = \mu^{-n}Y_n = Z_n.$$

Let $f(s) = E(e^s X)$ and q be the smallest nonnegative solution of $q = f(q)$. One can then show that

$$W_n = q^{Z_n}, n \geq 1$$

is a martingale.

Proof Exercise.   □

Doob Martingale
Let $\{X_n, n = 1, \ldots, N\}$ be random variables and $Y = f(X_1, \ldots, X_N)$, where f is some bounded measurable real-valued function. Then

$$Z_n := E[Y \mid X^n], n = 0, \ldots, N$$

is a martingale (by the smoothing property of conditional expectation, see Theorem 9.5) called a *Doob martingale*. Here are a two examples.

1. Throw N balls into M bins, and let Y be some function of the throws: the number of empty bins, the max load, the second-highly loaded bin, or some similar function. Let X_n be the index of the bin into which ball n lands. Then $Z_n = E[Y \mid X^n]$ is a martingale.
2. Suppose we have r red and b blue balls in a bin. We draw balls without replacement from this bin: what is the number of red balls drawn? Let X_n be the indicator for whether ball n is red, and let $Y = X_1 + \cdots + X_n$ be the number of red balls. Then Z_n is a martingale.

You Cannot Beat the House
To study convergence, we start by explaining a key property of martingales that says there is no winning recipe to play a fair game of chance.

Theorem 15.9 (You Cannot Win) *Let X_n be a martingale with respect to $\{X_n, Z_n, n \geq 0\}$ and V_n some bounded function of (X^n, Z^n). Then*

$$Y_n = \sum_{m=0}^n V_{m-1}(X_m - X_{m-1}), n \geq 1, \tag{15.10}$$

with $Y_0 := 0$ is a martingale.

■

Proof One has

$$E[Y_n - Y_{n-1} \mid X^{n-1}, Z^{n-1}]$$
$$= E[V_{n-1}(X_n - X_{n-1}) \mid X^{n-1}, Z^{n-1}]$$
$$= V_{n-1} E[X_n - X_{n-1} \mid X^{n-1}, Z^{n-1}] = 0.$$

$\square$

The meaning of Y_n is the fortune that you would get by betting V_{m-1} at time $m-1$ on the gain $X_m - X_{m-1}$ of the next round of the game. This bet must be based on the information (X^{m-1}, Z^{m-1}) that you have when placing the bet, not on the outcome of the next round, obviously. The theorem says that your fortune remains a martingale even after adjusting your bets in real time.

Stopping Times

When playing a game of chance, one may decide to stop after observing a particular sequence of gains and losses. The decision to stop is non-anticipative. That is, one cannot say "never mind, I did not mean to play the last three rounds." Thus, the random stopping time τ must have the property that the event $\{\tau \leq n\}$ must be a function of the information available at time n, for all $n \geq 0$. Such a random time is a *stopping time*.

Definition 15.4 (Stopping Time) A random variable τ is a stopping time for the sequence $\{X_n, Y_n, n \geq 0\}$ if τ takes values in $\{0, 1, 2, \ldots\}$ and

$$P[\tau \leq n \mid X_m, Y_m, m \geq 0] = \phi_n(X^n, Y^n), \forall n \geq 0$$

for some functions ϕ_n.

$\diamond$

For instance,

$$\tau = \min\{n \geq 0 \mid (X_n, Y_n) \in \mathscr{A}\},$$

where $\mathscr{A}$ is a set in $\mathfrak{R}^2$ is a stopping time for the sequence $\{X_n, Y_n, n \geq 0\}$. Thus, you may want to stop the first time that either you go broke or your fortune exceeds $1000.00.

One might hope that a smart choice of when to stop playing a fair game could improve one's expected fortune. However, that is not the case, as the following fact shows.

Theorem 15.10 (Optional Stopping) *Let $\{X_n, n \geq 0\}$ be a martingale and τ a stopping time with respect to $\{X_n, Y_n, n \geq 0\}$. Then*[6]

$$E[X_{\tau \wedge n} | X_0, Y_0] = X_0.$$

∎

In the statement of the theorem, for a random time σ one defines $X_\sigma := X_n$ when $\sigma = n$.

Proof Note that $X_{\tau \wedge n}$ is the fortune Y_n that one accumulates by betting $V_m = 1\{\tau \wedge n > m\}$ at time m in (15.10), i.e., by betting 1 until one stops at time $\tau \wedge n$. Since $1\{\tau \wedge n > m\} = 1 - \{\tau \wedge n \leq m\} = \phi(X^m, Y^m)$, the resulting fortune is a martingale.   □

You will note that bounding $\tau \wedge n$ in the theorem above is essential. For instance, let X_n correspond to the random walk described above with $P(Z_n = 1) = P(Z_n = -1) = 0.5$. If we define $\tau = \min\{n \geq 0 \mid X_n = 10\}$, one knows that τ is finite. (See the comments below Theorem 15.1.) Hence, $X_\tau = 10$, so that

$$E[X_\tau | X_0 = 0] = 10 \neq X_0.$$

However, if we bound the stopping time, the theorem says that

$$E[X_{\tau \wedge n} | X_0 = 0] = 0. \tag{15.11}$$

This result deserves some thought.

One might be tempted to take the limit of the left-hand side of (15.11) as $n \to \infty$ and note that

$$\lim_{n \to \infty} X_{\tau \wedge n} = X_\tau = 10,$$

because τ is finite. One then might conclude that the left-hand size of (15.11) goes to 10, which would contradict (15.11). However, the limit and the expectation do not interchange because the random variables $X_{\tau \wedge n}$ are not bounded. However, if they were, one would get $E[X_\tau | X_0] = X_0$, by the dominated convergence theorem. We record this observation as the next result.

Theorem 15.11 (Optional Stopping—2) *Let $\{X_n, n \geq 0\}$ be a martingale and τ a stopping time with respect to $\{X_n, Y_n, n \geq 0\}$. Assume that $|X_n| \leq V$ for some random variable V such that $E(V) < \infty$. Then*

$$E[X_\tau | X_0, Y_0] = X_0.$$

∎

---

[6] $\tau \wedge n := \min\{\tau, n\}$

L^1-Bounded Martingales

An L^1-bounded martingale cannot bounce up and down infinitely often across an interval $[a, b]$. For if it did, you could increase your fortune without bound by betting 1 on the way up across the interval and betting 0 on the way down. We will see shortly that this cannot happen. As a result, the martingale must converge. (Note that this is not true if the martingale is not L^1-bounded, as the random walk example shows.)

Theorem 15.12 (L^1-Bounded Martingales Convergence)   *Let $\{X_n, n \geq 0\}$ be a martingale such that $E(|X_n|) \leq K$ for all n. Then X_n converges almost surely to a finite random variable X_∞.*

■

Proof Consider an interval $[a, b]$. We show that X_n cannot up-cross this interval infinitely often. (See Fig. 15.16.) Let us bet 1 on the way up and 0 on the way down. That is, wait until X_n gets first below a, then bet 1 at every step until $X_n > b$, then stop betting until X_n gets below a, and continue in this way.

If X_m crossed the interval U_n times by time n, your fortune Y_n is now at least $(b - a)U_n + (X_n - a)$. Indeed, your gain was at least $b - a$ for every upcrossing and, in the last steps of your playing, you lose at most $X_n - a$ if X_n never crosses above b after you last resumed betting. But, since Y_n is a martingale, we have

$$E(Y_n) = Y_0 \geq (b - a)E(U_n) + E(X_n - a) \geq (b - a)E(U_n) - K - a.$$

(We used the fact that $X_n \geq -|X_n|$, so that $E(X_n) \geq -E(|X_n|) = -K$. This shows that $E(U_n) \leq B = (K + Y_0 + a)/(b - a) < \infty$. Letting $n \to \infty$, since $U_n \uparrow U$, where U is the total number of upcrossings of the interval $[a, b]$, it follows by the monotone convergence theorem that $E(U) \leq B$. Consequently, U is finite. Thus, X_n cannot up-cross any given interval $[a, b]$ infinitely often.

Consequently, the probability that it up-crosses infinitely often any interval with rational limits is zero (since there are countably many such intervals).

This implies that X_n must converge, either to $+\infty$, $-\infty$, or to a finite value. Since $E(|X_n|) \leq K$, the probability that X_n converges to $+\infty$ or $-\infty$ is zero.   □

Fig. 15.16 If X_n does not converge, there are some rational numbers $a < b$ such that X_n crosses the interval $[a, b]$ infinitely often

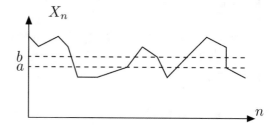

The following is a direct but useful consequence. We used this result in the proof of the convergence of the stochastic gradient projection algorithm (Theorem 12.2).

Theorem 15.13 (L^2-Bounded Martingales Convergence)   *Let X_n be a L^2-bounded martingale, i.e., such that $E(X_n^2) \leq K^2, \forall n \geq 0$, then $X_n \rightarrow X_\infty$, almost surely, for some finite random variable X_∞.*

∎

Proof We have

$$E(|X_n|)^2 \leq E(X_n^2) \leq K^2,$$

by Jensen's inequality. Thus, it follows that $E(|X_n|) \leq K$ for all n, so that the result of the theorem applies to this martingale.   □

One can also show that $E(|X_n - X_\infty|^2) \rightarrow 0$.

15.9.3   Law of Large Numbers

The SLLN can be proved as an application of the convergence of martingales, as Doob (1953) showed.

Theorem 15.14 (SLLN)   *Let $\{X_n, n \geq 1\}$ be i.i.d. random variables with $E(|X_n|) = K < \infty$ and $E(X_n) = \mu$. Then*

$$\frac{X_1 + \cdots + X_n}{n} \rightarrow \mu, \text{ almost surely as } n \rightarrow \infty.$$

∎

Proof Let

$$S_n = X_1 + \cdots + X_n, n \geq 1.$$

Note that

$$E[X_1 | S_n, S_{n+1}, \ldots] = \frac{1}{n} S_n =: Y_{-n}, \tag{15.12}$$

by symmetry. Thus,

$$E[Y_{-n} \mid S_{n+1}, \ldots] = E[E[X_1 \mid S_n, S_{n+1}, \ldots] \mid S_{n+1}, \ldots]$$
$$= E[X_1 \mid S_{n+1}, \ldots] = Y_{-n-1}.$$

Thus, $\{\ldots, Y_{-n-2}, Y_{-n-1}, Y_{-n}, \ldots\}$ is a martingale. (It is a Doob martingale.) This implies as before that the number U_n of upcrossings of an interval $[a, b]$ is such that $E(U_n) \leq B < \infty$. As before, we conclude that $U := \lim U_n < \infty$, almost surely. Hence, Y_n converges almost surely to a random variable $Y_{-\infty}$.

Now, since

$$Y_{-\infty} = \lim_{n \to \infty} \frac{X_1 + \cdots + X_n}{n},$$

we see that $Y_{-\infty}$ is independent of $(X_1, \ldots, X_n)$ for any finite n. Indeed, the limit does not depend on the values of the first n random variables. However, since $Y_{-\infty}$ is a function of $\{X_n, n \geq 1\}$, it must be independent of itself, i.e., be a constant. Since $E(Y_\infty) = E(Y_1) = \mu$, we see that $Y_\infty = \mu$. □

15.9.4 Wald's Equality

A useful application of martingales is the following. Let $\{X_n, n \geq 1\}$ be i.i.d. random variables. Let τ be a random variable independent of the X_n's that take values in $\{1, 2, \ldots\}$ with $E(\tau) < \infty$. Then

$$E(X_1 + \cdots + X_\tau) = E(\tau)E(X_1). \tag{15.13}$$

This expression is known as *Wald's Equality*.

To see this, note that $Y_n = X_1 + \cdots + X_n - nE(X_1)$ is a martingale. Also, τ is a stopping time. Thus,

$$E(Y_{\tau \wedge n}) = E(Y_1) = 0,$$

which gives the identity with τ replaced by $\tau \wedge n$. If $E(\tau) < \infty$, one can let n go to infinity and get the result. (For instance, replace X_i by X_i^+ and use MCT, similarly for X_i^-, then subtract.)

15.10 Summary

- General inference problems: guessing X given Y, Bayesian or not;
- Sufficient statistic: $h(Y)$ is sufficient for X;
- Infinite Markov Chains: PR, NR, T;
- Lyapunov–Foster Criterion;
- Poisson Process: independent stationary increments;
- Continuous-Time Markov Chain: rate matrix;
- Shannon Capacity of BSC: typical sequences and random codes;
- Bounds: Chernoff and Jensen;
- Martingales and Convergence;
- Strong Law of Large Numbers.

15.10.1 Key Equations and Formulas

Inference Problem	Guess X given Y: MAP, MLE, HT	S.15.1
Sufficient Statistic	$f_{Y\|X}[y\|x] = f(h(y), x)g(y)$	D.15.1
Infinite MC	Irreducible $\Rightarrow$ T, NR or PR	T.15.1
Poisson Process	Jumps w.p. $\lambda\epsilon$ in next ϵ seconds	D.15.2
Continuous-Time MC	Jumps from i to j w. rate $Q(i, j)$	D.6.1
Shannon Capacity C	Can transmit reliably at any rate $R < C$	S.15.7
" " of $BSC(p)$	$C = 1 + p\log_2(p) + (1 - p)\log_2(1 - p)$	(15.2)
Chernoff	$P(X > a) \le E(\exp\{\theta(X - a)\}), \forall \theta \ge 0$	(15.5)
Jensen	h convex $\Rightarrow E(h(X)) \ge h(E(X))$	(15.6)
Martingales	zero expected increase	D.15.3
MG Convergence	A.s. to finite RV if L^1 or L^2 bounded	T.15.12
Wald	$E(X_1 + \cdots + X_\tau) = E(\tau)E(X_1)$	(15.13)

15.11 References

For the theory of Markov chains, see Chung (1967). The text Harchol-Balter (2013) explains basic queueing theory and many applications to computer systems and operations research.

The book Bremaud (1998) is also highly recommended for its clarity and the breadth of applications. Information Theory is explained in the textbook Cover and

Thomas (1991). I learned the theory of martingales mostly from Neveu (1975). The theory of multi-armed bandits is explained in Cesa-Bianchi and Lugosi (2006). The text Hastie et al. (2009) is an introduction to applications of statistics in data science (Fig. 15.17).

15.12 Problems

Problem 15.1 Suppose that $y_1, \ldots, y_n$ are i.i.d. samples of $N(\mu, \sigma^2)$. What is a sufficient statistic for estimating μ given $\sigma = 1$. What is a sufficient statistic for estimating σ given $\mu = 1$?

Problem 15.2 Customers arrive to a store according to a Poisson process with rate 4 (per hour).

(a) What is the probability that exactly 3 customers arrive during 1 h?
(b) What is the probability that more than 40 min is required before the first customer arrives?

Problem 15.3 Consider two independent Poisson processes with rates λ_1 and λ_2. Those processes measure the number of customers arriving in stores 1 and 2.

(a) What is the probability that a customer arrives in store 1 before any arrives in store 2?
(b) What is the probability that in the first hour exactly 6 customers arrive at the two stores? (The total for both is 6)
(c) Given exactly 6 have arrived at the two stores, what is the probability all 6 went to store 1?

Problem 15.4 Consider the continuous-time Markov chain in Fig. 15.17.

(a) Find the invariant distribution.
(b) Simulate the MC and see that the fraction of time spent in state 1 converges to $\pi(1)$.

Fig. 15.17 CTMC

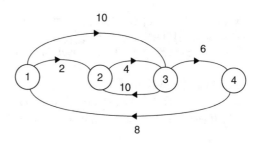

Problem 15.5 Consider a first-come-first-served discrete-time queuing system with a single server. The arrivals are Bernoulli with rate λ. The service times are i.i.d. and independent of the arrival times. Each service time Z takes values in $\{1, 2, \ldots, K\}$ such that $E(Z) = 1/\mu$ and $\lambda < \mu$.

(a) Construct the Markov chain that models the queue. What are the states and transition probabilities? [Hint: Suppose the head of the line task of the queue still requires z units of service. Include z in the state description of the MC.]
(b) Use Lyapunov–Foster argument to show the queue is stable or equivalently the MC is positive recurrent.

Problem 15.6 Suppose that random variable X takes value in the set $\{1, 2, \ldots, K\}$ such that $\Pr(X_1 = k) = p_k > 0$, and $\sum_{k=1}^{K} p_k = 1$. Suppose $X_1, X_2, \ldots, X_n$ is a sequence of n i.i.d. samples of X.

(a) How many possible sequences exist?
(b) How many typical sequences exist when n is large?
(c) Find a condition that answers to parts (a) and (b) are the same.

Problem 15.7 Let $\{N_t, t \geq 0\}$ be a Poisson process with rate λ. Let S_n denote the time of the n-th event. Find

(a) the pdf of S_n.
(b) $E[S_5]$.
(c) $E[S_4 | N(1) = 2]$.
(d) $E[N(4) - N(2) | N(1) = 3]$.

Problem 15.8 A queue has Poisson arrivals with rate λ. It has two servers that work in parallel. When there are at least two customers in the queue, two are being served. When there is only one customer, only one server is active. The service times are i.i.d. $Exp(\mu)$.

(a) Argue that the queue length is a Markov Chain.
(b) Draw the state transition diagram.
(c) Find the minimum value of μ so that the queue is positive recurrent and solve the balance equations.

Problem 15.9 Let $\{X_t, t \geq 0\}$ be a continuous-time Markov chain with rate matrix $Q = \{q(i, j)\}$. Define $q(i) = \sum_{j \neq i} q(i, j)$. Let also $T_i = \inf\{t > 0 | X_t = i\}$ and $S_i = \inf\{t > 0 | X_t \neq i\}$. Then (select the correct answers)

☐ $E[S_i | X_0 = i] = q(i)$;
☐ $P[T_i < T_j | X_0 = k] = q(k, i)/(q(k, i) + q(k, j))$ for i, j, k distinct;
☐ If $\alpha(k) = P[T_i < T_j | X_0 = k]$, then $\alpha(k) = \sum_s \frac{q(k,s)}{q(k)} \alpha(s)$ for $k \notin \{i, j\}$.

Problem 15.10 A continuous-time queue has Poisson arrivals with rate λ, and it is equipped with infinitely many servers. The servers can work in parallel on multiple customers, but they are non-cooperative in the sense that a single customer can only be served by one server. Thus, when there are k customers in the queue, k servers are active. Suppose that the service time of each customer is exponentially distributed with rate μ and they are i.i.d.

(a) Argue that the queue length is a Markov chain. Draw the transition diagram of the Markov chain.
(b) Prove that for all finite values of λ and μ the Markov chain is positive recurrent and find the invariant distribution.

Problem 15.11 Consider a Poisson process $\{N_t, t \geq 0\}$ with rate $\lambda = 1$. Let random variable S_i denote the time of the i-th arrival. [Hint: You recall that $f_{S_i}(x) = \frac{x^{i-1}e^{-x}}{(i-1)!}1\{x \geq 0\}$.]

(a) Given $S_3 = s$, find the joint distribution of S_1 and S_2. Show you work.
(b) Find $E[S_2|S_3 = s]$.
(c) Find $E[S_3|N_1 = 2]$.

Problem 15.12 Let $S = \sum_{i=1}^{N} X_i$ denote the total amount of money withdrawn from an ATM in 8 h, where:

(a) X_i are i.i.d. random variables denoting the amount withdrawn by each customer with $E[X_i] = 30$ and $Var[X_i] = 400$.
(b) N is a Poisson random variable denoting the total number of customers with $E[N] = 80$.

Find $E[S]$ and $Var[S]$.

Problem 15.13 One is given two independent Poisson processes M_t and N_t with respective rates λ and μ, where $\lambda > \mu$. Find $E(\tau)$, where

$$\tau = \max\{t \geq 0 \mid M_t \leq N_t + 5\}.$$

(Note that this is a max, not a min.)

Problem 15.14 Consider a queue with Poisson arrivals with rate λ. The service times are all equal to one unit of time. Let X_t be the queue length at time t $(t \geq 0)$.

(a) Is X_t a Markov chain? Prove or disprove.
(b) Let Y_n be the queue length just after the n-th departure from the queue $(n \geq 1)$. Prove that Y_n is a Markov chain. Draw a state diagram.
(c) Prove that Y_n is positive recurrent when $\lambda < 1$.

Problem 15.15 Consider a queue with Poisson arrivals with rate λ. The queue can hold N customers. The service times are i.i.d. $Exp(\mu)$. When a customer arrives, you can choose to pay him c so that he does not join the queue. You also pay c when a customer arrives at a full queue. You want to decide when to accept customers to minimize the cost of rejecting them, plus the cost of the average waiting time they spend in the queue.

(a) Formulate the problem as a Markov decision problem. For simplicity, consider a total discounted cost. That is, if x_t customers are in the system at time t, then the waiting cost during $[t, t + \epsilon]$ is $e^{-\beta t} x_t \epsilon$. Similarly, if you reject a customer at time t, then the cost is $ce^{-\beta t}$.

(b) Write the dynamic programming equations.

(c) Use Python to solve the equations.

Problem 15.16 The counting process $\mathbf{N} := \{N_t, 0 \le t \le T\}$ is defined as follows:

Given τ, $\{N_t, 0 \le t \le \tau\}$ and $\{N_t - N_\tau, \tau \le t \le T\}$ are independent Poisson processes with respective rates λ_0 and λ_1.

Here, λ_0 and λ_1 are known and such that $0 < \lambda_0 < \lambda_1$. Also, τ is exponentially distributed with known rate $\mu > 0$.

1. Find the MLE of τ given $\mathbf{N}$.
2. Find the MAP of τ given $\mathbf{N}$.

Problem 15.17 Figure 15.18 shows a system where a source alternates between the ON and OFF states according to a continuous-time Markov chain with the transition rates indicated. When the source is ON, it sends a fluid with rate 2 into the queue. When the source is OFF, it does not send any fluid. The queue is drained at constant rate 1 whenever it contains some fluid. Let X_t be the amount of fluid in the queue at time $t \ge 0$.

(a) Plot a typical trajectory of the random process $\{X_t, t \ge 0\}$.

(b) Intuitively, what are conditions on λ and μ that should guarantee the "stability" of the queue?

(c) Is the process $\{X_t, t \ge 0\}$ Markov?

Fig. 15.18 The system

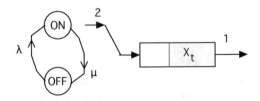

Problem 15.18 Let $\{N_t, t \geq 0\}$ be a Poisson process with rate λ that is exponentially distributed with rate $\mu > 0$.

(a) Find $MLE[\lambda|N_s, 0 \leq s \leq t]$;
(b) Find $MAP[\lambda|N_s, 0 \leq s \leq t]$;
(c) What is a sufficient statistic for λ given $\{N_s, 0 \leq s \leq t\}$;
(d) Instead of λ being exponentially distributed, assume that λ is known to take values in $[5, 10]$. Give an estimate of the time t required to estimate λ within 5% with probability 95%.

Problem 15.19 Consider two queues in parallel in discrete time with Bernoulli arrival processes of rates λ_1 and λ_2, and geometric service rates of μ_1 and μ_2, respectively. There is only one server that can serve either queue 1 and queue 2 at each time. Consider the scheduling policy that serves queue 1 at time n if $\mu_1 Q_1(n) > \mu_2 Q_2(n)$, and serve queue 2 otherwise, where $Q_1(n)$ and $Q_2(n)$ are queue lengths of the queues at time n. Use the Lyapunov function $V(Q_1(n), Q_2(n)) = Q_1^2(n) + Q_2^2(n)$ to show that the queues are stable if $\lambda_1/\mu_1 + \lambda_2/\mu_2 < 1$. This scheduling policy is known as Max-Weight or Back-Pressure policy.

Correction to: Probability in Electrical Engineering and Computer Science

---

Correction to:
J. Walrand, *Probability in Electrical Engineering and Computer Science*,
https://doi.org/10.1007/978-3-030-49995-2

The author of this book used Python materials and would like to acknowledge the source. The below mentioned text has been added to the copyright page of the updated version.

The link includes examples of Python demos and also Python labs used at Berkeley: https://sites.google.com/berkeley.edu/probabilityineecs/home

---

The online version of the original book can be found at
https://doi.org/10.1007/978-3-030-49995-2

© The Author(s) 2021
J. Walrand, *Probability in Electrical Engineering and Computer Science*,
https://doi.org/10.1007/978-3-030-49995-2_16

Correction to: Probability in Electrical Engineering and Computer Science (Funding Information)

Jean Walrand

Correction to:
J. Walrand, *Probability in Electrical Engineering and Computer Science,*
https://doi.org/10.1007/978-3-030-49995-2

This book was inadvertently published without the funding information on the "Ackowledgements" section. This should include:

This book is an open access publication. This title is freely available in open access edition with generous support from the Library of the University of California, Berkeley.

The online version of the original book can be found at
https://doi.org/10.1007/978-3-030-49995-2

© The Author(s) 2022
J. Walrand, *Probability in Electrical Engineering and Computer Science,*
https://doi.org/10.1007/978-3-030-49995-2_17

Elementary Probability

A

> **Topics:** Symmetry, conditioning, independence, expectation, law of large numbers, regression.

A.1 Symmetry

The simplest model of probability is based on *symmetry*. Picture a bag with 10 marbles that are identical, except that they are marked as shown in Fig. A.1.

You put the marbles in a bag that you shake thoroughly and you then pick a marble without looking. Out of the ten marbles, seven have a blue number equal to 1. We say that the *probability* that you pick a marble whose blue number is 1 is equal to $7/10 = 0.7$. The probability is the fraction of favorable outcomes (picking a marble with a blue number equal to 1) among all the equally likely possible outcomes (the different marbles). The notion of "equally likely" is defined by symmetry, so this definition of probability is not circular.

For ease of discussion, let us call B the blue number on the marble that you pick and R the red number on that marble. We write $P(B = 1) = 0.7$. Similarly, $P(B = 2) = 0.3$, $P(R = 1) = 0.2$, $P(B = 2, R = 3) = 0.1$, etc. We call R and B *random variables*. The justification for the terminology is that if we were to repeat the experiment of shaking the bag of ten marbles and picking a marble, the values of R and B would vary from experiment to experiment in an unpredictable way. Note that R and B are functions of the same outcome (the selected marble) of one experiment (picking a marble). Indeed, we do not pick one marble to read the value of B and then another one to read the value of R; we pick only one marble and the values of B and R correspond to that marble.

Let A be a subset of the marbles. We also write $P(A)$ for the probability that you pick a marble that is in the set A. Since all the marbles are equally likely to

© The Author(s) 2021
J. Walrand, *Probability in Electrical Engineering and Computer Science*,
https://doi.org/10.1007/978-3-030-49995-2

Fig. A.1 Ten marbles marked with a blue and a red number

be picked, $P(A) = |A|/10$, where $|A|$ is the number of marbles in the set A. For instance, if A is the set of marbles where $(B = 1, R = 3)$ or $(B = 2, R = 4)$, then $P(A) = 0.4$ since there are four such marbles out of 10.

It is clear that if A_1 and A_2 are disjoint sets (i.e., have no marble in common), then $P(A_1 \cup A_2) = P(A_1) + P(A_2)$. Indeed, when A_1 and A_2 are disjoint, the number of marbles in $A_1 \cup A_2$ is the number of marbles in A_1 plus the number of marbles in A_2. If we divide by ten, we conclude that the probability of picking a marble that is in $A_1 \cup A_2$ is the sum of the probabilities of picking one in A_1 or in A_2. We say that *probability is additive*. This property extends to any finite collection of events that are pairwise disjoint.

Note that if A_1 and A_2 are not disjoint, then $P(A_1 \cup A_2) < P(A_1) + P(A_2)$. For instance, if A_1 is the set of marbles such that $B = 1$ and A_2 is the set of marbles such that $R = 4$, then $P(A_1 \cup A_2) = 0.9$, whereas $P(A_1) + P(A_2) = 0.7 + 0.5 = 1.2$. What is happening is that $P(A_1) + P(A_2)$ is double-counting the marbles that are in both A_1 and A_2, i.e., the marbles such that $(B = 1, R = 4)$. We can eliminate this double-counting and check that $P(A_1 \cup A_2) = P(A_1) + P(A_2) - P(A_1 \cap A_2)$.

Thus, one has to be a bit careful when examining the different ways that something can happen. When adding up the probabilities of these different ways, one should make sure that they are exclusive, i.e., that they cannot happen together. For example, the probability that your car is red or that it is a Toyota is not the sum of the probability that it is red plus the probability that it is a Toyota. This sum double-counts the probability that your car is a red Toyota. Such double-counting mistakes are surprisingly common.

A.2   Conditioning

Now, imagine that you pick a marble and tell me that $B = 1$. How do I guess R?

Looking at the marbles, we see that there are 7 marbles with $B = 1$, among which two are such that $R = 1$. Thus, given that $B = 1$, the probability that $R = 1$ is 2/7. Indeed, given that $B = 1$, you are equally likely to have picked any one of the 7 marbles with $B = 1$. Since 2 out of these 7 marbles are such that $R = 1$, we conclude that the probability that $R = 1$ given that $B = 1$ is 2/7.

We write $P[R = 1 \mid B = 1] = 2/7$. Similarly, $P[R = 3 \mid B = 1] = 2/7$ and $P[R = 4 \mid B = 1] = 3/7$. We say that $P[R = 1 \mid B = 1]$ is the *conditional probability* that $R = 1$ given that $B = 1$.

So, we are not sure of the value of R when we are told that $B = 1$, but the information is useful. For instance, you can see that $P[R = 1 \mid B = 2] = 0$, whereas $P[R = 1 \mid B = 1] = 2/7$. Thus, knowing B tells us something about R.

Observe that $P(R = 1, B = 1) = N(1, 1)/10$ where $N(1, 1)$ is the number of marbles with $R = 1$ and $B = 1$. Also, $P[R = 1 \mid B = 1] = N(1, 1)/N(1, *)$ where $N(1, *)$ is the number of marbles with $B = 1$ and R taking an arbitrary value. Moreover, $P(B = 1) = N(1, *)/N$. It then follows that

$$P(B = 1, R = 1) = P(B = 1) \times P[R = 1 \mid B = 1].$$

Indeed,

$$\frac{N(1, 1)}{N} = \frac{N(1, *)}{N} \times \frac{N(1, 1)}{N(1, *)}.$$

To make the previous identity intuitive, we argue in the following way. For $(B = 1, R = 1)$ to occur, $B = 1$ must occur and then $R = 1$ must occur given that $B = 1$. Thus, the probability of $(B = 1, R = 1)$ is the probability of $B = 1$ times the probability of $R = 1$ given that $B = 1$.

The previous identity shows that

$$P[R = 1 \mid B = 1] = \frac{P(B = 1, R = 1)}{P(B = 1)}.$$

Intuitively, this expression says that the probability of $R = 1$ given $B = 1$ is the fraction of marbles with $B = 1$ and $R = 1$ among the marbles with $B = 1$.

More generally, for any two values b and r of B and R, one has

$$P[R = r \mid B = b] = \frac{P(B = b, R = r)}{P(B = b)} \tag{A.1}$$

and

$$P(B = b, R = r) = P(B = b)P[R = r \mid B = b]. \tag{A.2}$$

The most likely value of R given that $B = 1$ is 4. Indeed, $P[R = 4 \mid B = 1]$ is larger than $P[R = 1 \mid B = 1]$ and $P[R = 3 \mid B = 1]$. We say that 4 is the *maximum a posteriori (MAP) estimate* of R given that $B = 1$. Similarly, the MAP estimate of B given that $R = 4$ is 1. Indeed, $P[B = 1 \mid R = 4] = 3/5$, which is larger than $P[B = 2 \mid R = 4] = 2/5$.

A slightly different concept is the *maximum likelihood estimate (MLE)* of B given $R = 4$. By definition, this is the value of B that makes $R = 4$ most likely. We see that the MLE of B given that $R = 4$ is 2 because $P[R = 4 \mid B = 2] = 2/3 > P[R = 4 \mid B = 1]$.

For instance, the MLE of a disease of a person with high fever might be Ebola, but the MAP may be a common flu. To see this, imagine 100 marbles out of which one is marked (Ebola, High Fever), 15 are marked (Flu, High Fever), 5 are marked (Flu, Low Fever), and the others are marked (Something Else, No Fever). For this model, we see that P[High Fever | Ebola] = 1 > P[High Fever | Flu] = 0.75, and P[Flu | High Fever] = 15/16 > P[Ebola | High Fever] = 1/16.

A.3   Common Confusion

The discussion so far probably seems quite elementary. However, most of the confusion about probability arises with these basic ideas. Let us look at some examples.

You are told that Bill has two children and one of them is named Isabelle. What is the probability that Bill has two daughters? You might argue that Bill's other child has a 50% probability of being a girl, so that the probability that Bill has two daughters must be 0.5. In fact, the correct answer is 1/3. To see this, look at the four equally likely outcomes for the sex of the two children: $(M, M), (F, M), (M, F), (F, F)$ where (M, F) means that the first child is male and the second is female, and similarly for the other cases. Out of these four outcomes, three are consistent with the information that "one of them is named Isabelle." Out of these three outcomes, one corresponds to Bill having two daughters. Hence, the probability that Bill has two daughters given that one of his two children is named Isabelle is 1/3, not 50%.

This example shows that confusion in Probability is not caused by the sophistication of the mathematics involved. It is not a lack of facility with Calculus or Algebra that causes the difficulty. It is the lack of familiarity with the basic formalism: looking at the possible outcomes and identifying precisely what the given information tells us about these outcomes.

Another common source of confusion concerns chance fluctuations. Say that you flip a fair coin ten times. You expect about half of the outcomes to be tails and half to be heads. Now, say that the first six outcomes happen to be heads. Do you think the next four are more likely to be tails, to catch up with the average? Of course not. After 4 years of drought in California, do you expect the next year to be rainier than average? You should not.

Surprisingly, many people believe in the memory of purely random events. A useful saying is that "lady luck has no memory nor vengeance."

A related concept is "regression to the mean." A simple example goes as follows. Flip a fair coin twenty times. Say that eight of the first ten flips are heads. You expect the next ten flips to be more balanced. This does not mean that the next ten flips are more likely to be tails to compensate for the first ten flips. It simply means that the abnormal fluctuations in the first ten flips do not carry over to the next ten flips. More subtle scenarios of this example involve the stock market or the scores of sports teams, but the basic idea is the same. Of course, if you do not know whether the coin is fair or biased, observing eight heads out of the first ten flips suggests that the coin is biased in favor of heads, so that the next ten coin flips are likely to give more heads than tails. But, if you have observed many flips of that coin in the past, then you may know that it is fair, and in that case regression to the mean makes sense.

Now, you might ask "how can about half of the coin flips be heads if the coin does not make up for an excessive number of previous tails?" The answer is that among the $2^{10} = 1024$ equally likely strings of 10 heads and tails, a very large proportion have about 50% of heads. Indeed, 672 such strings have either 4, 5, or 6 heads. Thus the probability that the number of heads is between 40 and 60% is $672/1,024 = 65.6\%$. This probability gets closer to one as you flip more coins. For twenty coins, the probability that the fraction of heads is between 40 and 60 is 73.5%.

To avoid being confused, always keep the basic formalism in mind. What are the outcomes? How likely are they? What does the known information tell you about them?

A.4 Independence

Look at the marbles in Fig. A.2. For these marbles, we see that $P[R = 1 \mid B = 1] = P[R = 3 \mid B = 1] = 2/4 = 0.5$. Also, $P[R = 1 \mid B = 2] = P[R = 3 \mid B = 2] = 3/6 = 0.5$. Thus, knowing the value of B does not change the probability of the different values of R. We say that for this experiment, R and B are *independent*. Here, the value of R tells us something about which marble you picked, but that information does not change the probability of the different values of B.

In contrast, for the marbles in Fig. A.1, we saw that $P[R = 1 \mid B = 2] = 0$ and $P[R = 1 \mid B = 1] = 2/7$ so that, for that experiment, R and B are *not* independent: knowing the value of B changes the probability of $R = 1$. That is, B tells you something about R. This is rather common. The temperature in Berkeley tells us something about the temperature in San Francisco. If it rains in Berkeley, it is likely to rain in San Francisco.

This fact that observations tell us something about what do not observe directly is central in applied probability. It is at the core of data science. What information do we get from data? We explore this question later in this appendix.

Summarizing, we say that B and R are independent if $P[R = r \mid B = b] = P(R = r)$ for all values of r and b. In view of (A.2), B and R are independent if

$$P(B = b, R = r) = P(B = b)P(R = r), \quad \text{for all } b, r. \tag{A.3}$$

As a simple example of independence, consider ten flips of a fair coin. Let X be the number of heads in the first 4 flips and Y the number of heads in the last 6 flips. We claim that X and Y are independent. Intuitively, this is obvious. However, how

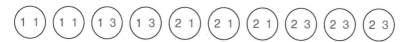

Fig. A.2 Ten other marbles marked with a blue and a red number

Fig. A.3 $A = \{X = x\}$ and
$B = \{Y = y\}$

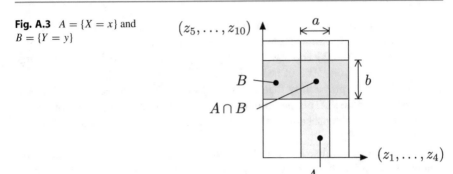

do we show this formally? In this experiment, an outcome is a string of ten heads and tails. These outcomes are all equally likely. Fix arbitrary integers x and y. Let A be the set of outcomes for which $X = x$ and B the set of outcomes for which $Y = y$. Figure A.3 illustrates the sets A and B. In the figure, the horizontal axis corresponds to the different strings of the first four flips; the vertical axis corresponds to the different strings of the last six flips.

The set A corresponds to a different strings of the first four flips that are such that $X = x$ and arbitrary values of the last six flips. Similarly, B corresponds to b different strings of the last six flips and arbitrary values of the first four flips. Note that A has $a \times 2^6$ outcomes since each of the a strings corresponds to 2^6 strings of the last 6 flips. Similarly, B has $2^4 \times b$ outcomes. Thus,

$$P(A) = \frac{a \times 2^6}{2^{10}} = \frac{a}{2^4} \text{ and } P(B) = \frac{2^4 \times b}{2^{10}} = \frac{b}{2^6}.$$

Moreover,

$$P(A \cap B) = \frac{a \times b}{2^{10}}.$$

Hence,

$$P(A \cap B) = P(A) \times P(B).$$

If you look back at the calculation, you will notice that it boils down to the area of a rectangle being the product of the sides. The key observation is then that the set of outcomes where $X = x$ and $Y = y$ is a rectangle. This is so because $X = x$ imposes a constraint on the first four flips and $Y = y$ imposes a constraint on the other flips.

A.5 Expectation

Going back to the marbles of Fig. A.1, what do you expect the value of B to be? How much would you be willing to pay to pick a marble given that you get the value of B in dollars? An intuitive argument is that if you were to repeat the experiment 1000 times, you should get a marble with $B = 1$ about 70% of the time, i.e., about 700 times. The other 300 times, you would get $B = 2$. The total amount should then be $1 \times 700 + 2 \times 300$. The average value per experiment is then

$$\frac{1 \times 700 + 2 \times 300}{1000} = 1 \times 0.7 + 2 \times 0.3.$$

We call this number the *expected value* of B and we write it $E(B)$. Similarly, we define

$$E(R) = 1 \times 0.2 + 3 \times 0.3 + 4 \times 0.5.$$

Thus, the expected value is *defined* as the sum of the values multiplied by their probability. The interpretation we gave by considering the experiment being repeated a large number of times is only an interpretation, for now.

Reviewing the argument, and extending it somewhat, let us assume that we have N marbles marked with a number X that takes the possible values $\{x_1, x_2, \ldots, x_M\}$ and that a fraction p_m of the marbles are marked with $X = x_m$, for $m = 1, \ldots, M$. Then we write $P(X = x_m) = p_m$ for $m = 1, \ldots, M$. We define the expected value of X as

$$E(X) = \sum_{m=1}^{M} x_m p_m = \sum_{m=1}^{M} x_m P(X = x_m). \tag{A.4}$$

Consider a random variable X that is equal to the same constant x for every outcome. For instance, X could be a number on a marble when all the marbles are marked with the same number x. In this case,

$$E(X) = x \times P(X = x) = x.$$

Thus, the expected value of a constant is the constant. For instance, if we designate by a a random variable that always takes the value a, then we have

$$E(a) = a.$$

There is a slightly different but very useful way to compute the expectation. We can write $E(X)$ as the sum over all the possible marbles we could pick of the product of X for that marble times the probability $1/N$ that we pick that particular marble. Doing this, we have

$$E(X) = \sum_{n=1}^{N} X(n)\frac{1}{N},$$

where $X(n)$ is the value of X for marble n. This expression gives the same value as the previous calculation. Indeed, in this sum there are $p_m N$ terms with $X(n) = x_m$ because we know that a fraction p_m of the N marbles, i.e., $p_m N$ marbles, are marked with $X = x_m$. Hence, the sum above is equal to

$$\sum_{m=1}^{M} (p_m N)x_m\frac{1}{N} = \sum_{m=1}^{M} p_m x_m,$$

which agrees with the previous expression for $E(X)$.

This latter calculation is useful to show that $E(B + R) = E(B) + E(R)$ in our example of Fig. A.1 with ten marbles. Let us examine this important property closely. You pick a marble and you get $B + R$. By looking at the marbles, you see that you get $1 + 1$ if you pick the first marble, and so on. Thus,

$$E(B + R) = (1 + 1)\frac{1}{10} + \cdots + (2 + 4)\frac{1}{10}.$$

If we decompose this sum by regrouping the values of B and then those of R, we see that

$$E(B + R) = \left[1\frac{1}{10} + \cdots + 2\frac{1}{10}\right] + \left[1\frac{1}{10} + \cdots + 4\frac{1}{10}\right].$$

The first sum is $E(B)$ and the second is $E(R)$. Thus, the expected value of a sum is the sum of the expected values. We say that *expectation is linear*. Notice that this is so even though the values B and R are not independent.

More generally, for our N marbles, if marble n is marked with two numbers $X(n)$ and $Y(n)$, then we see that

$$E(X + Y) = \sum_{n=1}^{N} (X(n) + Y(n))\frac{1}{N} = \sum_{n=1}^{N} X(n)\frac{1}{N} + \sum_{n=1}^{N} Y(n)\frac{1}{N} = E(X) + E(Y).$$

$$(A.5)$$

Linearity shows that if we get $5 + 3X^2 + 4Y^3$ when we pick a marble marked with the numbers X and Y, then

$$E(5 + 3X^2 + 4Y^3) = 5 + 3E(X^2) + 4E(Y^3).$$

Indeed,

$$E(5 + 3X^2 + 4Y^3)) = \frac{1}{N} \sum_{n=1}^{N} (5 + 3X^2(n) + 4Y^3(n)) = 5 + 3E(X^2) + 4E(Y^3).$$

As another example, we have

$$E(a + X) = E(a) + E(X) = a + E(X).$$

Similarly,

$$E((X - a)^2) = E(X^2 - 2aX + a^2) = E(X^2) - 2aE(X) + a^2.$$

Choosing $a = E(X)$ in the previous example, we find

$$E((X - E(X))^2) = E(X^2) - 2E(X)E(X) + [E(X)]^2 = E(X^2) - [E(X)]^2,$$

an example we discuss in the next section.

There is another elementary property of expectation that we use in the next section. Consider marbles where $B \leq R$, such as the marbles in Fig. A.1. Compare the sum over the marbles of $B(n) \times (1/10)$ with the sum of $R(n) \times (1/10)$. Term by term, the first sum is less than the second, so that $E(B) \leq E(R)$. Hence, if $B \leq R$ for every outcome, one has $E(B) \leq E(R)$. We say that *expectation is monotone*.

We will use yet another simple property of expectation in the next section. (Do not worry, there are not many such properties!) Assume that X and Y are independent. Recall that this means that $P(X = x_i, Y = y_j) = P(X = x_i)P(Y = y_j)$ for all possible pair of values (x_i, y_j) of X and Y. Then we claim that

$$E(XY) = E(X)E(Y). \tag{A.6}$$

To see this, we write

$$E(XY) = \sum_{n=1}^{N} X(n)Y(n)\frac{1}{N} = \sum_i \sum_j x_i y_j N(i, j)\frac{1}{N},$$

where $N(i, j)$ is the number of marbles marked with (x_i, y_j). We obtained the last term by regrouping the terms based on the values of $X(n)$ and $Y(n)$. Now, $N(i, j)/N = P(X = x_i, Y = y_j)$. Also, by independence, $P(X = x_i, Y = y_j) = P(X = x_i)P(Y = y_j)$. Thus, we can write the sum above as follows:

$$E(XY) = \sum_i \sum_j x_i y_j P(X = x_i, Y = y_j) = \sum_i \sum_j x_i y_j P(X = x_i)P(Y = y_j).$$

We now compute the sum on the right by first summing over j. We get

$$E(XY) = \sum_i \left[\sum_j x_i x_j P(X = x_i) P(Y = y_j) \right] = \sum_i x_i P(X = x_i)$$

$$\times \left[\sum_j x_j P(Y = y_j) \right].$$

We got the last expression by noticing that the factor $x_i P(X = x_i)$ is common to all the terms $x_i x_j P(X = x_i) P(Y = y_j)$ for different values of j when i is fixed. Now, the term between brackets is $E(Y)$. Hence, we find

$$E(XY) = \sum_i x_i P(X = x_i) E(Y) = E(X)E(Y),$$

as claimed. Thus, if X and Y are independent, the expected value of their product is the product of their expected values.

We can check that this property does not generally hold if the random variables are not independent. For instance, consider R and B in Fig. A.1. We find that

$$E(BR) = (1 + 1 + 3 + 3 + 4 + 4 + 4 + 6 + 8 + 8)/10 = 3.4,$$

whereas

$$E(B) = 1.3 \text{ and } E(R) = 3.1,$$

so that $E(BR) = 3.4 \neq E(B)E(R) = 1.3 \times 3.4 = 4.42$.

We invite you to construct an example of marbles where R and B are not independent and yet $E(BR) = E(B)E(R)$. This example will convince you that $E(XY) = E(X)E(Y)$ does not imply that X and Y are independent.

We have seen the following properties of expectation:

- expectation is linear
- expectation is monotone
- the expected value of the product of two independent random variables is the product of their expected values.

A.6   Variance

Let us define the *variance* of a random variable X as

$$\text{var}(X) = E((X - E(X))^2). \tag{A.7}$$

The variance measures the variability around the mean. If the variance is small, the random variable is likely to be close to its mean.

By linearity of expectation, we have

$$\text{var}(X) = E(X^2 - 2XE(X) + [E(X)]^2) = E(X^2) - 2E(X)E(X) + [E(X)]^2.$$

Hence,

$$\text{var}(X) = E(X^2) - [E(X)]^2, \tag{A.8}$$

as we saw already in the previous section.

Note also that

$$\text{var}(aX) = E((aX)^2) - [E(aX)]^2 = a^2 E(X^2) - a^2 [E(X)]^2 = a^2 [E(X^2) - [E(X)]^2].$$

Hence,

$$\text{var}(aX) = a^2 \text{var}(X). \tag{A.9}$$

Now, assume that X and Y are independent random variables. Then we find that

$$\text{var}(X + Y) = E((X + Y)^2) - [E(X + Y)]^2$$
$$= E(X^2 + 2XY + Y^2) - [E(X)]^2 - 2E(X)E(Y) - [E(Y)]^2$$
$$= E(X^2) - [E(X)]^2 + E(Y^2) - [E(Y)]^2 + 2E(XY) - 2E(X)E(Y).$$

Therefore, if X and Y are independent,

$$\text{var}(X + Y) = \text{var}(X) + \text{var}(Y), \tag{A.10}$$

where the last expression results from the fact that $E(XY) = E(X)E(Y)$ when the random variables are independent.

The square root of the variance is called the *standard deviation*.

Summing up, we saw the following results about variance:

- when one multiplies a random variable by a constant, its variance gets multiplied by the square of the constant
- the variance of the sum of independent random variables is the sum of their variances
- the standard deviation of a random variable is the square root of its variance.

A.7   Inequalities

The fact that expectation is monotone yields some inequalities that are useful to
bound the probability that a random variable takes large values. Intuitively, if a
random variable is likely to take large values, its expected value is large.

The simplest such inequality is as follows. Let X be a random variable that is
always non-negative, then

$$P(X \geq a) \leq \frac{E(X)}{a}, \text{ for } a > 0.$$

This is called *Markov's inequality*. To prove it, we define the random variable Y as
being 0 when $X < a$ and 1 when $X \geq a$. Hence,

$$E(Y) = 0 \times P(X < a) + 1 \times P(X \geq a) = P(X \geq a).$$

We note that $Y \leq X/a$. Indeed, that inequality is immediate if $X < a$ because then
$Y = 0$ and $X/a > 0$. It is also immediate when $X \geq a$ because then $Y = 1$ and
$X/a \geq 1$. Consequently, by monotonicity of expectation, $E(Y) \leq E(X/a)$. Hence,

$$P(X \geq a) = E(Y) \leq E\left(\frac{X}{a}\right) = \frac{E(X)}{a},$$

where the last equality comes from the linearity of expectation.

The second inequality is *Chebyshev's inequality*. It states that

$$P(|X - E(X)| \geq \epsilon) \leq \frac{\text{var}(X)}{\epsilon^2}, \text{ for } \epsilon > 0.$$

To derive this inequality, we define $Z = |X - E(X)|^2$. Markov's inequality says
that

$$P(Z \geq \epsilon^2) \leq \frac{E(Z)}{\epsilon^2} = \frac{\text{var} X}{\epsilon^2}.$$

Now, $Z \geq \epsilon^2$ is equivalent to $|X - E(X)| \geq \epsilon$. This proves the inequality.

A.8   Law of Large Numbers

Chebyshev's inequality is particularly useful when we consider a sum of independent random variables. Assume that $X_1, X_2, \ldots, X_n$ are independent random
variables with the same expected value μ and the same variance σ^2. Define $Y =
(X_1 + \cdots + X_n)/n$. Observe that

$$E(Y) = E\left(\frac{X_1 + \cdots + X_n}{n}\right) = \frac{n\mu}{n} = \mu.$$

Also,

$$\mathrm{var}(Y) = \frac{1}{n^2}\mathrm{var}(X_1 + \cdots + X_n) = \frac{1}{n^2} \times n\sigma^2 = \frac{\sigma^2}{n}.$$

Consequently, Chebyshev's inequality implies that

$$P(|Y - \mu| \geq \epsilon) \leq \frac{\sigma^2}{n\epsilon^2}.$$

This probability becomes arbitrarily small as n increases. Thus, if Y is the average of n random variables that are independent and have the same mean μ and the same variance, then Y is very close to μ, with a high probability, when n is large. This is called the *Weak Law of Large Numbers*.

Note that this result extends to the case where the random variables are independent, have the same mean, and have a variance bounded by some σ^2.

A.9 Covariance and Regression

Consider once again the N marbles with the numbers (X, Y). We define the *covariance* of X and Y as

$$\mathrm{cov}(X, Y) = E(XY) - E(X)E(Y).$$

By linearity of expectation, one can check that $\mathrm{cov}(XY) = E((X - E(X))(Y - E(Y))$. This expression suggests that $\mathrm{cov}(X, Y)$ is positive when X and Y tend to be large or small together. We make this idea more precise in this section.

Observe that if X and Y are independent, then $E(XY) = E(X)E(Y)$, so that $\mathrm{cov}(X, Y) = 0$. Two random variables are said to be *uncorrelated* if their covariance is zero. Thus, independent random variables are uncorrelated. The converse is not true.

Say that you observe X and you want to guess Y, as we did earlier with R and B. For instance, you observe the height of a person and you want to guess his weight. To do this, you choose two numbers a and b and you estimate Y by $\hat{Y}$ where

$$\hat{Y} = a + E(Y) + b(X - E(X)).$$

The goal is to choose a and b so that $\hat{Y}$ tends to be close to Y. Thus, $\hat{Y}$ can be an arbitrary linear function of X and we want to find the best linear function. We wrote the linear function in this particular form to simplify the subsequent algebra.

To make this precise, we look for the values of a and b that minimize $E((Y - \hat{Y})^2)$. That is, we want the error $Y - \hat{Y}$ to be small, on average. We consider the square of the error because this is much easier to analyze than choosing the absolute value of the error.

Now,

$$(Y - \hat{Y})^2 = [(Y - E(Y)) - a - b(X - E(X))]^2$$
$$= a^2 + (Y - E(Y))^2 + b^2(X - E(X))^2$$
$$- 2a(Y - E(Y)) - 2b(Y - E(Y))(X - E(X)) + 2ab(X - E(X)).$$

Taking the expected value, we get

$$E((Y - \hat{Y})^2) = a^2 + var(Y) + b^2 var(X) - 2b cov(X, Y).$$

To do the calculation, we used the linearity of expectation and the facts that the expected values of $X - E(X)$ and $Y - E(Y)$ are equal to zero. To minimize this expression over a, we should choose $a = 0$. To minimize it over b, we set the derivative with respect to b equal to zero and we find

$$2b var(X) - 2cov(X, Y) = 0,$$

so that $b = cov(X, Y)/var(X)$. Consequently,

$$\hat{Y} = E(Y) + \frac{cov(X, Y)}{var(X)}(X - E(X)).$$

We call $\hat{Y}$ the *Linear Least Squares Estimate (LLSE)* of Y given X. It is the linear function of X that minimizes the mean squared error with Y.

As an example, consider again the N marbles. There,

$$E(X) = \frac{1}{N}\sum_{n=1}^{N} X(n), \; E(Y) = \frac{1}{N}\sum_{n=1}^{N} Y(n)$$

$$var(X) = \frac{1}{N}\sum_{n=1}^{N} X^2(n) - [E(X)]^2$$

$$var(Y) = \frac{1}{N}\sum_{n=1}^{N} Y^2(n) - [E(Y)]^2$$

$$cov(X, Y) = \frac{1}{N}\sum_{n=1}^{N} X(n)Y(n) - E(X)E(Y).$$

In this case, one calls the resulting expression for $\hat{Y}$ the *linear regression* of Y against X. The linear regression is the same as the LLSE when one considers that (X, Y) are random variables that are equal to a given sample $(X(n), Y(n))$ with probability $1/N$ for $n = 1, \ldots, N$. That is, to compute the linear regression, one assumes that the sample values one has observed are representative of the random pair (X, Y) and that each sample is an equally likely value for the random pair.

A.10 Why Do We Need a More Sophisticated Formalism?

The previous sections show that one can get quite far in the discussion of probability concepts by considering a finite set of marbles, and random variables that can only take finitely many values. In engineering, one might think that this is enough. One may approximate any sensible quantity with a finite number of bits, so for all applications one may consider that there are only finitely many possibilities. All that is true, but results in clumsy models. For instance, try to write the equations of a falling object with discretized variables. The continuous versions are usually simpler than the discrete ones. As another example, a Gaussian random variable is easier to work with than a binomial or Poisson random variable, as we will see.

Thus we need to extend the model to random variables that have an infinite, even an uncountable set of possible values. Does this step cause formidable difficulties? Not at the intuitive level. The continuous version is a natural extension of the discrete case. However, there is a philosophical difficulty in going from discrete to continuous. Some thinkers do not accept the idea of making an infinite number of choices before moving on. That is, say that we are given an infinite collections $\{A_1, A_2, \ldots\}$ of nonempty sets. Can we reasonably define a new set B that contains one element from each A_n? We can define it, but if there is no finite way of building it, can we assume that it exists? A theory that does not rely on this *axiom of choice* is considerably more complex than those that do. The classical theory of probability (due to Kolmogorov) accepts the axiom of choice, and we follow that theory.

One key axiom of probability theory enables to define the probability of a set A of outcomes as a limit of that of simpler sets A_n that approach A. This is similar to approximating the area of a circle by the sum of the areas of disjoint rectangles that approach it from inside, or approximating an integral by a sum of rectangles. This key axiom says that if $A_1 \subset A_2 \subset A_3 \subset \cdots$ and $A = \cup_n A_n$, then $P(A) = \lim P(A_n)$. Thus, if sets A_n approximate A from inside in the sense that these sets grow and eventually contain every point of A, then the probability of A is equal to the limit of the probability of A_n. This is a natural way of extending the definition of probability of simple sets to more complex ones. The trick is to show that this is a consistent definition in the sense that different approximating sequences of sets must have the same limiting probability.

This key axiom enables to prove the *strong law of large numbers*. That law states that as you keep on flipping coins, the fraction of heads converges to the probability that one coin yields heads. Thus, not only is the fraction of heads very likely to be close to that probability when you flip many coins, but in fact the fraction gets closer and closer to that probability. This property justifies the *frequentist* interpretation of probability of an event as the long-term fraction of time that event occurs when one repeats the experiment. This is the interpretation that we used to justify the definition of expected value.

A.11   References

There are many useful texts and websites on elementary probability. Readers might find Walrand (2019) worthwhile, especially since it is free on Kindle.

A.12   Solved Problems

Problem A.1 You have a bag with 20 red marbles and 30 blue marbles. You shake the bag and pick three marbles, one at a time, without replacement. What is the probability that the third marble is red?

Solution *As is often the case, there is a difficult and an easy way to solve this problem. The difficult way is to consider the first marble, then find the probability that the second marble is red or blue given the color of the first marble, then find the probability that the third marble is red given the colors of the first two marbles.*

The easy way is to notice that, by symmetry, the probability that the third marble is red is the same as the probability that the first marble is red, which is 20/50 = 0.4.

It may be useful to make the symmetry argument explicit. Think of the marbles as being numbered from 1 to 50. Imagine that shaking the bag results in some ordering in which the marbles would be picked one by one out of the bag. All the orderings are equally likely. Now think of interchanging marble one and marble three in each ordering. You end up with a new set of orderings that are again equally likely. In this new ordering, the third marble is the first one to get out of the bag. Thus, the probability that the third marble is red is the same as the probability that the first marble is red.

Problem A.2 Your applied probability class has 275 students who all turn in their homework assignment. The professor returns the graded assignments in a random order to the students. What is the expected number of students who get their own assignment back?

Solution *The difficult way to solve the problem is to consider the first assignment, then the second, and so on, and for each to explore what happens if it is returned to its owner or not. The probability that one student gets her assignment back depends on what happened to the other students. It all seems very complicated.*

The easy way is to argue that, by symmetry, the probability that any given student gets his assignment back is the probability that the first one gets his assignment *back, which is 1/275. Let then $X_n = 1$ if student n gets his/her own assignment and $X_n = 0$ otherwise. Thus, $E(X_n) = 1/275$. The number of students who get their assignment back is $X_1 + \cdots + X_{275}$. Now, by linearity of expectation, $E(X_1 + \cdots + X_{275}) = E(X_1) + \cdots + E(X_{275}) = 275 \times (1/275) = 1$.*

Problem A.3 A monkey types a sequence of one million random letters on a typewriter. How many times does the name "walrand" appear in that sequence? Assume that the typewriter has 40 keys: the 26 letters and 14 other characters.

Solution *The easy solution uses the linearity of expectation. Let $X_n = 1$ if the name "walrand" appears in the sequence, starting at the n-th symbol of the string. The number of times that the name appears is then $Z = X_1 + \cdots + X_N$ with $N = 10^6 - 6$. By symmetry, $E(X_n) = E(X_1)$ for all n. Now, the probability that $X_1 = 1$ is equal to the probability that the first symbol is w, that the second symbol is a, and so on. Thus, $E(X_1) = P(X_1 = 1) = (1/40)^7$. Hence, the expected number of times that "walrand" appears is $E(Z) = (10^6 - 6) \times (1/40)^7 \approx 6 \times 10^{-6}$. So, it is true that a monkey could eventually type one of Shakespeare's plays, but he is likely to die before succeeding.*

Note that Markov's inequality implies that

$$P(Z \geq 1) \leq E(Z) \approx 6 \times 10^{-6}.$$

Problem A.4 You flip a fair coin n times and the fraction of heads is Y. How large does n have to be to be sure that $P(|Y - 0.5| \geq 0.05) \leq 0.05$?

Solution *We use Chebyshev's inequality that states*

$$P(|Y - 0.5| \geq \epsilon) \leq \frac{var(Y)}{\epsilon^2}.$$

We saw in our discussion of the weak law of large numbers that $var(Y) = var(X_1)/n$ where $X_1 = 1$ if the first coin yields heads and $X_1 = 0$ otherwise. Since $P(X_1 = 1) = P(X_1 = 0) = 0.5$, we find that $E(X_1) = 0.5$ and $E(X_1^2) = 0.5$. Hence, $var(X_1) = E(X_1^2) - [E(X_1)]^2 = 0.25$. Consequently,

$$P(|Y - 0.5| \geq 0.05) \leq \frac{0.25}{n \times 25 \times 10^{-4}} = \frac{100}{n}.$$

Thus, the right-hand side is 0.05 if $n = 2,000$. You have to be patient....

Problem A.5 What is the probability that two friends share a birthday? What about three friends? What about n? How large does n have to be for this probability to be 50%?

Solution *In the case of two friends, it is the probability that the second has the same birthday as the first, which is 1/365 (ignoring February 29).*

The case of three friends looks more complicated: two of the three or all of them could share a birthday. It is simpler to look at the probability that they do not share a birthday. This is the probability that the second friend does not have the same birthday as the first, which is 364/365 times the probability that the third does not share a birthday with the first two, which is 363/365. Let us explore this a bit further to make sure we fully understand this solution. First, we consider all the strings of three numbers picked in $\{1, 2, \ldots, 365\}$. There are 365^3 such strings because there are 365 choices for the first number, then 365 for the second, and finally 365 for the third. Second, consider the strings of three different numbers from the same set. There are 365 choices for the first, then 364 for the second, then 363 for the third. Hence, there are $365 \times 364 \times 363$ such strings. Since all the strings are equally likely to be picked (a reasonable assumption), the probability that the friends do not share a birthday is

$$\frac{365 \times 364 \times 363}{365 \times 365 \times 365}.$$

The case of n friends is then clear: they do not share a birthday with probability p where

$$p = \frac{365 \times 364 \times \cdots \times (365 - n + 1)}{365 \times 365 \times \cdots \times 365}$$

$$= 1 \times \left(1 - \frac{1}{365}\right) \times \left(1 - \frac{2}{365}\right) \times \cdots \times \left(1 - \frac{n-1}{365}\right).$$

To evaluate this expression, we use the fact that $1 - x \approx \exp\{-x\}$ when $|x| \ll 1$. We use this fact repeatedly in this book. Do not worry, there are not too many such tricks. In practice, this approximation is good for $|x| < 0.1$. For instance, $\exp\{-0.1\} \approx 0.90483$. Thus, assuming that $n/365 \leq 0.1$, i.e., $n \leq 36$, we find

$$p \approx 1 \times \exp\left\{-\frac{1}{365}\right\} \times \exp\left\{-\frac{2}{365}\right\} \times \cdots \times \exp\left\{-\frac{n-1}{365}\right\}$$

$$= \exp\left\{-\frac{1}{365} - \frac{2}{365} - \cdots - \frac{n-1}{365}\right\} = \exp\left\{-\frac{1 + 2 + \cdots + n - 1}{365}\right\}$$

$$= \exp\left\{-\frac{(n-1)n}{730}\right\}.$$

For instance, with $n = 24$, we find

$$p \approx \exp\left\{-\frac{23 \times 24}{730}\right\} \approx 0.5.$$

Hence, the probability that at least two friends in a group of 24 share a birthday is about 50%. This result is somewhat surprising because 24 is small compared to 365. One calls this observation the birthday paradox. *Many people think that it takes about 365/2 ≈ 180 friends for the probability that they share a birthday to be 50%. The paradox is less mysterious when you think of the many ways that friends can share birthdays.*

Problem A.6 You throw M marbles into B bins, each time independently and in a way that each marble is equally likely to fall into each bin. What is the expected number of empty bins? What is the probability that no bin contains more than one marble?

Solution *The first bin is empty with probability* $\alpha := [(B-1)/B]^M$, *and the same is true for every bin. Hence, if $X_b = 1$ when bin b is empty and $X_b = 0$ otherwise, we see that $E(X_b) = \alpha$. Hence, the expected value of the number $Z = X_1 + \cdots + X_B$ of empty bins is equal to*

$$E(Z) = BE(X_1) = B\alpha.$$

To evaluate this expression we use the following approximation:

$$\left(1 - \frac{a}{N}\right)^N \approx \exp\{-a\} \text{ for } N \gg 1.$$

This approximation is already quite good for $N = 10$ and $0 < a \leq 1$. For instance, $(1 - 1/10)^{10} \approx 0.35$ and $\exp\{-1\} \approx 0.37$. Hence, if $M = \beta B$, one can write

$$\alpha = (1 - 1/B)^M = (1 - \beta/M)^M \approx \exp\{-\beta\},$$

so that

$$E(Z) \approx B \exp\{-\beta\}.$$

For instance, with $M = 20$ and $B = 30$, one has $\beta = 2/3$ and $E(Z) \approx 30 \exp\{-2/3\} \approx 15$. That is, the 20 marbles are likely to fall into 15 of the 30 bins.

The probability that no bins contain more than one marble is the same as the probability that no two friends share a birthday when there are B different days and M friends. We saw in the last problems that this is given by

$$\exp\left\{-\frac{M(M-1)}{2B}\right\} \approx \exp\left\{-\frac{M^2}{2B}\right\}.$$

Problem A.7 As an error detection scheme, you compute a checksum of $b = 32$ bits from the bits of each of M files that you store in a computer and you attach the checksum to the file. When you read the file, you recompute the checksum and you compare with the one attached to the file. If the checksums agree, you assume that no storage/retrieval error occurred. How large can M be before the probability that two files share a checksum exceeds 10^{-6}. A similar scheme is used as a digital signature to make sure that files are not modified.

Solution *There are $B = 2^b$ possible checksums. Let us assume that each file is equally likely to get any one of the B checksums. In view of the previous problem, we want to find M such that*

$$\exp\left\{-\frac{M^2}{2B}\right\} = 10^{-6} = \exp\{-6\log(10)\} \approx \exp\{-14\}.$$

Thus, $M^2/(2B) = 14$, so that $M^2 = 28B = 28 \times 2^b$ and $M = 2^{b/2}\sqrt{28} \approx 5.3 \times 2^{b/2}$. With $b = 32$, we find $M \approx 350,000$.

Problem A.8 N people apply for a job with your company. You will interview them sequentially but you must either hire or decline a person right at the end of the interview. How should you proceed to maximize the chance of picking the best of all the candidates? Implicitly, we assume that the qualities of the candidates are all independent and equally likely to be any number in $\{1, \dots, Q\}$ where Q is very large.

Solution *The best strategy is to interview and decline about $M = N/e$ candidates and then hire the first subsequent candidate who is better that those M. Here, $e = \exp\{1\} \approx 2.72$. If no candidate among $\{M + 1, \dots, N\}$ is better than the first M, you hire the last candidate.*

To justify this procedure, we compute the probability that the candidate you select is the best, for a given value of M. By symmetry, the best candidate appears in position b with probability $1/N$. You then pick the best candidate if $b > M$ and if the best candidate among the first $b - 1$ is among the first M, which has probability $M/(b-1)$, by symmetry. Since probability is additive, the probability p that you pick the best candidate is given by

$$p = \frac{1}{N}\sum_{b=M+1}^{N}\frac{M}{b-1} = \frac{M}{N}\sum_{b=M}^{N-1}\frac{1}{b} \approx \frac{M}{N}\int_{M}^{N}\frac{1}{b}db = \frac{M}{N}[\log(N) - \log(M)].$$

To find the maximizing value of M, we set the derivative of this expression with respect to M equal to zero. This shows that $N/M \approx e$.

Basic Probability

B

Topics: General framework, conditional probability, independence, expectation, pdf, cdf, function of random variables, correlation, variance, transformation of jpdf.

B.1 General Framework

The general model of Probability Theory may seem a bit abstract and disconcerting. However, it unifies all the key ideas in a systematic framework and results in a great conceptual clarity. You should try to keep in mind this underlying framework when we discuss concrete examples.

B.1.1 Probability Space

To describe a *random experiment*, one first specifies the set Ω of all the possible outcomes. This set is called the *sample space*. For instance, when we flip a coin, the sample space is $\Omega = \{H, T\}$; when we roll a die, $\Omega = \{1, 2, 3, 4, 5, 6\}$, when one measures a voltage one may have $\Omega = \Re = (-\infty, +\infty)$; and so on.

Second, one specifies the probability that the outcome falls in subsets of Ω. That is, for $A \subset \Omega$, one specifies a number $P(A) \in [0, 1]$ that represents the likelihood that the random experiment yields an outcome in A. For instance, when rolling a die, the probability that the outcome is in a set $A \subseteq \{1, 2, 3, 4, 5, 6\}$ is given by $P(A) = |A|/6$ where $|A|$ is the number of elements of A. When we measure a voltage, the probability that it has any given value is typically 0, but the probability that it is less than 15 in absolute value may be 95%, which is why we specify the probability of subsets, not of specific outcomes.

© The Author(s) 2021
J. Walrand, *Probability in Electrical Engineering and Computer Science*,
https://doi.org/10.1007/978-3-030-49995-2

Of course, the specification of the probability of subsets of Ω cannot be arbitrary. For instance, if $A \subseteq B$, then one must have $P(A) \leq P(B)$. Also, $P(\Omega) = 1$. Moreover, if A and B are disjoint, i.e., if $A \cap B = \emptyset$, then $P(A \cup B) = P(A) + P(B)$. Finally, to be able to approximate a complex set by simple sets, one requires that if $A_1 \subseteq A_2 \subseteq A_3 \subseteq \cdots$ and if $A = \cup_n A_n$, then $P(A_n) \to P(A)$. Equivalently, if $A_1 \supseteq A_2 \supseteq A_3 \supseteq \cdots$ and if $A = \cap_n A_n$, then $P(A_n) \to P(A)$.

This property also implies the following result.

B.1.2   Borel–Cantelli Theorem

Theorem B.1 (Borel–Cantelli Theorem)   *Let A_n be events such that*

$$\sum_{n=1}^{\infty} P(A_n) < \infty.$$

Then

$$P(A_n, \ i.o.) = 0.$$

∎

Here, $\{A_n, \text{ i.o.}\}$ is defined as the set of outcomes ω that are in infinitely many sets A_n. So, stating that the probability of this set is equal to zero means that the probability that the events A_n occur for infinitely many n's is zero. So, the probability that the events A_n occur *infinitely often* is equal to zero. In other words, for any outcome ω that occurs, there is some m such that A_n does not occur for any n larger than m.

Proof   First note that

$$\{A_n, \ i.o.\} = \cap_n B_n =: B,$$

where $B_n = \cup_{m \geq n} A_m$ is a decreasing sequence of sets. To see this, note that the outcome ω is in infinitely many sets A_n, i.e., that $\omega \in \{A_n, \text{ i.o.}\}$, if and only if for every n, the outcome ω is in some A_m for $m \geq n$. Also, ω is in $\cup_{m \geq n} A_m = B_n$ for all n if and only if ω is in $\cap_n B_n = B$. Hence $\omega \in \{A_n, \text{ i.o.}\}$ if and only if $\omega \in B$.

Now, $B_1 \supseteq B_2 \supseteq \cdots$, so that $P(B_n) \to P(B)$. Thus,

$$P(A_n, \ i.o.) = P(B) = \lim_{n \to \infty} P(B_n)$$

and $P(B_n) \leq \sum_{m=n}^{\infty} P(A_m)$, so that[1]

$$P(B_n) \to 0 \text{ as } n \to \infty.$$

Consequently, $P(A_n, \text{ i.o.}) = 0$.   □

You may wonder whether $\sum_n P(A_n) = \infty$ implies that $P(A_n, \text{ i.o.}) = 1$. As a simple counterexample, imagine that you have an infinite collection of coins that you solder together in an infinite line, all heads up. Assume also that this long string is balanced and that you manage to flip it. Let A_n be the event that coin n yields heads. In this contraption, either all the coins yield heads, with probability 0.5, or all the coins yield tails. Also, $P(A_n) = 0.5$, so that $\sum_n P(A_n) = \infty$ and $P(A_n, \text{ i.o.}) = 0.5$. However, we show in the next section that the result holds if the events are mutually independent.

For the sake of completeness, we should mention that it is generally not possible to specify the probability of all the subsets of Ω. This does not really matter in applications. The terminology is that the subsets of Ω with a well-defined probability are *events*.

B.1.3   Independence

We say that the events A and B are independent if $P(A \cap B) = P(A)P(B)$. For instance, roll two dice. An outcome is a pair $(a, b) \in \{1, 2, \ldots, 6\}^2$ where a corresponds to the first die and b to the second. The event "the first die yields a number in $\{2, 4, 5\}$" corresponds to the set of outcomes $A = \{2, 4, 5\} \times \{1, \ldots, 6\}$. The event "the second die yields a number in $\{2, 4\}$" is the set of outcomes $B = \{1, \ldots, 6\} \times \{2, 4\}$. We can see that A and B are independent since $P(A) = 18/36$, $P(B) = 12/36$ and $P(A \cap B) = 6/36$.

A more subtle notion is that of *mutual independence*. We say that the events $\{A_j, j \in J\}$ are mutually independent if

$$P(\cap_{j \in K} A_j) = \Pi_{j \in K} P(A_j), \forall \text{ finite } K \subset J.$$

It is easy to construct events that are pairwise independent but not mutually independent. For instance, let $\Omega = \{1, 2, 3, 4\}$ where the four outcomes are equally likely and let $A = \{1, 2\}$, $B = \{1, 3\}$, and $C = \{1, 4\}$. You can check that these events are pairwise independent but not mutually independent since $P(A \cap B \cap C) = 1/4 \neq P(A)P(B)P(C)$.

---

[1]Recall that if the nonnegative numbers a_n are such that $\sum_{n=0}^{\infty} a_n < \infty$, then $\sum_{m=n}^{\infty} a_m$ goes to zero as $n \to \infty$.

B.1.4 Converse of Borel–Cantelli Theorem

Theorem B.2 (Converse of Borel–Cantelli Theorem) *Let $\{A_n, n \geq 1\}$ be a collection of mutually independent events with $\sum P(A_n) = \infty$. Then $P(A_n, \text{ i.o.}) = 1$.*

■

Proof Recall that

$$\{A_n, \text{ i.o.}\} = \cap_n B_n \text{ where } B_n = \cup_{m \geq n} A_m.$$

Hence,

$$\{A_n, \text{ i.o.}\}^c = \cup_n B_n^c \text{ where } B_n^c = \cap_{m \geq n} A_m^c.$$

Thus, to prove the theorem, it suffices to show that $P(B_n^c) = 0$ for all n. Indeed, if that is the case, then $P(\cup_{n=1}^N B_n^c) \leq \sum_{n=1}^N P(B_n^c) = 0$ and $\cup_{n=1}^N B_n^c$ are increasing with N and their union is $\cup_n B_n^c$, so that $P(\cup_n B_n^c) = \lim_{N \to \infty} P(\cup_{n=1}^N B_n^c) = 0$.

Now,

$$P(B_n^c) = P(\cap_{m \geq n} A_m^c) = \lim_{N \to \infty} P(\cap_{m=n}^N A_m^c)$$

$$= \lim_{N \to \infty} \Pi_{m=n}^N P(A_m^c) = \lim_{N \to \infty} \Pi_{m=n}^N [1 - P(A_m)]$$

$$\leq \lim_{N \to \infty} \Pi_{m=n}^N \exp\{-P(A_m)\} = \lim_{N \to \infty} \exp\left\{ -\sum_{m=n}^N P(A_m) \right\} = 0.$$

In this derivation we used the facts that $1 - x \leq \exp\{-x\}$ and $\sum_{m=n}^N P(A_m) \to \infty$ as $N \to \infty$.   □

B.1.5 Conditional Probability

Let A and B be two events. Assume that $P(B) > 0$. One defines the conditional probability $P[A|B]$ of A given B as follows:

$$P[A|B] := \frac{P(A \cap B)}{P(B)}.$$

The meaning of $P[A|B]$ is the probability that the outcome of the experiment is in A given that it is in B. As an example, say that a random experiment has 1000 equally likely outcomes. Assume that A contains $|A|$ outcomes and B contains $|B|$ outcomes. If we know that the outcome is in B, we know that it is equally likely to be any one of these $|B|$ outcomes. Given that information, the probability that the outcome is in A is then the fraction of outcomes in B that are also in A. This fraction is

$$\frac{|A \cap B|}{|B|} = \frac{|A \cap B|/1000}{|B|/1000} = \frac{P(A \cap B)}{P(B)}.$$

Note that the definition implies that if A and B are independent, then $P[A|B] = P(A)$, which makes intuitive sense. Also,

$$P(A \cap B) = P[A|B]P(B).$$

This expression extends to more than two events. For instance, with events $\{A_1, \ldots, A_n\}$ one has

$$P(A_1 \cap A_2 \cap \cdots \cap A_n) = P(A_1)P[A_2 \mid A_1]P[A_3 \mid A_1 \cap A_2]$$
$$\cdots P[A_n \mid A_1 \cap \cdots \cap A_{n-1}].$$

To verify this identity, note that the right-hand side is equal to

$$P(A_1)\frac{P(A_1 \cap A_2)}{P(A_1)}\frac{P(A_1 \cap A_2 \cap A_3)}{P(A_1 \cap A_2)} \cdots \frac{P(A_1 \cap \cdots A_n)}{P(A_1 \cap \cdots \cap A_{n-1})},$$

and this product is equal to the left-had side of the identity above.

B.1.6 Random Variable

A random variable X is a function $X : \Omega \to \Re$. Thus, one associates a real number $X(\omega)$ to every possible outcome ω of the random experiment.

For instance, when one flips a coin, with $\Omega = \{H, T\}$, one can define a random variable X by $X(H) = 1$ and $X(T) = 0$.

One then uses the notation $P(X \in B) = P(X^{-1}(B))$ for $B \subset \Re$ where

$$X^{-1}(B) := \{\omega \in \Omega | X(\omega) \in B\}.$$

The interpretation is the natural one: the probability that $X \in B$ is the probability that the outcome ω is such that $X(\omega) \in B$.

In particular, one defines the *cumulative distribution function (cdf)* of the random variable X as $F_X(x) = P(X \in (-\infty, x]) =: P(X \leq x)$. This function is nondecreasing and right-continuous; it tends to zero as $x \to -\infty$ and to one as $x \to +\infty$.

Figure B.1 summarizes this general framework for one random variable.

Fig. B.1 The random experiment is described by a set Ω of outcomes: the *sample space*. Subsets of Ω called *events* have a probability. A *random variable* is a real-valued function of the outcome ω of the random experiment

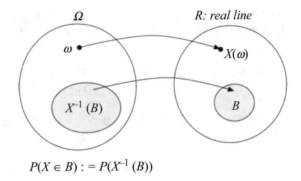

$$P(X \in B) := P(X^{-1}(B))$$

B.2   Discrete Random Variable

B.2.1   Definition

A *discrete random variable* X is defined by a list of distinct possible values and their probability:

$$X \equiv \{(x_n, p_n), n = 1, 2, \ldots, N\}. \tag{B.1}$$

Here, the x_n are real numbers and the p_n are positive and add up to one. By definition, p_n is the probability that X takes the value x_n and we write

$$p_n = P(X = x_n), n = 1, \ldots, N.$$

The number of values N can be infinite. This list is called the *probability mass function (pmf)* of the random variable X.

As an example,

$$V \equiv \{(1, 0.1), (2, 0.3), (3, 0.6)\}$$

is a random variable that has three possible values $(1, 2, 3)$ and takes these values with probability $0.1, 0.3,$ and 0.6, respectively. Equivalently, one can write

$$X = \begin{cases} 1, & \text{with probability } 0.1; \\ 2, & \text{with probability } 0.3; \\ 3, & \text{with probability } 0.6. \end{cases}$$

Note that the probabilities add up to one.

The connection with the general framework is the following. There is some probability space and some function $X : \Omega \to \Re$ that happens to take the values $\{x_1, \ldots, x_N\}$ and is such that

$$P(X = x_n) = P(\{\omega \in \Omega \mid X(\omega) = x_n\}) = p_n.$$

A possible construction is to define $\Omega = \{1, 2, 3\}$ with $P(\{1\}) = 0.1$, $P(\{2\}) = 0.3$, $P(\{3\}) = 0.6$, and $X(\omega) = \omega$. This construction is called the *canonical probability space*. It may not be the natural choice. For instance, say that you pick a marble out of a bag that has 10 identical marbles except that one is marked with the number 1, three with the number 2, and six with the number 3. Let then X be the number on the marble that you pick. A more natural probability space has ten outcomes (the ten marbles) and $X(\omega)$ is the number on marble ω for $\omega \in \Omega = \{1, 2, \ldots, 10\}$.

When one is interested in only one random variable X, one cares about its possible values and their probability, i.e., its pmf. The details of the random experiment do not matter. Thus, one may forget about the bag of marbles. However, if the marbles are marked with a second number Y, then one may have to go back to the description of the bag of marbles to analyze Y or to analyze the pair (X, Y).

B.2.2   Expectation

The *expected value*, or *mean*, of the random variable X is denoted $E(X)$ and is defined as (Fig. B.2)

$$E(X) = \sum_{n=1}^{N} x_n p_n.$$

In our example,

$$E(V) = 1 \times 0.1 + 2 \times 0.3 + 3 \times 0.6 = 2.5.$$

As another frequently used example, say that $X(\omega) = 1\{\omega \in A\}$ where A is an event in Ω. We say that X is the *indicator* of the event A. In this case, X is equal to one with probability $P(A)$ and to zero otherwise, so that $E(X) = P(A)$.

Fig. B.2 The expected value of a random variable

When N is infinite, the definition makes sense unless the sum of the positive terms and that of the negative terms are both infinite. In such a case, one says that X does not have an expected value.

It is a simple exercise to verify that the number a that minimizes $E((X - a)^2)$ is $a = E(X)$. Thus, the mean is the "least squares estimate" of X.

B.2.3   Function of a RV

Consider a function $h : \Re \to \Re$ and a discrete random variable X (Fig. B.3). Then $h(X)$ defines a new random variable with values and probabilities

$$\{(h(x_n), p_n), n = 1, \ldots, N\}.$$

Note that the values $h(x_n)$ may not be distinct, so that to conform to our definition of the pmf one should merge identical values and add their probabilities.

For instance, say that $h(1) = h(2) = 10$ and $h(3) = 15$. Then

$$h(V) \equiv \{(10, 0.4), (15, 0.6)\},$$

where we merged the two values $h(1)$ and $h(2)$ because they are equal to 10.

Thus,

$$E(h(V)) = 10 \times 0.4 + 15 \times 0.6 = 13.$$

Observe that

$$E(h(V)) = \sum_{n=1}^{N} h(v_n) p_n,$$

since

Fig. B.3 Function of a random variable

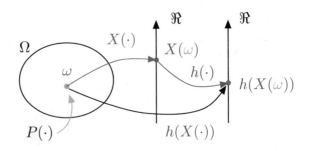

$$\sum_{n=1}^{3} h(v_n)p_n = h(1)0.1 + h(2)0.3 + h(3)0.6$$

$$= 10 \times 0.1 + 10 \times 0.3 + 15 \times 0.6$$

$$= 10 \times (0.1 + 0.3) + 15 \times 0.6,$$

which agrees with the previous expression.

Let us state that observation as a theorem.

Theorem B.3 (Expectation of a Function of a Random Variable) *Let X be a random variable with p.m.f. $\{(x_n, p_n), n = 1, \ldots, N\}$ and $h : \Re \rightarrow \Re$ some function. Then*

$$E(h(X)) = \sum_{n=1}^{N} h(x_n)p_n.$$

■

B.2.4   Nonnegative RV

We say that X is *nonnegative*, and we write $X \geq 0$, if all its possible values x_n are nonnegative. Observe that

$$\text{if } X \geq 0 \text{ and } E(X) = 0, \text{ then } P(X = 0) = 1.$$

Also,

$$\text{if } X \geq 0 \text{ and } E(X) < \infty, \text{ then } P(X < \infty) = 1.$$

B.2.5   Linearity of Expectation

Consider two functions $h_1 : \Re \rightarrow \Re$ and $h_2 : \Re \rightarrow \Re$ and define $h_1(X) + h_2(X)$ as follows:

$$h_1(X) + h_2(X) \equiv \{(h_1(x_n) + h_2(x_n), p_n), n = 1, \ldots, N\}.$$

As before,

$$E(h_1(X) + h_2(X)) = \sum_{n=1}^{N} (h_1(x_n) + h_2(x_n))p_n.$$

By regrouping terms, we see that

$$E(h_1(X) + h_2(X)) = E(h_1(X)) + E(h_2(X)).$$

We say that *expectation is linear*.

B.2.6   Monotonicity of Expectation

By $X \geq 0$ we mean that all the possible values of X are nonnegative, i.e., that $X(\omega) \geq 0$ for all ω. In that case, $E(X) \geq 0$ since $E(X) = \sum_n x_n P(X = x_n)$ and all the x_n are nonnegative.

We also write $X \leq Y$ if $X(\omega) \leq Y(\omega)$. The linearity of expectation then implies that $E(X) \leq E(Y)$ since $0 \leq E(Y - X) = E(Y) - E(X)$. Hence,

$$X \leq Y \text{ implies that } E(X) \leq E(Y). \tag{B.2}$$

One says that *expectation is monotone*.

B.2.7   Variance, Standard Deviation

The *variance* var(X) of a random variable X is defined as (Fig. B.4)

$$\text{var}(X) = E((X - E(X))^2),$$

By linearity, one has

$$\text{var}(X) = E(X^2 - 2XE(X) + E(X)^2)$$
$$= E(X^2) - 2E(X)E(X) + E(X)^2 = E(X^2) - E(X)^2.$$

With (B.1), one finds

$$\text{var}(V) = E(V^2) - E(V)^2 = 1^2 \times 0.1 + 2^2 \times 0.3 + 3^2 \times 0.6 - (2.5)^2 = 0.45.$$

Fig. B.4 The variance makes randomness interesting

The *standard deviation* σ_X of a random variable X is defined as the square root of its variance. That is,

$$\sigma_X := \sqrt{\mathrm{var}(X)}.$$

Note that a random variable W that is equal to $E(X) - \sigma_X$ or to $E(X) + \sigma_X$ with equal probabilities is such that $E(W) = E(X)$ and $\mathrm{var}(W) = \mathrm{var}(X)$. In that sense, σ_X is an "equivalent" deviation from the mean.

Observe that for any $a \in \Re$ and any random variable X one has

$$\mathrm{var}(aX) = a^2 \mathrm{var}(X). \tag{B.3}$$

Indeed,

$$\mathrm{var}(aX) = E((aX)^2) - [E(aX)]^2 = E(a^2 X^2) - [aE(X)]^2 = a^2 E(X^2)$$
$$- a^2 [E(X)]^2 = a^2 \mathrm{var}(X).$$

B.2.8   Important Discrete Random Variables

Here are a few important examples.

Bernoulli   We say that X is *Bernoulli with parameter* $p \in [0, 1]$, and we write $X =_D B(p)$, if[2]

$$X = \{(0, 1 - p), (1, p)\},$$

i.e., if

$$P(X = 0) = 1 - p \text{ and } P(X = 1) = p.$$

You should check that $E(X) = p$ and $\mathrm{var}(X) = p(1 - p)$. This random variable models a coin flip where 1 represents "heads" and 0 "tails."

Geometric   We say that X is *geometrically distributed with parameter* $p \in [0, 1]$, and we write $X =_D G(p)$, if

$$P(X = n) = (1 - p)^{n-1} p, n \geq 1.$$

You should check that $E(X) = 1/p$ and $\mathrm{var}(X) = (1 - p)p^{-2}$. This random variable models the number of coin flips until the first "heads" if the probability of

---

[2]The symbol $=_D$ means *equal in distribution*.

Fig. B.5 A geometric random variable models the number of coin flips until a first "heads"

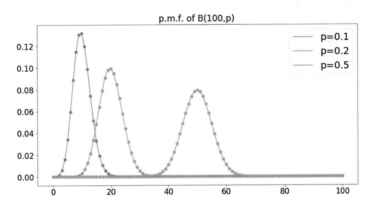

p.m.f. of B(100,p)

Fig. B.6 The probability mass function of the $B(100, p)$ distribution, for $p = 0.1, 0.2$, and 0.5

heads is p (Fig. B.5). (Sometimes, $X - 1$ is also called a geometric random variable on $\{0, 1, \ldots\}$. One avoids confusion by specifying the range. We will try to stick to our definition of X on $\{1, 2, \ldots\}$.).

Binomial We say that X is *binomial with parameters N and p*, and we write $X =_D B(N, p)$, if

$$P(X = n) = \binom{N}{n} p^n (1 - p)^{N-n}, n = 0, \ldots, N, \tag{B.4}$$

where

$$\binom{N}{n} = \frac{N!}{(N - n)! n!}.$$

You should verify that $E(X) = Np$ and $\text{var}(X) = Np(1 - p)$. This random variable models the number of heads in N coin flips; it is the sum of N independent Bernoulli random variables with parameter p. Indeed, there are $\binom{N}{n}$ strings of N symbols in $\{H, T\}$ with n symbols H and $N - n$ symbols T. The probability of each of these sequences is $p^n (1 - p)^{N-n}$ (Figs. B.6, and B.7).

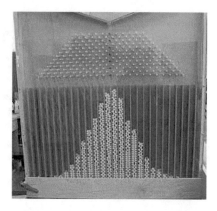

Fig. B.7 The binomial distribution as a sum of Bernoulli random variables. At each step, every steel ball moves to the left or to the right with equal probabilities, i.e., by $2X_n - 1$ where X_n is Bernoulli 0.5. The position after N steps is $Y = \sum n = 1^N (2X_n - 1) = 2B(N, 0.5) - N$. After M balls, the stacks show approximately the values of $M \times P(Y = y)$ for integer y's

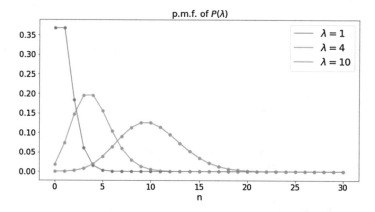

Fig. B.8 Poisson pmf, from Wikipedia

Poisson We say that X is *Poisson with parameter* λ, and we write $X =_D P(\lambda)$, if

$$P(X = n) = \frac{\lambda^n}{n!}e^{-\lambda}, n \geq 0. \tag{B.5}$$

You should verify that $E(X) = \lambda$ and $\mathrm{var}(X) = \lambda$. This random variable models the number of text messages that you receive in 1 day (Fig. B.8).

B.3 Multiple Discrete Random Variables

Quite often one is interested in multiple random variables. These random variables may be related. For instance, the weight and height of a person, the voltage that a

Fig. B.9 Height and weight are related

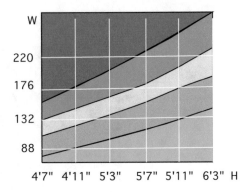

Fig. B.10 The jpmf of a pair of discrete random variables

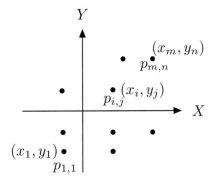

transmitter sends and the one that the receiver gets, and the backlog and delay at a queue are pairs of non-independent random variables (Fig. B.9).

B.3.1   Joint Distribution

To study such dependent random variables, one needs a description more complete than simply looking at the random variables individually. Consider the following example. Roll a die and let $X = 1$ if the outcome is odd and $X = 0$ otherwise. Let also $Y = 1$ if the outcome is in $\{2, 3, 4\}$ and $Y = 0$ if it is in $\{1, 5, 6\}$. Note that $P(X = 1) = P(X = 0) = 0.5$ and $P(Y = 1) = P(Y = 0) = 0.5$. Thus, individually, X and Y could describe the outcomes of flipping two fair coins. However, jointly, the pair (X, Y) does not look like the outcomes of two coin flips. For instance, $X = 1$ and $Y = 1$ only if the outcome is 3, which has probability 1/6. If X and Y were the outcomes of two flips of a fair coin, one would have $X = 1$ and $Y = 1$ in one out of four equally outcomes.

In the discrete case, one describes a pair (X, Y) of random variables by listing the possible values and their probabilities (see Fig. B.10):

$$p_{i,j} = P(X = x_i, Y = y_j), \forall (i, j) \in \{1, \ldots, m\} \times \{1, \ldots, n\},$$

where the $p_{i,j}$ are nonnegative and add up to one. Here, m and n can be infinite. This description specifies the *joint probability mass function (jpmf)* of the random variables (X, Y). (See Fig. B.10.)

From this description, one can in particular recover the probability mass of X and that of Y. For instance,

$$P(X = x_i) = \sum_{j=1}^{n} P(X = x_i, Y = y_j) = \sum_{j=1}^{n} p_{i,j}.$$

B.3.2   Independence

One says that X and Y are *independent* if

$$P(X = x, Y = y) = P(X = x)P(Y = y), \forall x, y.$$

In our die roll example, note that

$$P(X = 1, Y = 1) = \frac{1}{6} \neq P(X = 1)P(Y = 1) = \frac{1}{4},$$

so that X and Y are not independent (Fig. B.11).

B.3.3   Expectation of Function of Multiple RVs

For $h : \Re^2 \to \Re$, one then defines

Fig. B.11 Independence?

$$E(h(X, Y)) = \sum_{i=1}^{m} \sum_{j=1}^{n} h(x_i, y_j) p_{i,j}.$$

Note that if $h(x, y) = h_1(x, y) + h_2(x, y)$, then

$$\begin{aligned}
E(h(X, Y)) &= \sum_{i=1}^{m} \sum_{j=1}^{n} h(x_i, y_j) p_{i,j} \\
&= \sum_{i=1}^{m} \sum_{j=1}^{n} [h_1(x_i, y_j) + h_2(x_i, y_j)] p_{i,j} \\
&= \sum_{i=1}^{m} \sum_{j=1}^{n} h_1(x_i, y_j) p_{i,j} + \sum_{i=1}^{m} \sum_{j=1}^{n} h_2(x_i, y_j) p_{i,j} \\
&= E(h_1(X, Y)) + E(h_2(X, Y)).
\end{aligned}$$

so that *expectation is linear*.

B.3.4   Covariance

In particular, one defines the *covariance* of X and Y as

$$\text{cov}(X, Y) = E((X - E(X))(Y - E(Y))).$$

By linearity of expectation, one has

$$\text{cov}(X, Y) = E(XY - E(X)Y - XE(Y) + E(X)E(Y)) = E(XY) - E(X)E(Y).$$

One says that X and Y are *uncorrelated* if $\text{cov}(X, Y) = 0$. One says that X and Y are *positively correlated* if $\text{cov}(X, Y) > 0$ and that they are *negatively correlated* if $\text{cov}(X, Y) < 0$ (Fig. B.12).

In the die roll example, one finds

Fig. B.12 These random variables are positively correlated: if one is large, the other one tends to be large as well

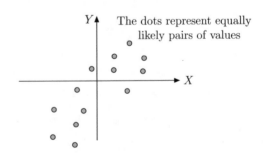

$$\mathrm{cov}(X, Y) = E(XY) - E(X)E(Y) = \frac{1}{6} - \frac{1}{4} < 0,$$

so that X and Y are negatively correlated. This negative correlation suggests that if X is larger than average, then Y tends to be smaller than average. In our example, we see that if $X = 1$, then the outcome is odd and Y is more likely to be 0 than 1.

Here is an important result:

Theorem B.4 (Independent Random Variables are Uncorrelated)

(a) *Independent random variables are uncorrelated.*
(b) *The converse is not true.*
(c) *The variance of a sum of uncorrelated random variables is the sum of their variances.*

■

Proof

(a) Let X, Y be independent. Then

$$E(XY) = \sum_{x,y} xy P(X = x, Y = y) = \sum_{x,y} xy P(X = x)P(Y = y)$$

$$= \left(\sum_{x} x P(X = x) \right) \left(\sum_{y} y P(Y = y) \right) = E(X)E(Y).$$

(b) As a simple example see Fig. B.13, say that (X, Y) is equally likely to take each of the following four values:

$$\{(-1, 0), (0, 1), (0, -1), (1, 0)\}.$$

Then one sees that $E(XY) = 0 = E(X)E(Y)$ so that X and Y are uncorrelated. However, $P(X = -1, Y = 1) = 0 \neq P(X = -1)P(Y = 1)$, so that X and Y are not independent.

(c) Let X and Y be uncorrelated random variables. Then

$$\begin{aligned}
\mathrm{var}(X + Y) &= E((X + Y - E(X + Y))^2) \\
&= E(X^2 + Y^2 + 2XY - E(X)^2 - E(Y)^2 - 2E(X)E(Y)) \\
&= E(X^2) - E(X)^2 + E(Y^2) - E(Y)^2 \\
&= \mathrm{var}(X) + \mathrm{var}(Y).
\end{aligned}$$

The third equality in this derivation comes from the fact that $E(XY) = E(X)E(Y)$.

□

Fig. B.13 The random
variables X and Y are
uncorrelated but not
independent

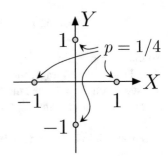

B.3.5   Conditional Expectation

Consider a pair (X, Y) of discrete random variables such that $P(X = x_i, Y = y_j) = p_{i,j}$ for $i = 1, \ldots, m$ and $j = 1, \ldots, n$. In particular, $P(X = x_i) = \sum_k P(X = x_i, Y = y_k) = \sum_k p_{i,k}$. Using the definition of conditional probability, we have

$$P[Y = y_j \mid X = x_i] = \frac{P(X = x_i, Y = y_j)}{P(X = x_i)}.$$

Thus, $P[Y = y_j \mid X = x_i]$ for $j = 1, \ldots, n$ is the *conditional distribution*, or conditional pmf, of Y given that $X = x_i$.

In particular, note that if X and Y are independent, then $P[Y = y_j \mid X = x_i] = P(Y = y_j)$.

We define $E[Y \mid X = x_i]$, the *conditional expectation* of Y given $X = x_i$, as follows:

$$E[Y \mid X = x_i] = \sum_j y_j P[Y = y_j \mid X = x_i].$$

We then define $E[Y \mid X]$ to be a new random variable that is equal to $E[Y \mid X = x_i]$ when $X = x_i$. That is, $E[Y \mid X]$ is a function $g(X)$ of X with $g(x_i) = E[Y \mid X = x_i]$.

The interpretation is that we observe $X = x_i$, which tells us that Y now has a new distribution: its conditional distribution given that $X = x_i$. Then $E[Y \mid X = x_i]$ is the expected value of Y for this conditional distribution.

Theorem B.5 (Properties of Conditional Expectation) *One has*

$$E(E[Y \mid X]) = E(Y) \tag{B.6}$$

$$E[h(X)Y \mid X] = h(X)E[Y \mid X] \tag{B.7}$$

$$E[Y \mid X] = E(Y), \text{ if } X \text{ and } Y \text{ are independent.} \tag{B.8}$$

■

Proof To verify (B.6), one notes that

$$E(E[Y \mid X]) = \sum_i P(X = x_i) E[Y \mid X = x_i] = \sum_i P(X = x_i)$$

$$\times \sum_j y_j P[Y = y_j \mid X = x_i]$$

$$= \sum_i \sum_j y_j P(X = x_i) P[Y = y_j \mid X = x_i]$$

$$= \sum_i \sum_j y_j P(X = x_i, Y = y_j) = \sum_j y_j P(Y = y_j) = E(Y).$$

For (B.7), we recall that, by definition, $E[h(X)Y \mid X]$ is a random variable that takes the value $E[h(X)Y \mid X = x_i]$ when $X = x_i$. Also, $E[h(X)Y \mid X = x_i]$ is the expected value of $h(X)Y$ given that $X = x_i$, i.e., of $h(x_i)Y$ given that $X = x_i$. By linearity of expectation, this is $h(x_i)E[Y \mid X = x_i]$.

Finally, (B.8) is immediate since the distribution of Y given $X = x_i$ is the original distribution of Y when X and Y are independent.   □

B.3.6   Conditional Expectation of a Function

In the same spirit as Theorem B.3, one has the following result:

Theorem B.6 (Conditional Expectation of a Function of a Random Variable)
One has

$$E[h(Y) \mid X = x_i] = \sum_j h(y_j) P[Y = y_j \mid X = x_i].$$

Also, conditional expectation is linear:

$$E[h_1(Y) + h_2(Y) \mid X] = E[h_1(Y) \mid X] + E[h_2(Y) \mid X].$$

                                                                                 ■

B.4   General Random Variables

Not all random variables have a discrete set of possible values. For instance, the voltage across a phone line, wind speed, temperature, and the time until the next customer arrives at a cashier have a continuous range of possible values.

In practice, one can always approximate values by choosing a finite number of bits to represent them. For instance, one can measure temperature in degrees, ignoring fractions, and fixing a lower and upper bound. Thus, discrete random variables suffice to describe systems with an arbitrary degree of precision. However, this discretization is rather artificial and complicates things. For instance, writing Newton's equation $F = ma$ where $a = dv(t)/dt$ with discrete variables is rather bizarre since a discrete speed does not admit a derivative. Hence, although computers perform all their calculations on discrete variables, the analysis and derivation of algorithms are often more natural with general variables. Nevertheless, the approximation intuition is useful and we make use of it.

We start with a definition of a general random variable.

B.4.1   Definitions

Definition B.1 (cdf and pdf)   Let X be a random variable.

(a) The *cumulative distribution function (cdf)* of X is the function $F_X(x)$ defined by

$$F_X(x) = P(X \leq x), x \in \Re.$$

(b) The *probability density function (pdf)* of X is

$$f_X(x) = \frac{d}{dx} F_X(x),$$

   if this derivative exists.

$\diamond$

Observe that, for $a < b$,

$$P(a < X \leq b) = P(X \leq b) - P(X \leq a) = F_X(b) - F_X(a) = \int_a^b f_X(x)dx,$$

where the last expression makes sense if the derivative exists. Also, if the pdf exists,

$$f_X(x)dx = F_X(x + dx) - F_X(x) = P(X \in (x, x + dx]).$$

This identity explains the term "probability density."

B.4.2   Examples

Example B.1 (U[a, b])   As a first example, we say that X is *uniformly distributed* in $[a, b]$, for some $a < b$, and we write $X =_D U[a, b]$ if

Fig. B.14 The pdf and cdf of
a $U[a, b]$ random variable

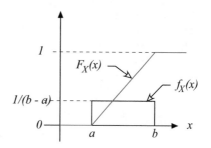

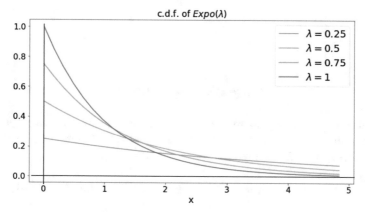

Fig. B.15 Density of exponential distribution

$$f_X(x) = \frac{1}{b-a}1\{a \le x \le b\}.$$

In this case, we see that

$$F_X(x) = \max\left\{0, \min\{1, \frac{x-a}{b-a}\}\right\}.$$

Figure B.14 illustrates the pdf and the cdf of a $U[a, b]$ random variable.

Example B.2 ($Exp(\lambda)$) As a second example, we say that X is *exponentially distributed* with rate $\lambda > 0$, and we write $X =_D Exp(\lambda)$, if

$$f_X(x) = \lambda e^{-\lambda x}1\{x \ge 0\}.$$

Figure B.15.
As before, you can verify that

$$F_X(x) = 1 - \exp\{-\lambda x\}, \text{ for } x \ge 0,$$

Fig. B.16 A discrete approximation of a continuous random variable

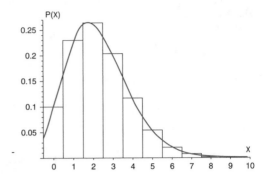

so that

$$P(X \geq x) = \exp\{-\lambda x\}, \forall x \geq 0.$$

It may help intuition to realize that a random variable X with cdf $F_X(\cdot)$ can be approximated by a discrete random variable Y that takes values in $\{\ldots, -2\epsilon, -\epsilon, 0, \epsilon, 2\epsilon, \ldots\}$ with

$$P(Y = n\epsilon) = F_X((n+1)\epsilon) - F_X(n\epsilon) = P(X \in (n\epsilon, (n+1)\epsilon]).$$

Figure B.16.

B.4.3 Expectation

For a function $h : \Re \to \Re$, one has

$$E(h(Y)) = \sum_n h(n\epsilon)[F_X((n+1)\epsilon) - F_X(n\epsilon)] \approx \int_{-\infty}^{\infty} h(x)dF_X(x),$$

where the last term is defined as the limit of the sum as $\epsilon \to 0$. If the pdf exists, one sees that

$$E(h(Y)) \approx \int_{-\infty}^{\infty} h(x)f_X(x)dx.$$

If ϵ is very small, the approximation of X by Y is very close, so that the expressions for $E(h(Y))$ should approach $E(h(X))$. We state these observations as a theorem.

Theorem B.7 (Expectation of a Function of a Random Variable) *Let X be a random variable with cdf $F_X(\cdot)$ and $h : \Re \to \Re$ some function. Then*

$$E(h(X)) = \int_{-\infty}^{\infty} h(x) dF_X(x).$$

If the pdf $f_X(\cdot)$ of X exists, then

$$E(h(X)) = \int_{-\infty}^{\infty} h(x) f_X(x) dx.$$

■

For example, if $X =_D U[0, 1]$, then

$$E(X^k) = \int_0^1 x^k dx = \frac{1}{k+1}.$$

In particular,

$$\mathrm{var}(X) = E(X^2) - E(X)^2 = \frac{1}{3} - \left(\frac{1}{2}\right)^2 = \frac{1}{12}.$$

As another example, if $X =_D Exp(\lambda)$, then

$$E(X) = \int_0^{\infty} x\lambda e^{-\lambda x} dx = -\int_0^{\infty} x de^{-\lambda x} = -[xe^{-\lambda x}]_0^{\infty} + \int_0^{\infty} e^{-\lambda x} dx$$
$$= -\lambda^{-1}[e^{-\lambda x}]_0^{\infty} = \lambda^{-1}.$$

Also,

$$E(X^2) = \int_0^{\infty} x^2 \lambda e^{-\lambda x} dx = -\lambda^{-1} \int_0^{\infty} x^2 de^{-\lambda x}$$
$$= -\lambda^{-1}[x^2 e^{-\lambda x}]_0^{\infty} + \lambda^{-1} \int_0^{\infty} e^{-\lambda x} dx^2$$
$$= 2\lambda^{-1} \int_0^{\infty} x e^{-\lambda x} dx = 2\lambda^{-2}.$$

In particular,

$$\mathrm{var}(X) = E(X^2) - E(X)^2 = 2\lambda^{-2} - (\lambda^{-1})^2 = \lambda^{-2}.$$

As a generally confusing example, consider the random variable that is equal to 0.3 with probability 0.4 and is uniformly distributed in [0, 1] with probability 0.6. That is, one flips a biased coins that yields "head" with probability 0.4. If the

Fig. B.17 The pdf and cdf of
the mixed random variable X

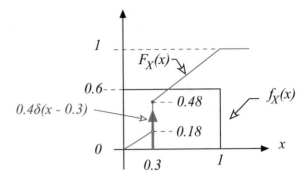

outcome of the coin flip is head, then $X = 0.3$. If the outcome is tail, then X is
picked uniformly in $[0, 1]$. Then,

$$F_X(x) = P(X \leq x) = 0.4 \times 1\{x \geq 0.3\} + 0.6x, \, x \in [0, 1].$$

This cdf is illustrated in Fig. B.17. We can define the derivative of $F_X(x)$ formally
by using the Dirac impulse as the formal derivative of a step function.

For this random variable, one finds that[3]

$$E(X^k) = \int_{-\infty}^{\infty} x^k f_X(dx)$$

$$= \int_{-\infty}^{\infty} x^k 0.4\delta(x - 0.3)dx + \int_0^1 x^k 0.6dx$$

$$= 0.4(0.3)^k + 0.6\frac{1}{k+1}.$$

In particular, we find that

$$\text{var}(X) = E(X^2) - E(X)^2$$

$$= 0, 4(0.3)^2 + 0.6\frac{1}{3} - \left[0.4(0.3) + 0.6\frac{1}{2}\right]^2 = 0.0596.$$

---

[3]Recall that the Dirac impulse is defined by

$$\int_{-\infty}^{\infty} g(x)\delta(x - a)dx = g(a)$$

for any function $g(\cdot)$ that is continuous at a.

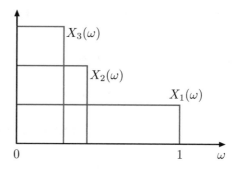

Fig. B.18 The random variables X_n converge to zero but their expectation does not

B.4.4 Continuity of Expectation

We state without proof two useful technical properties of expectation. They address the following question. Assume that $X_n \to X$ as $n \to \infty$. Can we conclude that $E(X_n) \to E(X)$? In other words, is expectation "continuous"?

The following counterexample shows that some conditions are needed (see Fig. B.18). Say that ω is chosen uniformly in $[0, 1]$, so that $P([0, a]) = a$ for $a \in \Omega := (0, 1]$. Define $X_n(\omega) = n \times 1\{\omega \leq 1/n\}$ for $n \geq 1$. That is, $X_n(\omega) = n$ if $\omega \leq 1/n$ and $X_n(\omega) = 0$ otherwise. Then $P(X_n = n) = 1/n$ and $P(X_n = 0) = 1 - 1/n$, so that $E(X_n) = 1$ for all n. Also, $X_n(\omega) \to 0$ as $n \to \infty$, for all $\omega \in \Omega$. Indeed, $X_n(\omega) = 0$ for all $n > 1/\omega$. Thus, $X_n \to X = 0$ but $E(X_n) = 1$ does not converge to $0 = E(X)$.

Theorem B.8 (Dominated Convergence Theorem (DCT)) *Assume that* $|X_n(\omega)| \leq Y(\omega)$ *for all* $\omega \in \Omega$ *where* $E(Y) < \infty$. *Assume also that, for all* $\omega \in \Omega$, $X_n(\omega) \to X(\omega)$ *as* $n \to \infty$. *Then* $E(X_n) \to E(X)$ *as* $n \to \infty$.

∎

Theorem B.9 (Monotone Convergence Theorem (MCT)) *Assume that* $0 \leq X_n(\omega) \leq X_{n+1}(\omega)$ *for all* ω *and* $n = 1, 2, \ldots$. *Assume also that* $X_n(\omega) \to X(\omega)$ *as* $n \to \infty$ *for all* $\omega \in \Omega$. *Then* $E(X_n) \to E(X)$ *as* $n \to \infty$.

∎

One also has the following useful fact.

Theorem B.10 (Expectation as Integral of Complementary cdf) *Let* $X \geq 0$ *be a nonnegative random variable with* $E(X) < \infty$. *Then*

$$E(X) = \int_0^\infty P(X > x)dx.$$

∎

Proof Recall the *integration by parts* formula:

$$\int_a^b u(x)dv(x) = [u(x)v(x)]_a^b - \int_a^b v(x)du(x)$$

that follows from the fact that

$$\frac{d}{dx}[u(x)v(x)] = u'(x)v(x) + u(x)v'(x).$$

Using that formula, one finds

$$E(X) = \int_0^\infty x \, dF_X(x) = -\int_0^\infty x \, d(1 - F_X(x))$$

$$= -[x(1 - F_X(x))]_0^\infty + \int_0^\infty (1 - F_X(x))dx = \int_0^\infty P(X > x)dx.$$

For the last equality we use the fact that $x(1 - F_X(x)) = xP(X > x)$ goes to zero as $x \to \infty$. This fact follows from *DCT*. To see this, define $X_n = n1\{X > n\}$. Then $|X_n| \leq X$ for all n. Also, $X_n \to 0$ as $n \to \infty$. Since $E(X) < \infty$, DCT then implies that $nP(X > n) = E(X_n) \to 0$.

□

The function $P(X > x) = 1 - F_X(x)$ is called the *complementary cdf*. For instance, if $X =_D Exp(\lambda)$, then

$$E(X) = \int_0^\infty P(X > x)dx = \int_0^\infty \exp\{-\lambda x\}dx = \frac{1}{\lambda}.$$

As another example, if $X =_D G(p)$, then $P(X > x) = (1-p)^n$ for $x \in [n, n+1)$ and

$$E(X) = \int_0^\infty P(X > x)dx = \sum_{n \geq 0}(1 - p)^n = \frac{1}{p}.$$

B.5   Multiple Random Variables

B.5.1   Random Vector

A random vector $\mathbf{X} = (X_1, \ldots, X_n)'$ is a vector whose components are random variables defined on the same probability space. That is, it is a function $\mathbf{X} : \Omega \to \mathfrak{R}^n$. One then defines the *joint cumulative distribution function (jcdf)* $F_{\mathbf{X}}$ as follows:

$$F_{\mathbf{X}}(\mathbf{x}) = P(X_1 \leq x_1, \ldots, X_n \leq x_n), \mathbf{x} \in \mathfrak{R}^n.$$

The derivative of this function, if it exists, is the *joint probability density function (jpdf)* $f_{\mathbf{X}}(\mathbf{x},)$. That is,

$$F_{\mathbf{X}}(\mathbf{x},) = \int_{-\infty}^{x_1} \cdots \int_{-\infty}^{x_n} f_{\mathbf{X}}(\mathbf{u}) du_1 \cdots du_n.$$

The interpretation of the jpdf is that

$$f_{\mathbf{X}}(\mathbf{x}) dx_1 \cdots dx_n = P(X_m \in (x_m, x_m + dx_m) \text{ for } m = 1, \ldots, n).$$

For instance, let

$$f_{X,Y}(x, y) = \frac{1}{\pi} 1\left\{x^2 + y^2 \leq 1\right\}, x, y \in \Re.$$

Then, we say that (X, Y) is picked uniformly at random inside the unit circle.

One intuitive way to look at these random variables is to approximate them by points on a fine grid with size $\epsilon > 0$. For instance, an ϵ-approximation of a pair (X, Y) is $(\tilde{X}, \tilde{Y})$ defined by

$$(\tilde{X}, \tilde{Y}) = (m\epsilon, n\epsilon) \text{ w. p. } f_{X,Y}(m\epsilon, n\epsilon)\epsilon^2.$$

This approximation suggests that

$$E(h(X, Y)) = \sum_{m,n} h(m\epsilon, n\epsilon) f_{X,Y}(m\epsilon, n\epsilon)\epsilon^2$$

$$\approx \int_{-\infty}^{\infty} \int_{-\infty}^{\infty} h(x, y) f_{X,Y}(x, y) dx dy.$$

We take this as a definition.

Definition B.2 Let (X, Y) be a pair of random variables and $h : \Re^2 \to \Re$. If the jpdf exists, then

$$E(h(X, Y)) := \int_{-\infty}^{\infty} \int_{-\infty}^{\infty} h(x, y) f_{X,Y}(x, y) dx dy.$$

More generally,

$$E(h(\mathbf{X})) = \int_{-\infty}^{\infty} \cdots \int_{-\infty}^{\infty} h(\mathbf{x}) dx_1 \cdots dx_n.$$

◇

This definition guarantees that expectation is linear. The covariance of X and Y is defined as before.

Definition B.3 (Independence) Two random variables X and Y are independent if

$$P(X \in A, Y \in B) = P(X \in A)P(Y \in B)$$

for all sets A and B in $\mathfrak{R}$.

◇

It is a simple exercise to show that, if the jpdf exists, the random variables are independent if and only if

$$f_{X,Y}(x, y) = f_X(x)f_Y(y), \forall x, y \in \mathfrak{R}.$$

If X is a random variable and $g : \mathfrak{R} \to \mathfrak{R}$ is some function, then $g(X)$ is a random variable. Note that

$$g(X) \in A \text{ if and only if } X \in g^{-1}(A) := \{x \in \mathfrak{R} \mid g(x) \in A\}.$$

Of course, this is a tautology.

Here is a very useful observation.

Theorem B.11 (Functions of Independent Random Variables are Independent)
Let X, Y be two independent random variables and $g, h : \mathfrak{R} \to \mathfrak{R}$ be two functions. Then $g(X)$ and $h(Y)$ are two independent random variables.

∎

Proof Note that

$$P(g(X) \in A, h(Y) \in B) = P(X \in g^{-1}(A), Y \in h^{-1}(B))$$
$$= P(X \in g^{-1}(A))P(Y \in h^{-1}(B))$$
$$= P(g(X) \in A)P(h(Y) \in B).$$

□

B.5.2   Minimum and Maximum of Independent RVs

One is often led to considering the minimum or the maximum of independent random variables. The basic observation is as follows. Let X, Y be independent random variables. Let $V = \min\{X, Y\}$ and $W = \max\{X, Y\}$. Then,

$$P(V > v) = P(X > v, Y > v) = P(X > v)P(Y > v).$$

Also,

$$P(W \leq w) = P(X \leq w, Y \leq w) = P(X \leq w)P(Y \leq w).$$

These observations often suffice to do useful calculations.

For example, assume that $X = Exp(\lambda)$ and $Y = Exp(\mu)$. Then

$$P(V > v) = P(X > v)P(Y > v) = \exp\{-\lambda v\}\exp\{-\mu v\} = \exp\{-(\lambda + \mu)v\}.$$

Thus, the minimum of two exponentially distributed random variables is exponentially distributed, with a rate equal to the sum of the rates.

Let X, Y be i.i.d. $U[0, 1]$. Then,

$$P(W \leq w) = P(X \leq w)P(Y \leq w) = w^2, \text{ for } w \in [0, 1].$$

B.5.3   Sum of Independent Random Variables

Let X, Y be independent random variables and let $Z = X + Y$. We want to calculate $f_Z(z)$ from $f_X(x)$ and $f_Y(y)$. The idea is that

$$P(Z \in (z, z + dz)) = \int_{-\infty}^{+\infty} P(X \in (x, x + dx), Y \in (z - x, z - x + dz)).$$

Hence,

$$f_Z(z)dz = \int_{-\infty}^{+\infty} f_X(x)f_Y(z - x)dxdz.$$

We conclude that

$$f_Z(z) = \int_{-\infty}^{+\infty} f_X(x)f_Y(z - x)dx = f_X * f_Y(z),$$

where $g * h$ indicates the convolution of two functions. If you took a class on signals and systems, you learned the "flip and drag" graphical method to find a convolution.

B.6   Random Vectors

In many situations, one is interested in a collection of random variables.

Definition B.4 (Random Vector) A random vector $\mathbf{X} = (X_1, \ldots, X_n)'$ is a vector whose components are random variables. It is characterized by the *Joint Cumulative Probability Distribution Function (jcdf)*

$$F_{\mathbf{X}}(x_1, \ldots, x_n) := P(X_1 \leq x_1, \ldots, X_n \leq x_n), x_i \in \mathfrak{R}, i = 1, \ldots, n.$$

The *Joint Probability Density Function (jpdf)* is the function $f_{\mathbf{X}}(\mathbf{x})$ such that

$$F_{\mathbf{X}}(x_1, \ldots, x_n) = \int_{-\infty}^{x_1} \cdots \int_{-\infty}^{x_n} f_{\mathbf{X}}(u_1, \ldots, u_n) du_1 \ldots du_n,$$

if such a function exists. In that case,

$$f_{\mathbf{X}}(\mathbf{x}) dx_1 \ldots dx_n = P(X_i \in [x_i, x_i + dx_i], i = 1, \ldots, n).$$

◇

Thus, the jcdf and the jpdf specify the likelihood that the random vector takes values in given subsets of $\mathfrak{R}^n$.

As in the case of two random variables, one has

$$E(h(\mathbf{X})) = \int \cdots \int h(\mathbf{x}) f_{\mathbf{X}}(\mathbf{u}) du_1 \ldots du_n,$$

if the jpdf exists.

The following definitions are used frequently.

Definition B.5 (Mean and Covariance) Let $\mathbf{X}, \mathbf{Y}$ be random vectors. One defines

$$E(\mathbf{X}) = (E(X_1), \ldots, E(X_n))'$$
$$\Sigma_{\mathbf{X}} = E((\mathbf{X} - E(\mathbf{X}))(\mathbf{X} - E(\mathbf{X}))')$$
$$\text{cov}(\mathbf{X}, \mathbf{Y}) = E((\mathbf{X} - E(\mathbf{X}))(\mathbf{Y} - E(\mathbf{Y}))').$$

We say that $\mathbf{X}$ and $\mathbf{Y}$ are uncorrelated if $\text{cov}(\mathbf{X}, \mathbf{Y}) = \mathbf{0}$, i.e., if X_i and Y_j are *uncorrelated* for all i, j.

◇

Thus, the mean value of a vector is the vector of mean values. Similarly, the mean value of a matrix is defined as the matrix of mean values. Also, the covariance of $\mathbf{X}$ and $\mathbf{Y}$ is the matrix of covariances. Indeed,

$$\text{cov}(\mathbf{X}, \mathbf{Y})_{i,j} = E((X_i - E(X_i))(Y_j - E(Y_j))) = \text{cov}(X_i, Y_j).$$

Note also that $\Sigma_{\mathbf{X}} = \text{cov}(\mathbf{X}, \mathbf{X}) =: \text{cov}(\mathbf{X})$.

As a simple exercise, note that

$$\text{cov}(A\mathbf{X} + \mathbf{a}, B\mathbf{Y} + \mathbf{b}) = A\text{cov}(\mathbf{X}, \mathbf{Y})B'.$$

Fig. B.19 A geometric view
of orthogonality

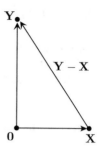

B.6.1 Orthogonality and Projection

The notions of orthogonality and of projection are essential when studying estimation.

Let $\mathbf{X}$ and $\mathbf{Y}$ be two random vectors. We say that $\mathbf{X}$ and $\mathbf{Y}$ are orthogonal and we write $\mathbf{X} \perp \mathbf{Y}$ if

$$E(\mathbf{XY'}) = 0.$$

Thus, $\mathbf{X}$ and $\mathbf{Y}$ are orthogonal if and only if each X_i is orthogonal to every Y_j.

Note that if $E(\mathbf{X}) = \mathbf{0}$, then $\mathbf{X} \perp \mathbf{Y}$ if and only if $\text{cov}(\mathbf{X}, \mathbf{Y}) = 0$. Indeed,

$$\text{cov}(\mathbf{X}, \mathbf{Y}) = E(\mathbf{XY'}) - E(\mathbf{X})E(\mathbf{Y})' = E(\mathbf{XY'}),$$

since $E(\mathbf{X}) = \mathbf{0}$.

The following fact is very useful (see Fig. B.19).

Theorem B.12 (Orthogonality) *If* $\mathbf{X} \perp \mathbf{Y}$*, then*

$$E(||\mathbf{Y} - \mathbf{X}||^2) = E(||\mathbf{X}||^2) + E(||\mathbf{Y}||^2).$$

This statement is the equivalent of Pythagoras' theorem.

∎

Proof One has

$$E(||\mathbf{Y} - \mathbf{X}||^2) = E((\mathbf{Y} - \mathbf{X})'(\mathbf{Y} - \mathbf{X})) = E(\mathbf{Y'Y}) - 2E(\mathbf{X'Y}) + E(\mathbf{X'X})$$
$$= E(||\mathbf{Y}||^2) - 2E(\mathbf{X'Y}) + E(||\mathbf{X}||^2).$$

Now, if $\mathbf{X} \perp \mathbf{Y}$, then $E(X_i Y_j) = 0$ for all i, j. Consequently, $E(\mathbf{X'Y}) = \sum_i E(X_i Y_i) = 0$. This proves the result. $\square$

B.7   Density of a Function of Random Variables

Assume that $\mathbf{X}$ has a known p.d.f. $f_{\mathbf{X}}(\mathbf{x})$ on $\mathfrak{R}^n$ and that $g : \mathfrak{R}^n \to \mathfrak{R}^n$ is a differentiable function. Let $\mathbf{Y} = g(\mathbf{X})$. How do we find $f_{\mathbf{Y}}(\mathbf{y})$?

We start with the linear case and then explain the general case.

B.7.1   Linear Transformations

Assume that X has p.d.f. $f_X(x)$. Let $Y = aX + b$ for some $a > 0$. How do we calculate $f_Y(y)$?

As we see in Fig. B.20, we have

$$P(Y \in (y, y + dy)) = P(aX + b \in (y, y + dy))$$
$$= P(X \in (a^{-1}(y - b), a^{-1}(y + dy - b))).$$

Recall that $P(Z \in (z, z + dz)) = f_Z(z)dz$. Accordingly,

$$f_Y(y)dy = f_X(a^{-1}(y - b)) \times a^{-1}dy,$$

so that

$$f_Y(y) = \frac{1}{a} f_X(x) \text{ where } ax + b = y. \tag{B.9}$$

The case $a < 0$ is not that different. Repeating the argument above, one finds

$$f_Y(y) = \frac{1}{|a|} f_X(x) \text{ where } ax + b = y.$$

What about a pair of random variables? Assume that $\mathbf{X}$ is a random vector that takes values in $\mathfrak{R}^2$ with p.d.f. $f_{\mathbf{X}}(\mathbf{x})$. Let

Fig. B.20 The linear
transformation $Y = aX + b$

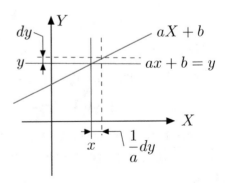

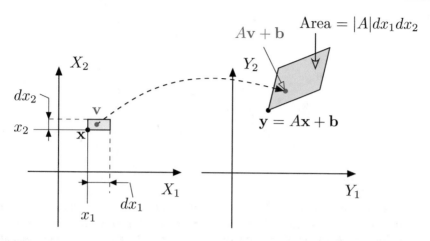

Fig. B.21 The linear transformation $\mathbf{Y} = A\mathbf{X} + \mathbf{b}$

$$\mathbf{Y} = A\mathbf{X} + \mathbf{b},$$

where $A \in \Re^{2 \times 2}$ and $\mathbf{b} \in \Re^{2}$.

Figure B.21 shows that, under the linear transformation, the rectangle $[x_1, x_1 + dx_1] \times [x_2, x_2 + dx_2]$ gets mapped into a parallelogram with area $|A|dx_1dx_2$ where $|A|$ is the absolute value of the determinant of the matrix A. Hence, the probability that $\mathbf{Y}$ falls in this parallelogram with area $|A|dx_1dx_2$ is $f_{\mathbf{X}}(\mathbf{x})dx_1dx_2$. Since the probability that $\mathbf{Y}$ takes value in a small area is proportional to that area, the probability that $\mathbf{Y}$ falls in this parallelogram with area $|A|dx_1dx_2$ is also given by $f_{\mathbf{Y}}(\mathbf{y})|A|dx_1dx_2$ where $\mathbf{y} = A\mathbf{x} + \mathbf{b}$. Thus,

$$f_{\mathbf{Y}}(\mathbf{y})|A|dx_1dx_2 = f_{\mathbf{X}}(\mathbf{x})dx_1dx_2, \text{ with } \mathbf{y} = A\mathbf{x} + \mathbf{b}.$$

Hence,

$$f_{\mathbf{Y}}(\mathbf{y}) = \frac{1}{|A|}f_{\mathbf{X}}(\mathbf{x}) \text{ where } A\mathbf{x} + \mathbf{b} = \mathbf{y}.$$

In fact, this result holds for n random variables.

Given the importance of this result, we state it as a theorem.

Theorem B.13 (Change of Density Through Linear Transformation) *Let* $\mathbf{Y} = A\mathbf{X} + \mathbf{b}$ *where* A *is an* $n \times n$ *nonsingular matrix. Then*

$$f_{\mathbf{Y}}(\mathbf{y}) = \frac{1}{|A|}f_{\mathbf{X}}(\mathbf{x}) \text{ where } A\mathbf{x} + \mathbf{b} = \mathbf{y}. \tag{B.10}$$

∎

Fig. B.22 A singular transformation $\mathbf{Y} = (X_1, X_1)'$

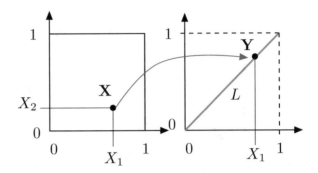

When the matrix A is singular, the random vector $\mathbf{Y} = A\mathbf{X} + \mathbf{b}$ takes values in a set of dimension less than n. In that case, the vector $\mathbf{Y}$ does not have a density in $\mathfrak{R}^n$. As a simple example of this situation, assume that X_1 and X_2 are independent and uniformly distributed in $[0, 1]$. (See Fig. B.22.)

Let

$$\mathbf{Y} = \begin{bmatrix} X_1 \\ X_1 \end{bmatrix} = A\mathbf{X} \text{ where } A = \begin{bmatrix} 1 & 0 \\ 1 & 0 \end{bmatrix}.$$

Then $\mathbf{Y}$ has no density in $\mathfrak{R}^2$. Indeed, if it had one, one would find that, with

$$L = \{\mathbf{y} \mid y_1 = y_1 \text{ and } 0 \le y_1 \le 1\},$$

$$P(\mathbf{Y} \in L) = \int \int_L f_{\mathbf{Y}}(\mathbf{y}) dy = 0$$

since L has measure 0 in $\mathfrak{R}^2$. But $P(\mathbf{Y} \in L) = 1$.

B.7.2   Nonlinear Transformations

The case when $Y = g(X)$ for a nonlinear function $g(\cdot)$ is slightly more tricky. Let us look at one example first.

First Example
Say that $X =_D U[0, 1]$ and $Y = X^2$, as shown in Fig. B.23.

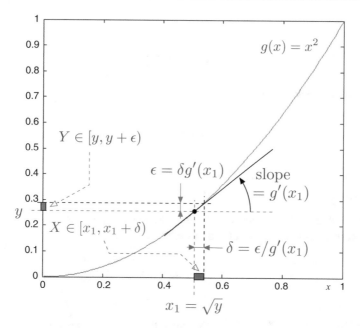

Fig. B.23 The transformation $Y = X^2$ with $X =_D U[0, 1]$

As the figure shows, for $0 < \epsilon \ll 1$, one has $Y \in [y, y + \epsilon)$ if and only if $X \in [x_1, x_1 + \delta)$ where

$$\delta = \frac{\epsilon}{g'(x_1)} = \frac{\epsilon}{2x_1} \text{ where } g(x_1) = x_1^2 = y.$$

Now,[4]

$$P(Y \in [y, y + \epsilon)) = f_Y(y)\epsilon + o(\epsilon)$$

and

$$P(X \in [x_1, x_1 + \delta)) = f_X(x_1)\delta + o(\delta).$$

Also, $o(\delta) = o(\epsilon)$. Hence,

$$f_Y(y)\epsilon + o(\epsilon) = f_X(x_1)\delta + o(\epsilon) = \frac{1}{g'(x_1)} f_X(x_1)\epsilon + o(\epsilon),$$

---

[4]Recall that $o(\epsilon)$ designates a function of ϵ such that $\frac{o(\epsilon)}{\epsilon} \to 0$ as $\epsilon \to 0$.

so that

$$f_Y(y) = \frac{1}{g'(x_1)} f_X(x_1) \text{ where } g(x_1) = y.$$

In this example, we see that

$$f_Y(y) = \frac{1}{2\sqrt{y}},$$

because $g'(x_1) = 2x_1 = 2\sqrt{y}$ and $f_X(x_1) = 1$.

Second Example
We now look at a slightly more complex example. Assume that $Y = g(X) = X^2$ where X takes values in $[-1, 1]$ and has p.d.f.

$$f_X(x) = \frac{3}{8}(1+x)^2, x \in [-1, 1].$$

Figure B.24.
Consider one value of $y \in (0, 1)$. Note that there are now two values of x, namely $x_1 = \sqrt{y}$ and $x_2 = -\sqrt{y}$ such that $g(x) = y$. Thus,

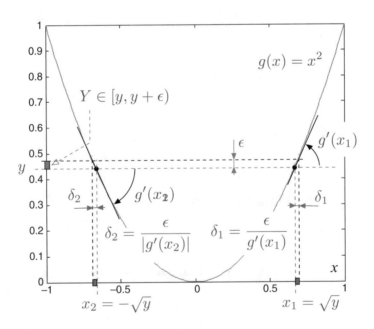

Fig. B.24 The transformation $Y = X^2$ with $X \in [-1, 1]$

$$P(Y \in (y, y + \epsilon)) = P(X \in (x_1, x_1 + \delta_1)) + P(X \in (x_2 - \delta_2, x_2)),$$

where

$$\delta_1 = \frac{\epsilon}{g'(x_1)} \text{ and } \delta_2 = \frac{\epsilon}{|g'(x_2)|}.$$

Hence,

$$f_Y(y)\epsilon + o(\epsilon) = \frac{\epsilon}{g'(x_1)} f_X(x_1) + \frac{\epsilon}{|g'(x_2)|} f_X(x_2) + o(\epsilon)$$

and we conclude that

$$f_Y(y) = \frac{1}{g'(x_1)} f_X(x_1) + \frac{1}{|g'(x_2)|} f_X(x_2).$$

For this specific example, we find

$$f_Y(y) = \frac{1}{2\sqrt{y}} \frac{3}{8} (1 + \sqrt{y})^2 + \frac{1}{2\sqrt{y}} \frac{3}{8} (1 - \sqrt{y})^2 = \frac{3}{8} \frac{1+y}{\sqrt{y}}.$$

Third Example

Our next example is a general differentiable function $g(\cdot)$. From the second example, we can see that if $Y = g(X)$, then

$$f_Y(y) = \sum_i \frac{1}{|g'(x_i)|} f_X(x_i), \tag{B.11}$$

where the sum is over all the x_i such that $g(x_i) = y$.

Fourth Example

What about the multi-dimensional case? The key idea is that, locally, the transformation from $\mathbf{x}$ to $\mathbf{y}$ looks linear. Observe that

$$g_i(\mathbf{x} + d\mathbf{x}) \approx g_i(\mathbf{x}) + \sum_j \frac{\partial}{\partial x_j} g_i(\mathbf{x}) dx_j \approx g(\mathbf{x}) + J(\mathbf{x}) d\mathbf{x},$$

where $J(\mathbf{x})$ is the matrix defined by

$$J_{i,j}(\mathbf{x}) = \frac{\partial}{\partial x_j} g_i(\mathbf{x}).$$

This matrix is called the *Jacobian* of the function $g : \Re^n \to \Re^n$. Thus, locally, the transformation looks like $\mathbf{Y} = A\mathbf{X} + \mathbf{b}$ where $\mathbf{b} = g(\mathbf{x})$ and $A = J(\mathbf{x})$.

Consequently, the density of $f_{\mathbf{X}}$ around $\mathbf{x}$ such that $g(\mathbf{x}) = \mathbf{y}$ gets transformed as if the transformation were linear: it is stretched by the determinant of $J(\mathbf{x})$. Consequently, we have the following theorem.

Theorem B.14 (Density of Function of Random Variables) *Assume that* $\mathbf{Y} = g(\mathbf{X})$ *where* $\mathbf{X}$ *has density* $f_{\mathbf{X}}$ *in* $\mathfrak{R}^n$ *and* $g : \mathfrak{R}^n \to \mathfrak{R}^n$ *is differentiable. Then*

$$f_{\mathbf{Y}}(\mathbf{y}) = \sum_i \frac{1}{|J(\mathbf{x}_i)|} f_{\mathbf{X}}(\mathbf{x}_i),$$

where the sum is over all the $\mathbf{x}_i$ *such that* $g(\mathbf{x}_i) = \mathbf{y}$ *and* $|J(\mathbf{x}_i)|$ *is the absolute value of the determinant of the Jacobian evaluated at* $\mathbf{x}_i$.

∎

Here is an example to illustrate this result. Assume that $\mathbf{X} = (X_1, X_2)$ where the X_i are i.i.d. $U[0, 1]$. Consider the transformation

$$Y_1 = X_1^2 + X_2^2 \text{ and } Y_2 = 2X_1 X_2.$$

Then

$$J(\mathbf{x}) = \begin{bmatrix} 2x_1 & 2x_2 \\ 2x_2 & 2x_1 \end{bmatrix}.$$

Hence,

$$|J(\mathbf{x})| = 4|x_1^2 - x_2^2|.$$

There are two values of $\mathbf{x}$ that correspond to each value of $\mathbf{y}$. These values are

$$x_1 = \frac{1}{2} \left[\sqrt{y_1 + y_2} + \sqrt{y_1 - y_2} \right] \text{ and } x_2 = \frac{1}{2} \left[\sqrt{y_1 + y_2} - \sqrt{y_1 - y_2} \right]$$

and

$$x_1 = \frac{1}{2} \left[\sqrt{y_1 + y_2} - \sqrt{y_1 - y_2} \right] \text{ and } x_2 = \frac{1}{2} \left[\sqrt{y_1 + y_2} + \sqrt{y_1 - y_2} \right].$$

For these values,

$$|J(\mathbf{x})| = \sqrt{y_1^2 - y_2^2}.$$

Hence,

$$f_{\mathbf{Y}}(\mathbf{y}) = \frac{2}{\sqrt{y_1^2 - y_2^2}}$$

for all possible values of $\mathbf{y}$.

B.8 References

Mastering probability theory requires curiosity, intuition, and patience. Good books are very helpful. Personally, I enjoyed Pitman (1993). The home page of David Aldous (2018) is a source of witty and inspiring comments about probability. The textbooks Bertsekas and Tsitsiklis (2008), Grimmett and Stirzaker (2001), and Billingsley (2012) are very useful. The text Wong and Hajek (1985) provides a deeper discussion of the topics in this book. The books Gallager (2014) and Hajek (2017) are great resources and are highly recommended to complement this course.

Wikipedia and YouTube are cool sources of information about everything, including probability. I like to joke, "Don't take notes, it's all on the web."

B.9 Problems

Problem B.1 You have a collection of coins and that the probability that coin n yields heads is p_n. Show that, as you keep flipping the coins, the flips yield a finite number of heads if and only if $\sum p_n < \infty$.

Hint This is a direct consequence of the Borel–Cantelli Theorem and its converse.

Problem B.2 Indicate whether the following statements are true or false:

(a) Disjoint events are independent.
(b) The variance of a sum of random variables is always the sum of their variances.
(c) The expected value of a sum of random variables is the sum of their expected values.

Problem B.3 Provide examples of events A, B, C such that

$$P[A|C] < P(A), P[A|B] > P(A) \text{ and } P[B|A] > P(B).$$

Problem B.4 Roll two balanced dice. Let A be the event "the sum of the faces is less than or equal to 8." Let B be the event "the face of the first die is larger than or equal to 3."

- What is the probability space $(\Omega, \mathcal{F}, P)$?
- Calculate $P[A|B]$ and $P[B|A]$.

Problem B.5 You flip a fair coin repeatedly, forever.

- What is the probability that out of the first 1000 flips the number of heads is even?
- What is the probability that the number of heads is always ahead of the number of tails in the first 4 flips?

Problem B.6 Let X, Y be i.i.d. $Exp(1)$, i.e., exponentially distributed with rate 1. Derive the p.d.f. of $Z = X + Y$.

Problem B.7 You pick four cards randomly from a perfectly shuffled 52-card deck. Assume that the four cards you got are all numbered between 2 and 10. For instance, you got a 2 of diamonds, a 10 of hearts, a 6 of clubs, and a 2 of spades. Write a MATLAB script to calculate the probability that the sum of the numbers on the black cards is exactly twice the sum of the numbers on the red cards.

Problem B.8 Let $X =_D G(p)$, i.e., geometrically distributed with parameter p. Calculate $E(X^3)$.

Problem B.9 Let X, Y be i.i.d. $U_D[0, 1]$. Calculate $E(\max\{X, Y\} - \min\{X, Y\})$.

Problem B.10 Let $X =_D P(\lambda)$ (i.e., Poisson distributed with mean λ). Find $P(X \text{ is even})$.

Problem B.11 Consider $\Omega = [0, 1]$ with the uniform distribution. Let $X(\omega) = 1\{a < \omega < b\}$ and $Y = 1\{c < \omega < d\}$ for some $0 < a < b < 1$ and $0 < c < d < 1$. Assume that X and Y are uncorrelated. Are they necessarily independent?

Problem B.12 Let X and Y be i.i.d. $U[-1, 1]$ and define $Z = XY$. Are X and Z uncorrelated? Are they independent?

Problem B.13 Let $X =_D U[-1, 3]$ and $Y = X^3$. Calculate $f_Y(\cdot)$.

Problem B.14 You are given a one meter long stick. You choose two points X and Y independently and uniformly along the stick and cut the stick at those two points. What is the probability that you can make a triangle with the three pieces?

Problem B.15 Two friends go independently to a bar at times that are uniformly distributed between 5:00 pm and 6:00 pm. They wait for ten minutes when they get there. What is the probability that they meet?

Problem B.16 Choose $V \geq 0$ so that $V^2 =_D Exp(2)$. Now choose $\theta =_D U[0, 2\pi]$, independent of V. Define $X = V \cos(\theta)$ and $Y = V \sin(\theta)$. Calculate $f_{X,Y}(x, y)$.

Problem B.17 Assume that Z and $1/Z$ are random variables with the same probability distribution and such that $E(|Z|)$ is well-defined. Show that $E(|Z|) \geq 1$.

Problem B.18 Let $\{X_n, n \geq 1\}$ be i.i.d. with mean 0 and variance 1. Define $Y_n = (X_1 + \cdots + X_n)/n$.

(a) Calculate $\text{var}(Y_n)$.
(b) Show that $P(|Y_n| \geq \epsilon) \to 0$ as $n \to \infty$, for all $\epsilon > 0$.

Problem B.19 Let X, Y be i.i.d. $U[0, 1]$ and $\mathbf{Z} = A(X, Y)^T$ where A is a given 2×2 matrix. What is the p.d.f. of Z?

Problem B.20 Let $X =_D U[1, 7]$ and $Y = \ln(X) + 3\sqrt{X}$. Show that $E(Y) \leq 7.4$.

Problem B.21 Pick two points X and Y independently and uniformly in $[0, 1]^2$. Calculate $E(||X - Y||^2)$.

Problem B.22 Let (X, Y) be picked uniformly in the triangle with corners $(-1, 0)$, $(1, 0)$, $(0, 1)$. Find $\text{cov}(X, Y)$.

Problem B.23 Let X be a random variable with mean 1 and variance 0.5. Show that

$$E(2X + 3X^2 + X^4) \geq 8.5.$$

Problem B.24 Let X, Y, Z be i.i.d. and uniformly distributed in $\{-1, +1\}$ (i.e., equally likely to be -1 or $+1$). Define $V_1 = XY$, $V_2 = YZ$, $V_3 = XZ$.

(a) Are $\{V_1, V_2, V_3\}$ pairwise independent? Prove.
(b) Are $\{V_1, V_2, V_3\}$ mutually independent? Prove.

Problem B.25 Let A and B be events with probabilities $P(A) = 3/4$ and $P(B) = 1/3$. Show that $\frac{1}{12} \leq P(A \cap B) \leq 1/3$, and give examples to show that both upper and lower bound are tight. Find corresponding bounds for $P(A \cup B)$.

Problem B.26 A power system supplies electricity to a city from N plants. Each power plant fails with probability p independently of the other plants. The city will experience a blackout if fewer than k plants are supplying it, where $0 < k < N$. What is the probability of blackout?

Fig. B.25 Reliability graph
of a system

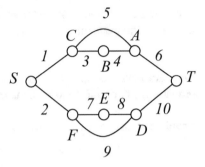

Fig. B.26 A circuit used as a
simple timer. An external
circuit detects when the
voltage $V(t)$ drops below $1V$

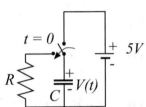

Problem B.27 Figure B.25 is the reliability graph of a system. The links of the
graph represent components of the system. Each link i is working with probability
p_i and defective with probability $1 - p_i$, independently of the other links. The system
is operational if the nodes S and T are connected. Thus, the system is built of two
redundant subsystems. Each subsystem consists of a number of components.

(a) Calculate the probability that the system is operational.
(b) Assume now the reliability graph is a binary tree with n levels and that the links
    fail independently with probability $1 - p$. What is the probability $g(n)$ that there
    is a working path from the root to a leaf?
(c) Show that $g(n) \rightarrow 0$ as $n \rightarrow \infty$ if $p < 0.5$. Also, prove that $g(n) \rightarrow q > 0$ if
    $p > 0.5$. What is the limit q?

Problem B.28 Figure B.26 illustrates an RC-circuit used as a timer. Initially, the
capacitor is charged by the power supply to $5\,V$. At time $t = 0$, the switch is flipped
and the capacitor starts discharging through the resistor. An external circuit detects
the time τ when $V(t)$ first drops below $1\,V$.

(a) Calculate τ in terms of R and C.
(b) Assume now that R and C are independent random variables that are uniformly
    distributed in $[R_0(1 - \epsilon), R_0(1 + \epsilon)]$ and $[C_0(1 - \epsilon), C_0(1 + \epsilon)]$, respectively.
    Calculate the variance of τ.
(c) Let τ_0 be the value of τ that corresponds to $R = R_0$ and $C = C_0$. Find an upper
    bound on the probability that $|\tau - \tau_0| > \delta\tau_0$ for some small δ.

Fig. B.27 Alice and Bob
play the game "matching
pennies"

Problem B.29 Alice and Bob play the game of *matching pennies*. In this game, they both choose the side of the penny to show. Alice wins if the two sides are different and Bob wins otherwise (Fig. B.27).

(a) Assume that Alice chooses to show "head" with probability $p_A \in [0, 1]$. Calculate the probability p_B with which Bob should show "head" to maximize his probability of winning.
(b) From your calculations, find the best choices of p_A and p_B for Alice and Bob. Argue that those choices are such that Alice cannot improve her chance of winning by modifying p_A and similarly for Bob. A solution with that property is called a *Nash equilibrium*.

Problem B.30 You find two old batteries in a drawer. They produce the voltages X and Y. Assume that X and Y are i.i.d. and uniformly distributed in $[0, 1.5]$.

(a) What is the probability that if you put them in series they produce a voltage larger than 2?
(b) What is the probability that at least one of the two batteries has a voltage that exceeds $1V$?
(c) What is the probability that both batteries have a voltage that exceeds $1 V$?
(d) You find more similar batteries in that drawer. You test them one by one until you find one whose voltage exceeds $1.2 V$. What is the expected number of batteries that you have to test?
(e) You pick three batteries. What is the probability that at least two of them have voltages that add up to more than 2? (Fig. B.28).

Problem B.31 You want to sell your old iPhone 4S. Two friends, Alice and Bob, are interested. You know that they value the phone at X and Y, respectively, where X and Y are i.i.d. $U[50, 150]$. You propose the following *auction*. You ask for a price R. If Alice bids A and Bob B, then the phone goes to the highest bidder, provided that it is larger than R, and the highest bidder pays the maximum of the second bid and R. Thus, if $A < R < B$, then Bob gets the phone and pays R. If $R < A < B$, then Bob gets the phone and pays A (Fig. B.29).

(a) What is the expected payment that you get for the phone if $A = X$ and $B = Y$?
(b) Find the value of R that maximizes this expected payment.

Fig. B.28 Batteries

Fig. B.29 Alice and Bob
have private valuations X and
Y for the phone and they bid
A and B, respectively

(c) The surplus of Alice is $X - P$ if she gets the phone and pays P for it; it is
zero if she does not get the phone. Bob's surplus is defined similarly. Show that
Alice maximizes her expected surplus by bidding $A = X$ and similarly for Bob.
We say that this auction is incentive compatible. Also, this auction is revenue
maximizing.

Problem B.32 Recall that the trace $\text{tr}(S)$ of a square matrix S is the sum of its
diagonal elements. Let A be an $m \times n$ matrix and B an $n \times m$ matrix. Show that
$\text{tr}(AB) = \text{tr}(BA)$.

Problem B.33 Let Σ be the covariance of some random vector $\mathbf{X}$. Show that
$\mathbf{a}'\Sigma\mathbf{a} \geq 0$ for all real vector $\mathbf{a}$.

Fig. B.30 What size solar panels should you buy for your house?

Problem B.34 You want to buy solar panels for your house. Panels that deliver a maximum power K cost αK per unit of time, after amortizing the cost over the lifetime of the panels. Assume that the actual power Z that such panels deliver is $U[0, K]$ (Fig. B.30).

The power X that you need is $U[0, A]$ and we assume it is independent of the power that the solar panels deliver. If you buy panels with a maximum power K, your cost per unit time is

$$\alpha K + \beta \max\{0, X - Z\},$$

where the last term is the amount of power that you have to buy from the grid. Find the maximum power K of the panels you should buy to minimize your expected cost per unit time.

References

N. Abramson, The ALOHA System – Another Alternative for Computer Communications, in *Proceedings of 1970 Fall Joint Computer Conference* (1970)

S. Agrawal, N. Goyal, Analysis of Thompson sampling for the multi-armed bandit problem, in *Proceedings of the 21st Annual Conference on Learning Theory (COLT)*, PMLR, vol. 23 (2012), pp. 39.1–39.26

D. Aldous, David Aldous's Home Page (2018). http://www.stat.berkeley.edu/~aldous/

D. Bertsekas, *Dynamic Programming and Optimal Control* (Athena, Nashua, 2005)

D. Bertsekas, J. Tsitsiklis, *Distributed and Parallel Computation: Numerical Methods* (Prentice-Hall, Upper Saddle River, 1989)

D. Bertsekas, J. Tsitsiklis, *Introduction to Probability* (Athena, Nashua, 2008)

P. Billingsley, *Probability and Measure, Third Edition* (Wiley, Hoboken, 2012)

P. Bremaud, *An Introduction to Probabilistic Modeling* (Springer, Berlin, 1998)

P. Bremaud, *Markov Chains: Gibbs Fields, Monte Carlo Simulation, and Queues* (Springer, Berlin, 2008)

P. Bremaud, *Discrete Probability Models and Methods: Probability on Graphs and Trees, Markov Chains and Random Fields, Entropy and Coding (Probability Theory and Stochastic Modelling)* (Springer, Berlin, 2017)

R.G. Brown, P.Y.C. Hwang, *Introduction to Random Signals and Applied Kalman Filtering* (Wiley, Hoboken, 1996)

E.J. Candes, B. Recht, Exact matrix completion via convex optimization. Found. Comput. Math. **9**, 717–772 (2009)

E.J. Candes, J. Romberg, Sparsity and incoherence in compressive sampling. Inverse Prob. **23**, 969 (2007)

N. Cesa-Bianchi, G. Lugosi, *Prediction Learning and Games* (Cambridge University Press, Cambridge, 2006)

H. Chernoff, Some reminiscences of my friendship with Herman Rubin. *Institute of Mathematical Statistics Lecture Notes – Monograph Series*, vol. 4 (2004), pp. 1–4

K.L. Chung, *Markov Chains with Stationary Transition Probabilities* (Springer, Berlin, 1967)

T.M. Cover, J.A. Thomas, *Elements of Information Theory* (Wiley-Interscience, Hoboken, 1991)

J.L. Doob, *Stochastic Processes* (Wiley, Hoboken, 1953)

D. Easley, J. Kleinberg, *Networks, Crowds, and Markets: Reasoning About a Highly Connected World* (2012). http://www.cs.cornell.edu/home/kleinber/networks-book/

R.G. Gallager, *Low Density Parity Check Codes* (M.I.T. Press, Cambridge, 1963)

R.G. Gallager, *Stochastic Processes: Theory for Applications* (Cambridge University Press, Cambridge, 2014)

G.C. Goodwin, K.S. Sin, *Adaptive Filtering Prediction and Control* (Dover, New York, 2009)

G.R. Grimmett, D.R. Stirzaker, *Probability and Random Processes* (Oxford University Press, Oxford, 2001)

B. Hajek, *Random Processes for Engineers* (Cambridge University Press, Cambridge, 2017)

© The Author(s) 2021

J. Walrand, *Probability in Electrical Engineering and Computer Science*,

https://doi.org/10.1007/978-3-030-49995-2

B. Hajek, T. Van Loon, Decentralized dynamic control of a multiple access broadcast channel. IEEE Trans. Autom. Control, **AC-27**(3), 559–569 (1982)

M. Harchol-Balter, *Performance Modeling and Design of Computer Systems: Queueing Theory in Action.* (Cambridge University Press, Cambridge, 2013)

T. Hastie, R. Tibshirani, J. Friedman, *The Elements of Statistical Learning: Data Mining, Inference, and Prediction*, 2nd edn. (Springer, Berlin, 2009)

D.A. Huffman, A method for the construction of minimum-redundancy codes. *Proceeding of the IRE*, pp. 1098–1101 (1952)

R.E. Kalman, A new approach to linear filtering and prediction problems. Trans. ASME J. Basic Eng. **82**(Series D), 35–45 (1960)

F. Kelly, *Reversibility and Stochastic Networks* (Wiley, Hoboken, 1979)

F. Kelly, E. Yudovina, *Lecture Notes in Stochastic Networks* (2013). http://www.statslab.cam.ac.uk/~frank/STOCHNET/LNSN/book.pdf

L. Kleinrock, *Queueing Systems*, vols.1 and 2 (J. Wiley, Hoboken, 1975–1976)

P.R. Kumar, P.P. Varaiya, *Stochastic Systems: Estimation, Identification and Adaptive Control* (Prentice-Hall, Upper Saddle River, 1986)

E.L. Lehmann, *Testing Statistical Hypotheses*, 3d edn. (Springer, Berlin, 2010)

J.D.C. Little, A proof for the queuing formula: $L = \lambda W$. Oper. Res. **9**(3), 383–401 (1961)

R. Lyons, Y. Perez, *Probability on Trees and Networks.* Cambridge Series in Statistical and Probabilistic Mathematics (2017)

S.G. Mallat, Z. Zhang, Matching pursuits with time-frequency dictionaries. IEEE Trans. Signal Process. **41**, 3397–3415 (1993)

M.J. Neely, *Stochastic Network Optimization with Application to Communication and Queueing Systems* (Morgan & Claypool, San Rafael, 2010)

J. Neveu, *Discrete Parameter Martingales* (American Elsevier, North-Holland, 1975)

J. Neyman, E.S. Pearson, On the problem of the most efficient tests of statistical hypotheses. Phil. Trans. R. Soc. Lond. **231**, 289–337 (1933)

L. Page, Method for node ranking in a linked database. United States Patent and Trademark Office, 6,285,999 (2001)

J. Pitman, *Probability.* Springer Texts in Statistics (Springer, New York, 1993)

J. Proakis, *Digital Communications*, 4th edn. (McGraw-Hill Science/Engineering/Math, New York, 2000)

T. Richardson, R. Urbanke. *Modern Coding Theory* (Cambridge University Press, Cambridge, 2008)

E. Roche, EM algorithm and variants: an informal tutorial (2012). arXiv:1105.1476v2 [stat.CO]

S.M. Ross, *Introduction to Stochastic Dynamic Programming* (Academic Press, Cambridge, 1995)

D. Russo, B. Van Roy, A. Kazerouni, I. Osband, Z. Wen. A Tutorial on Thompson Sampling problem. IEEE Trans. Signal Process. **11**, 1–96 (2018)

D. Shah, Gossip algorithms. Found. Trends Netw. **3**, 1–25 (2009)

R. Srikant, L. Ying, *Communication Networks: An Optimization, Control, and Stochastic Networks Perspective* (Cambridge University Press, Cambridge, 2014)

E.L. Strehl, M.L. Littman, Online linear regression and its application to model-based reinforcement learning, in *In Advances in Neural Information Processing Systems 20 (NIPS-07*, pp. 737–744 (2007)

D. Tse, P. Viswanath, *Fundamentals of Wireless Communication* (Cambridge University Press, Cambridge, 2005)

M.J. Wainwright, M. Jordan, *Graphical Models, Exponential Families, and Variational Inference* (Now Publishers, Boston, 2008)

J. Walrand, *An Introduction to Queueing Networks* (Prentice-Hall, Upper Saddle River, 1988)

J. Walrand, *Uncertainty: A User Guide* (Amazon, Seattle, 2019)

E. Wong, B. Hajek, *Stochastic Processes in Engineering Systems* (Springer, Berlin, 1985)

Index

© The Author(s) 2021
J. Walrand, *Probability in Electrical Engineering and Computer Science*,
https://doi.org/10.1007/978-3-030-49995-2

Printed in the United States
by Baker & Taylor Publisher Services